Dynamics of Fibre Formation and Processing

Roland Beyreuther
Harald Brünig

Dynamics of Fibre Formation and Processing

Modelling and Application in Fibre and Textile Industry

With 171 Figures

Springer

Prof. Dr.-Ing. habil. Roland Beyreuther
Dr. rer. nat. Harald Brünig
Leibniz-Institut für Polymerforschung Dresden e. V.
Hohe Straße 6
01069 Dresden, Germany
RBeyreuther@t-online.de
bruenig@ipfdd.de

ISBN 978-3-540-46221-7 Springer Berlin Heidelberg New York
DOI 10.1007/978-3-540-46223-1

Library of Congress Control Number: 2006933731

Springer is a part of Springer Science+Business Media
springer.com

Typesetting: By the authors
Production: LE-TEX, Jelonek, Schmidt & Vöckler GbR, Leipzig
Coverdesign: design&production, Heidelberg

Printed on acid-free paper 2/3100/YL – 5 4 3 2 1 0

Dedicated to our wives

Ute Beyreuther

and

Dietlinde Brünig

with a warm thanks for many hours of appreciative patience

Preface

The special subject *"dynamics"* calls forth mostly mixed feelings to students and graduates of technical fields. The first foresee and the second know the problems to reach a good or very good mark before or after an appropriate leaving examination.
The reasons for this are well known and even partly understandable. Dynamics – the science of changes or motions which are caused by means of energetic impulses or forces – requires namely a certain amount of applicable knowledge of mathematics as well as logical abstraction and intellectual power to its understanding. These are not within reach by means of schematic memorising.

But basic knowledge and field specific applications of special branches of knowledge are absolutely necessary for engineers, physicists, chemists or mathematicians, if they in practice and research intend to treat determined tasks and operation fields successfully. Such knowledges and considerations are unconditional prerequisites specifically to the analysis of technological processes, to the suppression of disturbances and to the process stabilisation by means of qualitative high-class automatic control systems.

This is especially valid for many textile-technological processes. The mastery of their dynamics decisively influences the process economy as well as the product quality because these processes are often connected with a high manufacturing velocity and/or with a special disturbance sensitivity.

The manufacture of textile products is one of the oldest machine technologies of mankind because it satisfies one of the elementary basic necessities, namely clothing. Many additional technical applications of textile products came along in the nearest past. The processes of manufacturing, the treatment and the processing of fibre threads have a special importance in this context because the fibres (in their multiform kind) were and are the most important basic elements for textile fabrics.

Several ten-thousands of parallel single fibre formations and processing stages are realised simultaneously in a textile or man-made fibre plant. These processes are to be organised with the lowest possible disturbance (e. g. fibre break) levels. Additionally, a strongly increased production velocity has been realised in the last decades for the natural as well as for the man made fibre manufacturing. Both – the multitude of single work positions and the high production and processing velocities – require an increasing knowledge and reliable mastery of the process dynamics from engineers.

This presented book collects the results of industry orientated research which the authors have been carrying out during their work at the Leibniz-Institute of Polymer Research Dresden, Germany (former Institute of Technology of Fibres, until 1990). It deals with dynamic-analytical investigations of different basic processes of the yarn formation and processing for natural as well as chemical (man made) and also glass fibres. The carding processes (roller top card and stationary flat card) and also the false twist texturing process are included (Chaps. 4 and 5, Beyreuther). The main point of these investigations is the dynamic transfer and step response functions and their methodical fundamentals will be fully explained before hand in the introductory Chaps. 1 and 2 (Beyreuther). Therefore the explanations should be understandable for experts who are not skilled in the process-dynamical thinking, too.

A large scale occupies the engineering modelling of the steady state man made fibre process (Chap. 3, Brünig). The presented results are based on the current knowledge of the theory of fibre formation but also on own developed description statements. They include the single filament and multifilament melt spinning processes as well as the fibre formation of the spunbonded nonwoven process.

Some representations about the importance of the tensile force time function for the process stability, its measurement and evaluation conclude the book (Chap. 6, Beyreuther).

All theoretical investigations and results are discussed and verified by means of numerous examples within the industrial practice. With the representation of these complex subjects the book should be qualified for natural and engineer scientists of research and education as well as of textile and man made fibre industry to extend their know-how and know-why knowledge about the processes of fibre formation and processing. The book is also recommendable for lecturers and students of appropriate technical and chemical special branches at technical universities and colleges. The authors also considered didactic experiences in the present subject representation which they have obtained during the last three decades at lectures and seminars at the Technische Universität Dresden/Germany to graduate students of textile engineering and macromolecular chemistry.

We thank all collaborators from the Leibniz-Institute of Polymer Research Dresden/Germany, especially our fellow-workers of the former department *"Fibre Formation"*, for the given support and assistance with numerous experiments, for their evaluation and for many scientific discussions which were very helpful to the conception and planning of the presented book.

Furthermore we are very grateful to our longterm research partners from industry, research institutes and research organisations for the financial support and many fruitful discussions to application related projects which stimulated us to the shaping of general examples into single book chapters. This particularly concerns the companies EMS Inventa AG Domat/Switzerland, DSM Research Geleen/Netherlands, Barmag AG Remscheid/Germany, Saxon Textile Research Institute Chemnitz/Germany and the German Research Association (DFG) Bonn/Germany.

We warmly thank Prof. Dr. P. Offermann, Technische Universität Dresden/Germany (Institute of Textiles and Clothing), and Prof. Dr. T. Kikutani, Tokyo Institute of Technology/Japan (Department of Organic and Polymeric Materials), for their suggestions and helpful comments to the manuscript.

We are very thankful last but not least to Springer Publishers for the appreciative cooperation and the possibility of publication.

Dresden,
October 2006

Roland Beyreuther
Harald Brünig

Contents

1. Introduction

Technologies for the production of textiles are amongst the oldest technologies developed and used by mankind. Firstly, there are clothing textiles which protect against cold and heat, and have the largest share of the manufactured textile industry; then there is the increasing contribution of fabrics for home and technical use. But of course, the classical textiles represent just one part of a wide range of products and there has been an increasing growth in textile intermediary substances which include fibres, yarns and nonwovens, substances that are employed as reinforcing and/or filling elements in thermoplastic or duromeric composites, yarns. The textile yarn or fibre is the most important basic component of most textile final products; and in the past several years yarnless and more efficient alternative technologies, for instance those based on porous sheets, have been developed. But despite the fact that in certain fields they meet the requirements, there is no doubt among the experts that in the near future, no new technology can really substitute textile yarns as the basic component for the production of textile goods. Therefore, all efforts aimed at increasing efficiency and process rationalisation will be focused on the processes of fibre formation and processing; improvements which normally only result from a substantial scientific-analytic description of the essential cause-effect relations of the processes involved.

The fundamental goals and motives for such a process analysis, which within the scope of this book are limited to the dynamic aspects, can be summarised as follows:

1. A generally accepted rule says that a process can only be considered controllable if there exists a sophisticated model by which it can be described mathematically [1]. Only when the analytical work coupled with the development of a satisfactory mathematical model has reached this level, is the *know-how* finally complemented by the control of the *know-why*. The latter, which can never be obtained from any equipment supplier, leads to the path of stable and optimised process control. Furthermore it is then possible to estimate the limits of the process (that is the maximal speed or throughput) realistically.
2. Already on the level of planning and execution of the theoretical and experimental investigations on a process, the dynamic process analysis itself

can lead to a decisive gain of knowledge. The process analysis always accompanies an analytical way of thinking. Such theoretical considerations supply useful know-how for the future work with the technical equipment.

3. The development of different types of technological processes, especially those for fibre formation and processing, can be characterised by the integration of single process stages at simultaneously higher process speeds. Some typical examples are the several processes of melt spinning and texturing of endless yarns. Here, the drawing process, which had been a separate process in the past, was technologically and technically integrated into the spinning or texturing process. One of the consequences was the increasing need for automation, for instance by means of microprocessor control. It is well known that the efficiencies of all process stages have to be multiplied to get the total efficiency of the integrated process. Therefore, even a small disturbance can result in a high loss of production; and hence, planning of a process automation and rationalisation needs a thorough knowledge of the dynamic and steady behaviour of the process under consideration. Then, all the questions about the number, type and quantitative and qualitative effects of the appearing disturbances, as well as their influence on fundamental process and product qualities, can be answered. Furthermore, the meaningful choice of the measurement and control parameters, as well as stable and high quality production, can be guaranteed. Fig. 1.1 (after [2]) shows the various functional levels in automation. The analytic model of the technological process, at the zero level (that means the process level itself), must be seen as the base for all other levels.
4. To sum up, it can be said that the quality of the analytical model is directly related to the efficiency of the process and the quality of its final products.

According to the aims and motives discussed above, process analysis in general is a continuous task within the process maintenance.

This present monograph aims at a description of the dynamic cause-effect relations between a technological machine and its product, the fibre, at several fibre formation and fibre processing stages. Following the long-standing experiences of the authors, the main, but not exclusive focus of this book, is laid on the fibre formation and fibre processing stages of melt spun fibres. Firstly, the content is based on strongly revised parts of a former monograph in the German language [3] as well as on extensive and exhaustive investigations of the modelling of the fibre formation in melt spinning processes. These investigations, which are still in progress, have been carried out at the Leibniz-Institute of Polymer Research Dresden, Germany, over the last twenty five years.

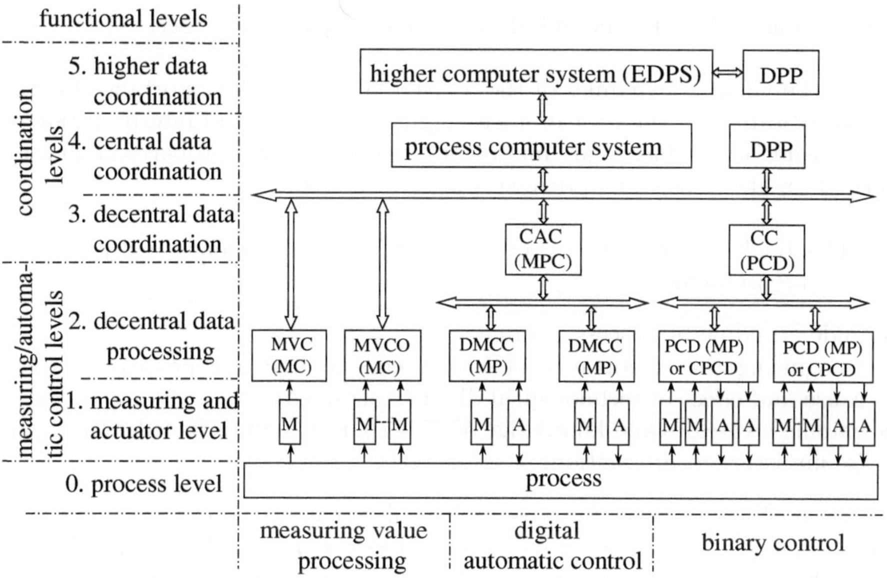

Fig. 1.1. Functional levels of a complex process automatic control system by use of a process computer according to [2]:
M measuring device, A actuator device, MVC measuring value computer, MC micro computer, MP micro processor, MPC micro processor controller, MVCO measuring value concentrator, DMCC digital multi channel controller, PCD programmable controller device, CPCD connection programmed controller device, CAC coordination automatic controller, CC coordination controller, DPP data processing peripherals, EDPS electronic data processing system

The Chap. "Steady State and Non-Steady State Technological Processes", deals with the fundamentals of modelling steady state technological processes on the one hand and non-steady state (dynamic) processes, on the other. The ideas are expounded for both the fibre formation and fibre processing process.

In the Chap. "Modelling of Steady State Fibre Formation Process in Melt Spinning" a model is developed which describes the fibre formation of thermoplastic polymers for different melt spinning processes, which include the monofilament, multifilament, staple fibre and spunbonded nonwoven processes.

The Chap. "Dynamics of Fibre Formation Processes", focuses on the effect of process disturbances on the essential fibre properties during the melt spinning process of polymers and glass, as well as during carding and drafting processes.

The Chap. "Dynamics of Fibre Processing Processes", presents dynamic models for the description of different fundamental fibre processing processes and demonstrates their potential for some exemplary processes.

In the Chap. "Dynamics of the Tensile Force and its Importance for the Process Stability", the measurement, signal evaluation and interpretation of the tensile force (which is an important dynamic product property in all fibre formation and processing processes) are considered.

The book is aimed at experts and students of technical subjects and should be useful for:

- understanding of process parameters,
- recognising the importance of occurrences in dynamic processes as the cause and effect of technological disturbances, and
- gaining strategies and knowledge of disturbance analysis in order to organise undisturbed production.

As is the case with most technological research, the current subject is highly specific and requires a union of various classical fields. As such, fundamentals from mathematics, physics, technical mechanics, automated control engineering, textile technology and textile testing were employed to develop the physical-analytic connections.

2. Steady State and Non-Steady State Technological Processes

2.1 Definitions

In this section the meaning of some technical terms, which are necessary for the understanding of the following parts, will be explained. Besides the terms "fibre formation" and "fibre processing" characterising the *textile technological process* the title of the book contains the terms of *dynamics* and *modelling*. The latter are to be defined first.

2.1.1 The Technological Process

Before the consideration of some specific points, the technological process itself is defined:

The technological process is an organised series of scientifically determined changes of the treated product in order to fulfill a certain manufacturing task. Those changes are initiated by the employment of production instruments and machines.

The succession of each technological process is coupled with the appearance of certain energetic processes:

Each technological process is a meaningful coupling of different energetic processes, which are applied to the treated product. Their combination serves the realisation of a goal-directed procedure.

The example of the drawing process, a basic process in the production of chemical fibres, shall clarify the given definitions.

Figure 2.1 shows a series of important parameters of the manufacturing task of drawing. During the process they are connected by varied cause-effect relations. The drawing process is put into reality by input and output godet pairs driven with different velocities. The parameters can be divided into two groups:

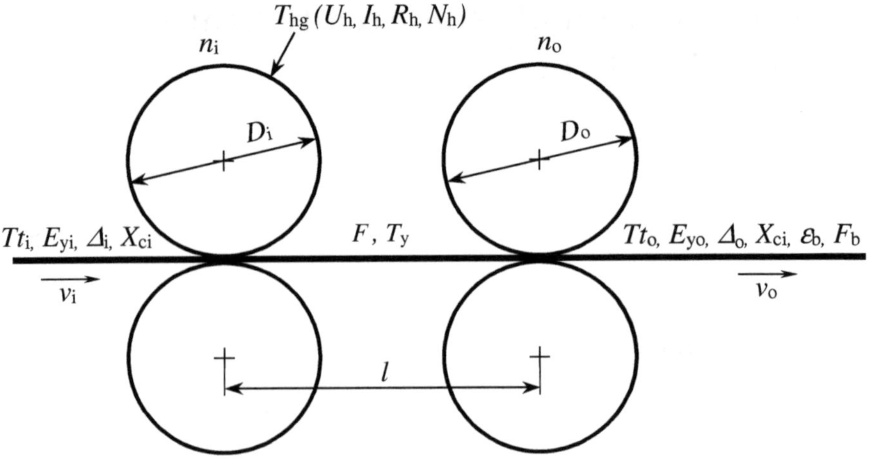

Fig. 2.1. Technological scheme of the drawing process, essential process control and product variables

1. the *process characteristics*, which define the mechanical, geometric and energetic, in other words, the technical-physical set point values characterising a certain technological operating point;
2. the *product characteristics*, which define the quality or state of the input, the processed and the output materials of the process. Thereby the input materials can either be raw materials or semifinished products, whereas the processed materials are semifinished or finished products.

According to Fig. 2.1 the group of the process characteristics contains the input velocity v_i and the output velocity v_o of the drawing godets, which are characterised by the diameters D_i and D_o and the motive revolutions per minute n_i and n_o. Furthermore, the length of the drawing zone l, as well as the godet temperature T_{hg} (only if a hot drawing process, for instance for the polyester production, is considered), which is given by the heating voltage U_h, the heating current I_h, the heating resistance R_h and the heating power N_h, belong to this group.

The product characteristics group comprises for instance the finesses Tt_i and Tt_o, the elastic moduli E_{yi} and E_{yo}, the birefringences Δ_i and Δ_o and the crystallinities X_{ci} and X_{co} of the incoming undrawn and the outcoming drawn yarns, the breaking elongation ε_b of the drawn yarn, the breaking force F_b of the drawn yarn, the tensile force F of the yarn in the drawing zone and the temperature T_y of the yarn at a certain point of the drawing zone.

As the above definitions show, the modification of the properties of the involved materials is characteristic for each technological process. Those modifications can either be intentional or disturbing. The materials' properties, which partly serve as a measure of the quality of the raw materials and the semifinished or finished products, are the product characteristics.

Just as well, the process characteristics are not constant during a certain period of the process. Compared with those of the product characteristics the modifications of the process characteristics are quite small but in general undesirable. In order to emphasise the dynamic character it seems to be meaningful to employ the terms *product variable* and *process variable* instead. As already mentioned, the product and process variables are only constant in some very special cases. In the quality characterisation and classification it is therefore usual to permit tolerances around a desired mean value of a product property.

This fact allows a further conclusion. As the process variables vary around their operating point, the product variables of interest are oscillating around the defined set point value, as well. In other words, the quantities fluctuate with the dynamics caused by the cause-effect relations of the single process lines. Thus, the process dynamics are an inherent component of the process itself.

2.1.2 Dynamics, Process Dynamics

In technical mechanics, dynamics mean the theory of motions caused by forces. In connection with the technological process the term *dynamics* has to be modified and adjusted. It must be restricted to:

Dynamics relating to the technological process, which will be called process dynamics in the following sections, mean the behaviour of a process run during a transition (transition behaviour) from one technologically adjusted operating point to another one. This transition is initiated by a determined or a stochastic disturbance and can be described in its quantitative and temporary progress.

The above definition can be easily explained for the drawing process. Let us take a yarn, which is, according to its macromolecular structure, relatively unoriented.

As shown in Fig. 2.2 the yarn is classically spun between the drawing godets and consequently continuously formed by the elongation $\varepsilon(t)$. The elongation results in a higher oriented form of the yarn which is connected with some desired physical properties. The realised elongation $\varepsilon(t)$ depends on the input and output velocities of the drawing godets $v_\mathrm{i}(t)$ and $v_\mathrm{o}(t)$ in the following way:

$$\varepsilon(t) = \frac{v_\mathrm{o}(t) - v_\mathrm{i}(t)}{v_\mathrm{i}(t)} \tag{2.1}$$

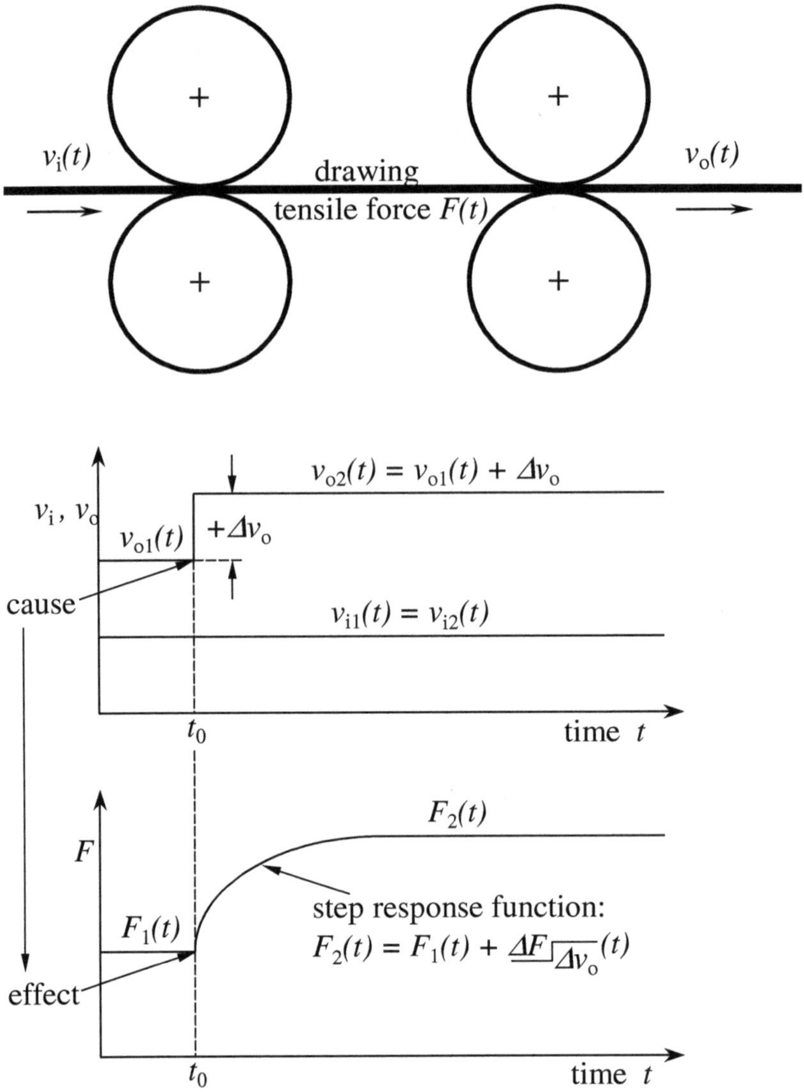

Fig. 2.2. Technological scheme of the drawing process, step-like disturbance of the output velocity Δv_o (symbol ⌐, see Sect. 2.4.2 too), effect $\Delta F(t)$

The raised elongation causes a reaction force in the drawn yarn, the drawing yarn tensile force $F(t)$. Let the process run at the technological operating point 1 characterised by the input velocity $v_{i1}(t) = const.$ and the drawing yarn tensile force $F_1(t) = const.$ At the moment $t = 0$ the output velocity $v_o(t)$ of the drawing godets is step-like increased by Δv_o and consequently reaches the new level $v_{o2}(t) = v_{o1}(t) + \Delta v_o$, which shall characterise the operating point 2.[1] Let level 2 be constant for $t > t_0$. The input velocity shall remain constant. Caused by the step-like disturbance $+\Delta v$, which can be classified as determined aperiodic disturbance (Fig. 2.2), the drawing tensile force (as effect quantity) will leave the level $F_1(t)$ for $t > t_0$. This transition is continuous and not step-like! Thus, the tensile drawing force reaches a level $F_2(t)$ after a transition period. Then, $F_2(t)$ remains constant if there is no further change of the new godet velocities $v_{i2}(t)$ and $v_{o2}(t)$.

The quantitative chronological description of the transition behaviour of the effect quantity drawing yarn tensile force from $F_1(t)$ to $F_2(t)$ caused by a step-like disturbance of a cause quantity (here: step-like increase of $v_{o1}(t)$ to $v_{o2}(t)$) is called step response function or simply *step response*. The response function describes the transitory behaviour of an effect caused by a step-like disturbance and is thus an expression for the *dynamics* of the process under consideration. Already the evaluation of such a simple dynamic standard function allows some fundamental statements regarding the characterisation and the assessment of the dynamic properties of technological processes or specific parts of them. This will be explained in detail in further sections, general facts are given in Sect. 2.4.2.

Let us assume we have a stable technological operating point. Now let the output velocity $v_o(t)$ periodically (for instance sinusoidal) oscillate with the amplitude Δv_{o1} and the circular frequency ω_1 around its mean value v_{om} (see Fig. 2.3).

This periodic disturbance makes the drawing tensile force $F(t)$ oscillate with the amplitude ΔF_1 and the same circular frequency ω_1 around its mean value F_m. The initial oscillation (v_o-disturbance) and the response oscillation (F-disturbance) are not synchronous for the effect runs behind the cause. This behaviour is reflected in the phase shift angle φ_1. A change of the circular frequency of the periodic disturbance from ω_1 to ω_2 at a constant amplitude Δv_{o1} results in a modification of the response oscillation. Then, $F(t)$ oscillates at ω_2 with the modified amplitude ΔF_2 and a changed phase shift angle φ_2. The magnitude of the ratio $\Delta F/\Delta v_o$ and the change of the phase shift angle φ dependent upon the disturbing frequency are, similarly to the step response, an expression of the dynamic characteristics of the drawing process. They lead to the terms *transfer function, frequency response, amplitude frequency response* (simply *amplitude response*) and *phase frequency response* (simply *phase response*). Explanations will follow later in Sect. 2.4.3.

[1] A step-like change in mass effected technical systems is normally not possible to realise. Nevertheless, this model image will be used here and in the following, because this affords an insight in the dynamic process behaviour on the basis of a well developed mathematical signal and system theory.

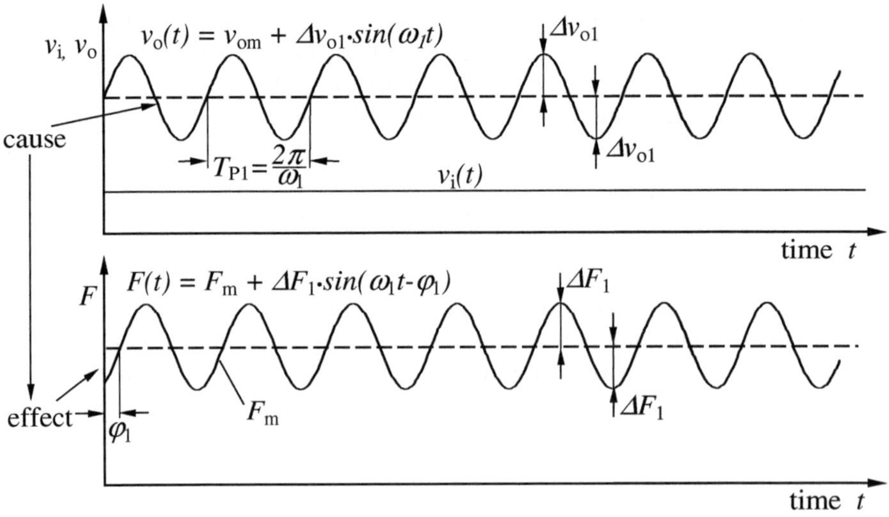

Fig. 2.3. Drawing process, periodic disturbances of the output velocity Δv_{o1}, effect ΔF_1, circular frequency of the disturbances ω_1, phase shift angle φ_1

2.2 Modelling of the Steady State Melt Spinning Process

2.2.1 Goal of Modelling

The goal of each modelling procedure is to obtain a mathematical description of the technological process for better understanding of the main relations between process parameters, material behaviour and product properties. An entire theory of melt spinning should also take into account the history of the polymer, thermal and deformational (rheological) behaviour, and the non-equilibrium conditions for the transfer processes. Such a theory would be quite complicated and is only marginally realised with respect to rheology. But, stationarity in melt spinning means that there are stable conditions in time, that there are no dependences of the process variables, no changes, disturbances or drifts in time. The values describing the fibre formation process change only with respect to space and describe therefore a *steady state* process, but it is noticeable that the process in reality is not one of equilibrium. A reasonable model for this process involves the dynamics of melt spinning and the resulting fibre properties. The best model then is one which is simple enough for handling but good enough for answering the questions which the fibre producers and developers are interested in. Early investigations to the melt spinning process, and based on this knowledge the development of the fundamentals of the model of fibre formation were done by Ziabicki [14–23], Andrews [24], Kase and Matsuo [25–28], Hamana [29,30], Han [31–37],

George [38–40], Shimizu [41–53, 56–60], Yasuda [61–67], and many other authors [78–99, 176–183, 256].

The research on this interesting topic is still actively persued, there are many efforts made by scientists, engineers and producers to get a deeper understanding of the process. Recently, a multitude of contributions to the theory of fibre formation and melt spinning have been published. A major review about the literature is given in two books by Ziabicki [184, 274].

The following two sections only give a short and very simplified introduction to the model of fibre formation, more details to the steady state model will be discussed in Chap. 3 later. The goal of this section is only to give an impression of what the model analysis is capable of doing, therefore, as an example, a first and simple estimation to the fibre cooling process is made.

2.2.2 Balance Equations

To form filaments, the molten polymer is extruded through capillaries and is drawn down by a take-up unit which applies the necessary force. The take-up unit is often realised by godets but in principle it is also possible to wind up the fibres directly onto bobbins. Another common procedure for taking up the filaments is by using special air suction devices (for example in the spunbonded nonwoven process). At their path from the spinneret to the take-up unit, the filaments cool down, become accelerated to their final take-up speed, solidify and at last they can be partly oriented and crystallised. The engineering analysis of this process is made by the application of the physical balance equations of mass, energy and momentum to the fibre forming process in combination with material behaviour. In the following, the three basic equations are briefly summarised.

The most important (and simplest) relation is the continuity equation, it describes the *mass balance* in melt spinning:

$$Q = Tt \cdot v \,. \tag{2.2}$$

The filament cooling, i. e. the heat loss by heat transfer from the fibre surface to the surrounding air, is described by the *energy balance*. If we only regard the convective heat transfer (see later, Eq. 3.10) it is given in the following form:

$$\frac{\mathrm{d}T}{\mathrm{d}x} = -(T - T_{\mathrm{air}}) \cdot \frac{1}{L_{\mathrm{c}}} \,. \tag{2.3}$$

Then, the *momentum balance* describes the forces acting at the fibres:

$$F = F_0 + F_{\mathrm{surf}} + F_{\mathrm{inert}} + F_{\mathrm{drag}} - F_{\mathrm{grav}} \,. \tag{2.4}$$

Using Eq. 2.2 above, the fineness (titre) Tt_{L} of the as-spun filament with take-up velocity v_{L} then is given by

$$Tt_{\mathrm{L}} = \frac{Q}{v_{\mathrm{L}}} \,. \tag{2.5}$$

2.2.3 Example: Heat Transfer

The differential equation 2.3 can easily be integrated for constant parameter L_c and constant temperature T_{air} of surrounding air. The solution is a simple exponential function

$$T(x) = T_{air} + (T_0 - T_{air})e^{-x/L_c} \,. \tag{2.6}$$

with the cooling length L_c proportional to the primary variable mass throughput: $L_c \propto Q$ (for more details see Sect. 3.1.2). The formula (2.6) can be used as a rough estimation for how fast the filaments will cool down. For poly(ethylene terephthalate) (PET) as typical melt spinning polymer the cooling length L_c per mass throughput becomes $L_c/Q \approx 0.2\,\mathrm{m}/(\mathrm{g}\cdot\mathrm{min}^{-1})$ (see Sect. 3.1.2). If the exponential course of the filament temperature is taken into consideration it means for PET, to cool down a filament with a mass throughput of about 1 g/min from an initial melt temperature of 290°C to a temperature of 50°C where solidification is surely reached, one needs about 0.5 m cooling length, for 2 g/min about 1.0 m, etc. (see Fig. 2.4).

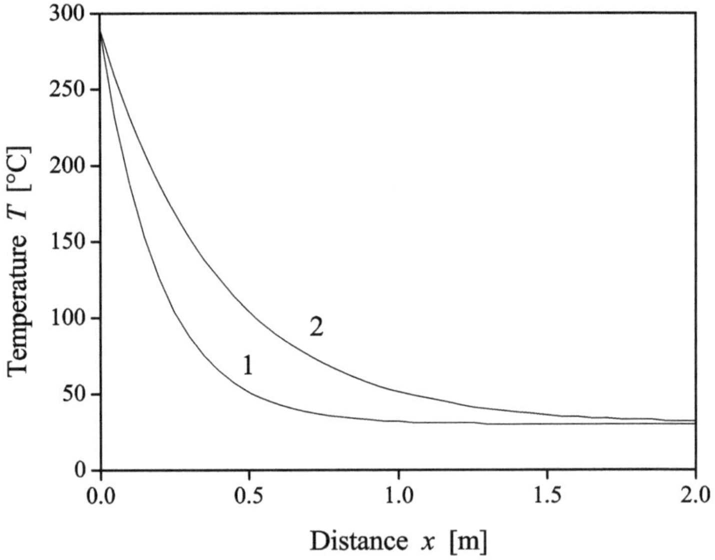

Fig. 2.4. Estimated course of PET filament temperature vs. distance, 1 – mass throughput $Q = 1$ g/min, 2 – $Q = 2$ g/min

2.3 Modelling of Non-Steady State Dynamic Process

2.3.1 System and Signal

In view of the closing remarks of Sect. 2.1.2 a second definition of the technological process is necessary. The definition can be derived from the following way of thinking, which is the norm in the fields of automatic control engineering, system engineering or information technology:
The realisation of each technological process requires an arrangement of certain mechanical, electrical, electronic, pneumatic or hydraulic devices and instruments, which are assembled in a machine or a part of a machine. They are the material basis for the fulfillment of the manufacturing task for the materials passing the process.

The machine and the passing and processed materials together are called the *system.*

As soon as the system, which is assumed to be at standstill and is not performing any material or energy transfer at first, begins to run, one can speak about a *process.* Only then the previously defined process and product variables begin to interact according to their varied cause-effect relations. Those interactions imply that all process and product variables are reflected in their (desired) mean values *and* their (in most cases undesired) fluctuations around the mean values. Those lapses of time are called *time functions* of the process and product variables, for instance $Tt_{\mathrm{o}}(t)$, $F(t)$, $\varepsilon_{\mathrm{b}}(t)$, $v_{\mathrm{i}}(t)$, $T_{\mathrm{hg}}(t)$ and so on. The information content of a time function, in other words the mean value and the fluctuations, are called their *signal.* If those signals are missing (for instance process interruption after a fibre break) the interaction between the process and product variables will not exist any longer. The process will fall back into its static (unproductive) state. Thus, the following process definition, which is the most contensed one, seen under the system technical viewpoint, is possible:

$$\text{process} = \text{system} + \text{signal(s)}$$

Consequently, the scientific analysis of a process with respect to its dynamics includes both, at least one signal analysis and one system analysis; in other words:

$$\text{process analysis} = \text{system analysis} + \text{signal analysis}$$

This situation is once more illustrated in Fig. 2.5 (after [4]), some terms appearing in this figure will be explained in the following section within the presentation of specific examples.

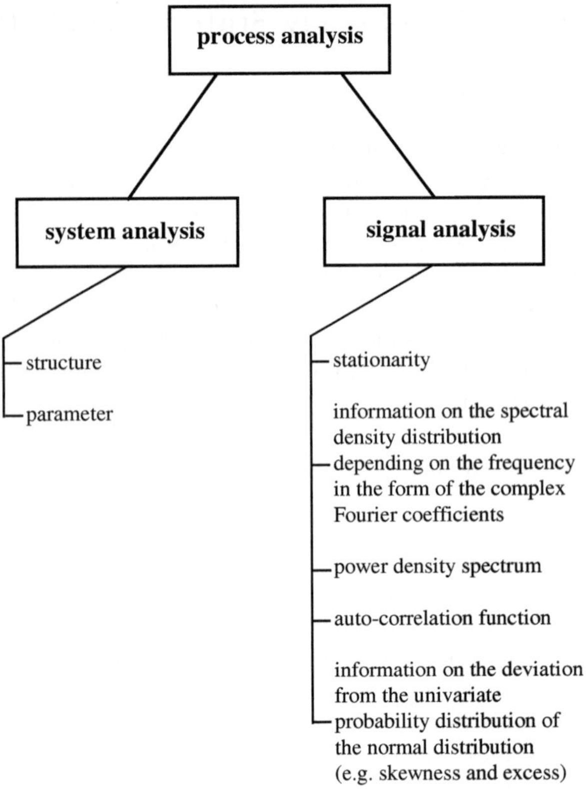

Fig. 2.5. Elements of a process analysis under system-technical aspect according to [4]

2.3.2 Model

The term *model* is another important expression being introduced now. Dealing with a given process it always has a certain reason, for instance:

- The process runs too instable. The prescribed tolerances of the product qualities cannot be maintained. The causes have to be investigated and eliminated.
- There are too many disturbances leading to process interruptions (= interruptions of the signal exchange). The causes must be determined and removed, too.
- A better machine shall be designed (= construction and design task).

Each of those exemplary mentioned tasks means that, at first, an analysis of the given state has to be made. This fact leads to the theoretical and/or experimental attempt to investigate the quantitative relations between the cause-effect relations of the process and product variables and their time-dependent behaviour. In most cases the results of this work have the shape of an image of the real constellation. This can be a formula, a graph, a regression equation, a DEq., a machine on a reduced scale, which is more transparent in its working, or an electronically aided simulation of the process or its parts. Those representations are designated as a more or less complete and sufficient *model*. Thus, relating to the technological process,

the model is an image of a process in its most significant parts, which describes the essential aspects of and the relations between the process and product variables with reference to a certain question.

Within the scope of this book the following classification is chosen: The term *model* is divided into the *steady state* model on the one hand and the *dynamic* model on the other. The first describes the relations between the constant mean values of the process and product variables whereas the latter includes the relations between the changes and fluctuations of the process and product variables. Thus, the dynamic model represents the time dependence of the cause-effect relations.

As both models will be presented mathematically, it has to be mentioned that in most cases the steady state model does not require DEqs. in time whereas the dynamic model *always* leads to such DEqs.. This must be explained with the fact that the time-dependent behaviour can only be described with differentials of time.

As in the former section, the above definitions shall be clarified by an easy example. Again, the drawing process according to Fig. 2.1 is employed. With some fundamental technological knowledge we can write for the fineness of the drawn yarn Tt_{o}:

$$Tt_{\mathrm{o}} = Tt_{\mathrm{i}} \cdot \frac{v_{\mathrm{i}}}{v_{\mathrm{o}}} \tag{2.7}$$

Here, Tt_{o}, Tt_{i}, v_{o} and v_{i} are the mean values of the corresponding process and product variables. In this case, Eq. 2.7 would be the quite simple steady state mathematical model expressing the relations between the mean values of the target quantity Tt_{o} and its determining variables Tt_{i}, v_{o} and v_{i}.

The dynamic mathematical model describing the modification behaviour of the same target quantity Tt_{o} at the same process level is given by the mass balance equation. In our case, the balance equation is realised by the following DEq.:

$$v_{\mathrm{o}} \cdot Tt_{\mathrm{o}} + l \cdot \frac{\mathrm{d}Tt_{\mathrm{o}}}{\mathrm{d}t} - v_{\mathrm{i}} \cdot Tt_{\mathrm{i}} = 0 \tag{2.8}$$

The meaning of the differential element $l \cdot \dfrac{\mathrm{d}Tt_\mathrm{o}}{\mathrm{d}t}$ in Eq. 2.8 will be explained below the Eq. 2.23 in Sect. 2.5.2.

For a step-like disturbance of the input velocity v_i of the magnitude Δv_i the solution of the DEq. 2.8 would be the following step response function:

$$\underline{\Delta Tt_\mathrm{o}}|\overline{\Delta v_\mathrm{i}} = \Delta v_\mathrm{i} \cdot \frac{Tt_\mathrm{i}}{v_\mathrm{o}} \left[1 - \exp\left(-\frac{v_\mathrm{o}}{l} \cdot t\right)\right] \quad , \tag{2.9}$$

which describes the lapse of time of the effect of this disturbance on the modification of the output fineness ΔTt_o.

The DEq. 2.8 and its solutions (for the case of an aperiodic step-like disturbance such a solution is given by the Eq. 2.9) are a *dynamic mathematical model* of the drawing process.

2.4 Characterisation of the Dynamic Process Behaviour

2.4.1 Differential Equation

Restricting the dynamic mathematical modelling of processes or process steps on the clear representation in the time range and/or the frequency range, further terms, which accompany the methodical approach, have to be introduced.

For the determination of the dynamics of technological processes the DEq., describing the time behaviour of the system, plays a fundamental role. The DEq. is the mathematical reproduction of the cause-effect relations of those processes and product variables which are taken into account within the scope of the model. The DEqs. formation for specific technical systems is based on the dynamic mass-, energy- and momentum balance relations. Hereby, not only the steady state, but the dynamic (related to the general case of a running process) balances (processing mode) must be employed. Different a-priori knowledge of the process of interest, mathematical-methodical knowledge as well as some basic scientific laws are required for the DEqs. development.

2.4.2 Description in the Time Range; Step Response

If the DEq. shall supply facts about the dynamic properties of the system forming its basis, it must be solved first. According to the type of the time function of the independent variables of the DEq., which correspond to the causes for the system modifications, several solutions for the dependent variable, corresponding to the effect, are possible.

If the time function of the independent (cause-)variable, which disturbs the system, initiates a single (aperiodic) change of the previous (steady state) mean value or technological operating point, the arising new time function of the dependent (effect-)variable will be called *response function.* For special time functions of the independent (disturbance- or cause-) variables the *response time-* (or simple only *time-*) *function* can be designated more exactly. If the cause-time function is a single step (symbol $\_|\overline{\ }$, see also Fig. 2.2), the response time function will be called *step response*. If the cause time function is a single impulse (symbol $\perp$), which means that the cause variable leaves its steady state value only for an infinitely short time $t = t_0$, say impulse-like, and comes back to this value instantaneously, the response time function will be called *impulse response function*. Additionally, it has to be mentioned that, besides those two standard types of a cause time function, arbitrary signal types can appear as disturbances. The mathematical algorithm for the solution of the DEq. for step-like and impulse-like disturbances is well-known and ready for application. But at technological process steps the impulse-like disturbance has almost no practical significance, for in an experiment this type cannot be realised with sufficient exactness. Therefore, all common descriptions are based on the *step response* or simply *response function.* As the response functions are solutions of the DEq. reflecting the time behaviour of the dependent effect variables, we also speak about a representation in the time range. The following equation shows the general form:

$$\underline{\Delta y}|\overline{\Delta x} = f(t) \tag{2.10}$$

with Δx as cause, Δy as effect and t as running time.

2.4.3 Description in the Frequency Range

Dynamic Transfer Function; Complex Frequency Response

The cause variable can also be assumed to be a periodic function of time. In other words, the variable is disturbed frequently and not only at one time as considered in the above section. Now, in contrast to the case of a single disturbance, the solution of the DEq. does not reveal the exact time behaviour of the effect variable after a defined cause disturbance, but the solution specifies:

a) how the ratio of the amplitudes of the dependent and the periodic independent variable, and
b) how the phase shift angle φ between the cause-and the effect disturbance

will depend on the frequency ω of the periodic disturbance.

In principle, similar to the case of aperiodic disturbances, arbitrary types of disturbance signals are possible. In most practical cases the sine function is used because its mathematical treatment is quite easy.

However, the amplitude- and phase shift angle dependence on the disturbing frequency are the solution of the DEq., which models the dynamics of the system under investigation, in the *frequency range.* This solution is called *dynamic transfer function* or *complex frequency response.* In the common literature of system control or automatic control engineering the dynamic transfer function is given as the following complex function:

$$G(\mathrm{j}\omega) = \frac{\widetilde{\Delta y}(\omega)}{\widetilde{\Delta x}(\omega)} \cdot \mathrm{e}^{\mathrm{j}\varphi(\omega)} \tag{2.11}$$

with the following symbols[2]:

- $\widetilde{\Delta x}(\omega)$ vector of the sinusoidal disturbance of the independent cause variable $\Delta x \sin(\omega t)$
- $\widetilde{\Delta y}(\omega)$ vector of the sinusoidal response of the dependent response (effect) variable $\Delta y \sin(\omega t + \varphi)$
- $\mathrm{e}^{\mathrm{j}\varphi(\omega)}$ factor, which gives the phase shift between cause- and effect oscillation in the complex plane
- j imaginary unit, $\mathrm{j}^2 = -1$

As detailed examples later in the text will show, the dynamic transfer function or complex frequency response are directly obtained as the solution of the DEq. for periodic sinusoidal disturbances of the independent variable.

The complex frequency response Eq. 2.11 can, as common with complex quantities, be separated into its real and imaginary part:

$$G(\mathrm{j}\omega) = \mathrm{Re}(\omega) + \mathrm{j} \cdot \mathrm{Im}(\omega) \quad , \tag{2.12}$$

whereas $\mathrm{Re}(\omega)$ represents the real part and $\mathrm{Im}(\omega)$ corresponds to the imaginary part of the complex frequency response.

The representation of the complex frequency response Eq. 2.11 or Eq. 2.12 in the complex plane (x axis = real axis; y axis = imaginary axis) marks the end points of all those vectors, which can be drawn from the origin of the coordinate system dependent upon the excitation frequency ω. This curve is called *transfer locus* of the complex frequency response (Fig. 2.6).

Equivalent to the response function in the time range, the transfer locus describes the dynamic behaviour of a system in the frequency range, because the amplitude of the dependent effect variable is now plotted against the frequency.

---

[2] In mathematics the imaginary unit is usually designated with i. As in technical context the symbol i is often used for electric currents, the symbol j is used for the imaginary unit here.

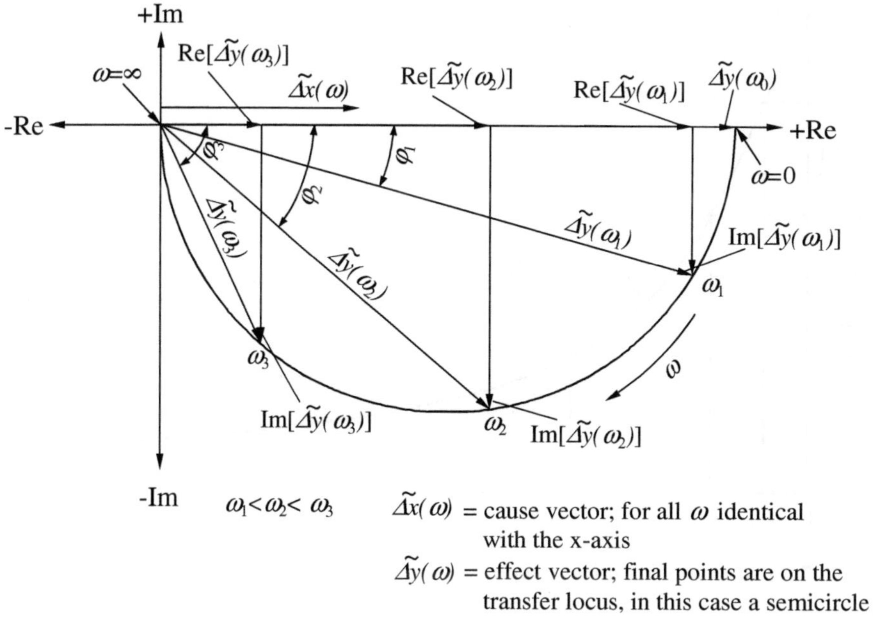

Fig. 2.6. Transfer locus of a complex frequency response; general example

It can be easily seen that for the excitation frequency $\omega = 0$, the amplitude vector of the independent cause variable (which always remains on the real axis) and the amplitude vector of the dependent variable have the same direction and melt together on the real axis. This represents the initial point of the transfer locus defining a kind of steady state excitation for the very special case of $\omega = 0$. For this exemplary situation there is no phase shift between cause and effect oscillation: $\varphi = 0$.

Amplitude Frequency Response

Proceeding from the complex frequency response Eq. 2.12 the ratio of the amplitudes of the cause- and the effect oscillation, $\widetilde{\Delta y}(\omega)/\widetilde{\Delta x}(\omega)$, can be plotted against the excitation frequency ω. This amplitude ratio corresponds to the absolute value of the single complex number of the frequency response $G(\mathrm{j}\omega)$, that is $|G(\mathrm{j}\omega)|$, or to the ratio of the mentioned vectors of the transfer locus of the complex frequency response. Those absolute values are calculated as follows:

$$|G(\mathrm{j}\omega)| = \sqrt{\mathrm{Re}(\omega)^2 + \mathrm{Im}(\omega)^2} \tag{2.13}$$

Equation 2.13 gives the *amplitude frequency response*, which is one part of the complex frequency response. Consequently it is just another form of

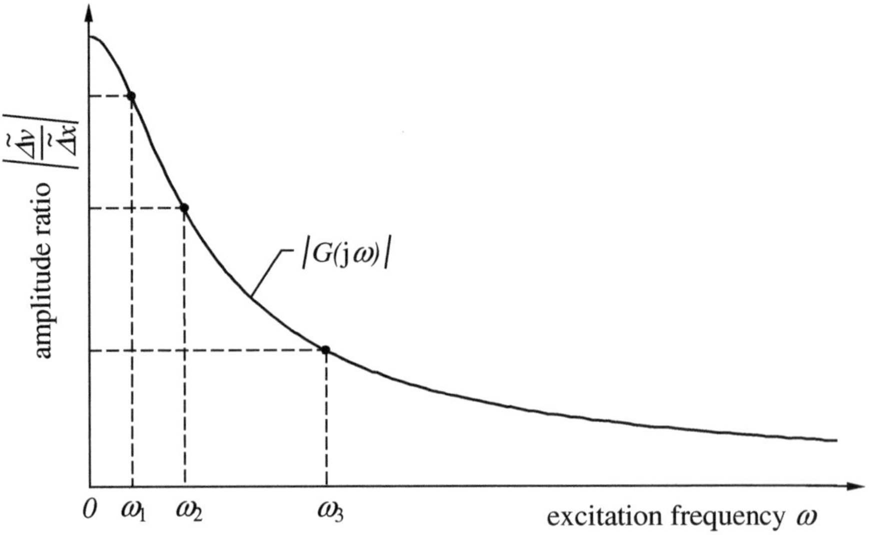

Fig. 2.7. Amplitude frequency response of the transfer locus according to Fig. 2.6

representation in the frequency range, as the solution is plotted against the frequency ω, again (Fig. 2.7).

Phase Frequency Response

The second part which can be extracted from the complex frequency response, is the *phase frequency response*. This function describes how the phase shift angle φ between cause- and effect oscillation depends on the excitation frequency ω. In the transfer response representation it is exactly that angle included by the real axis and each vector beginning at the origin of the coordinate system and ending on the transfer locus curve. The tangent of the phase shift angle φ is equal to the ratio of the imaginary and the real part of the complex frequency response 2.12, namely:

$$\tan[\varphi(\omega)] = \frac{\mathrm{Im}(\omega)}{\mathrm{Re}(\omega)} \tag{2.14}$$

According to Eq. 2.14, the phase shift angle φ can be calculated directly:

$$\varphi(\omega) = \arctan\left[\frac{\mathrm{Im}(\omega)}{\mathrm{Re}(\omega)}\right] \tag{2.15}$$

For the general case the phase frequency response is depicted in Fig. 2.8.

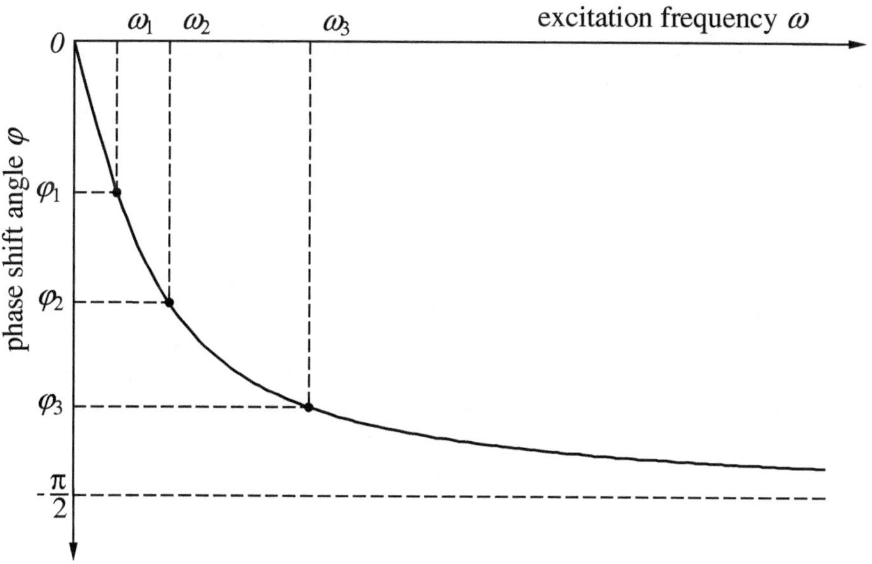

Fig. 2.8. Phase frequency response of the transfer locus according to Fig. 2.6

While the complex frequency response or its transfer locus, respectively, contain the information from the DEqs. solution for a certain sinusoidal disturbance of the independent cause variable, i.e. the dependence of the amplitude ratio *and* the phase shift angle on ω, the amplitude frequency response and the phase frequency response provide only the dependence of *either* the amplitude ratio *or* the phase shift angle on ω.

For more detailed information, didactically reasoned descriptions and analytical proofs see for instance [5], [6] or [7]. Within the scope of this book, only the basic knowledge which seems to be inherently necessary for the understanding of the following sections, has been sketched. A very detailed example, which is thought to clarify the above mathematical terms and connections, is presented in Sect. 2.5.2.

2.4.4 Correlation and Power Density Spectrum Functions

In the previous sections only such functions characterising the dynamic system behaviour have been considered, which are based on the transfer description of *determined* disturbances or changes of the system. But in principle it is also possible to develop a mathematical description for the transfer of *non-determined* or *stochastic* disturbances of the cause variables. This becomes necessary as soon as it is experimentally impossible to produce determined disturbances or if the process of interest should not be faced with higher disturbance amplitudes, which are in most cases needed for determined disturbances, for reasons of safety or any other reasons. In those cases the arbitrary (stochastic) fluctuations of the cause and effect variables around their mean values, which are generally always present and which determine their technological operating point, can be used for the dynamic system analysis. Consequently, this method of analysis implies the possibility to gain information during the normal process run. In doing so, some very strict conditions and standards have to be fulfilled. Those conditions and standards specifically concern the parts of system engineering and system analysis of such an investigation. This methodical procedure for the system analysis will not be shown in this book, the approach to this specific theoretical and experimental discipline can be studied for instance in [5–8]. Within the scope of stochastic system investigations statistic characteristic functions of the time functions of the process and product variables are used as they allow a so far hardly practised approach to the system analysis with a higher gain of information in the field of advanced, process related disturbance analysis of textile products. Therefore the fundamental equations with some short explications are now given and will be applied in later sections. These equations are the *correlation function* and the *power density spectrum function.*

The *correlation function* $K(\tau)$ of *one* time function $x(t)$ is defined in the following way.

Integral representation:

$$K(\tau) = \lim_{\mathrm{t}\to\infty} \frac{1}{T} \int_0^t [x(t) - \bar{x}][x(t+\tau) - \bar{x}]\mathrm{d}t \tag{2.16}$$

Sum representation:

$$K(k \cdot \Delta t) = \frac{1}{n-k} \sum_{\mathrm{i}=0}^{n-k} (x_\mathrm{i} - \bar{x})(x_{\mathrm{i+k}} - \bar{x}) \tag{2.17}$$

Boundary condition: $k_{\mathrm{max}} \leq \frac{n}{5}$, with $k = 0, 1, 2, 3, \ldots k_{\mathrm{max}}$.

$x(t)$	time function
$\bar{x}$	mean value of the time function in the evaluation range $0 \leq t \leq T$ or $0 \leq k \cdot \Delta t \leq n-k$, respectively
x_i	discrete value of time function $x(t)$, taken in steps of Δt
T	time period of the integration range
$\tau, k \cdot \Delta t$	time shift
k, n, m	running (sequence) indices
$K(\tau), K(k \cdot \Delta t)$	single values of the correlation function for τ or $k \cdot \Delta t$, respectively

As can be seen from Eqs. 2.16 and 2.17 all values of the time function are multiplied continuously or at certain points with the values of the same time function after a time shift of τ or $k \cdot \Delta t$, namely $x(t+\tau)$ or $x_{\mathrm{i+k}}$, and finally added up to get $K(\tau)$ or $K(k \cdot \Delta t)$ (see Fig. 2.9 too).

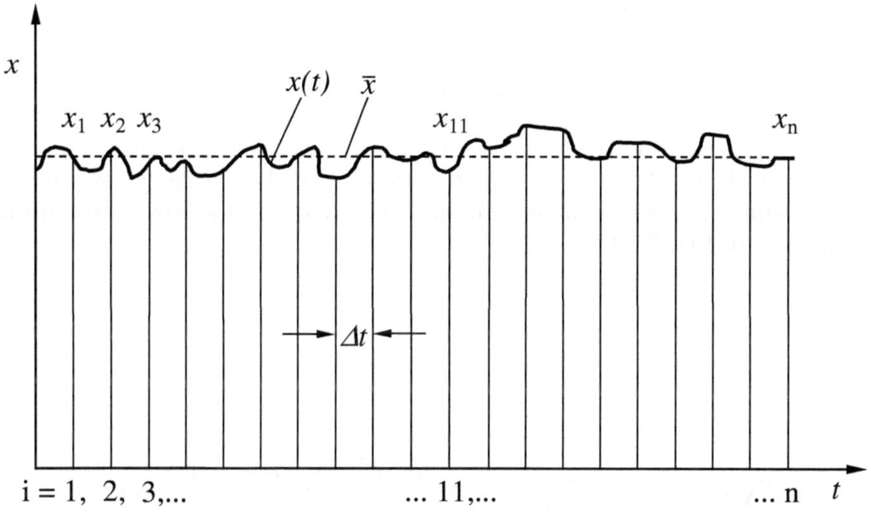

Fig. 2.9. Time function $x(t)$, distributed into n single values for the design of the auto-correlation function

The result of this summation divided by the length of the addition- or interpretation interval T represents the single value of the correlation function $K(\tau)$ or $K(k \cdot \Delta t)$. The repetition of this calculation for different τ or $k \cdot \Delta t$ yields the complete correlation function.

As $x(t)$ is correlated with itself according to Eqs. 2.16 and 2.17, the corresponding correlation function is called *auto-correlation function.*

Without an extended discussion of the efficiency and the interpretation possibilities (more about this topic in Sect. 6.4.2), it should be mentioned that the initial value of the auto-correlation function $K(0)$ is equal to the square spread of the time function in the interpretation- and integration range.

The algorithm for the calculation of the correlation function can also be applied to *two* different time functions $x(t)$ and $y(t)$, which could be for instance coupled by certain cause-effect relations. Analogous to Eqs. 2.16 and 2.17 the corresponding equations are:

$$K(\tau) = \lim_{t \to \infty} \frac{1}{T} \int_0^t [x(t) - \bar{x}][y(t+\tau) - \bar{y}] \mathrm{d}t \quad , \tag{2.18}$$

$$K(k \cdot \Delta t) = \frac{1}{n-k} \sum_{\mathrm{i}=1}^{n-k} (x_{\mathrm{i}} - \bar{x})(y_{\mathrm{i+k}} - \bar{y}) \tag{2.19}$$

Boundary condition: $k_{\max} \leq \frac{n}{5}$, with $k = 0, 1, 2, 3, \ldots k_{\max}$.

As the partial products of two cross-wisely analyzed time functions are summed up, the resulting correlation function is called *cross-correlation function.* To support a better understanding the situation of a cross-correlation is visualised in Fig. 2.10.

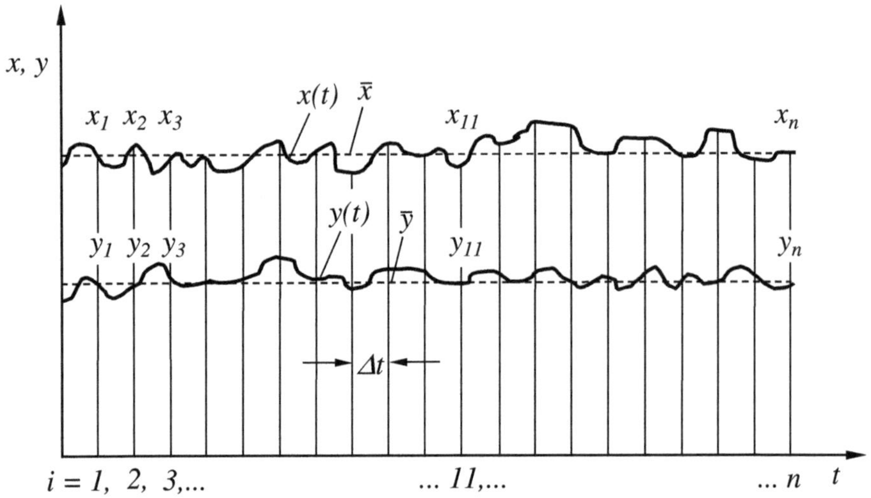

Fig. 2.10. Time functions $x(t)$ and $y(t)$, distributed into n single values for the design of the cross-correlation function

The independent variable of the auto- or cross-correlation function is the time shift τ or $k \cdot \Delta t$. Thus, this form of representation is a representation in the time range (see also Sect. 2.4.2).

The statistic characteristic function, which supplies the equivalent information about the signal shape of the time function in the frequency range is the *power density spectrum function.* This function can also be written as auto- (APSF) or cross-power density spectrum function (CPSF).

Following, the calculation rule for the power density spectrum function is given, as Sect. 6.4.2 will exclusively refer to it. Furthermore, the principle is commercially available in the form of modules for yarn and fibre uniformity testing devices or spectrometers. These instruments allow the automatised calculation of the amplitude spectrum, which is similar to the power density spectrum function, of measured signal shapes of the fibre fineness.[3]

Integral representation to the auto-power density spectrum function:

$$S(\omega) = \lim_{\mathrm{T}\to\infty} \frac{1}{T}\left\{ \left[\int_0^T (x(t)-\bar{x})\cos(\omega t)\mathrm{d}t\right]^2 + \left[\int_0^T (x(t)-\bar{x})\sin(\omega t)\mathrm{d}t\right]^2 \right\} \tag{2.20}$$

Sum representation:

$$S(\omega) = \frac{1}{n}\left\{ \left[\sum_{\mathrm{i}=1}^{n}(x(t)-\bar{x})\cos(\omega\cdot i\cdot \Delta t)\right]^2 + \left[\sum_{\mathrm{i}=1}^{n}(x(t)-\bar{x})\sin(\omega\cdot i\cdot \Delta t)\right]^2 \right\} \tag{2.21}$$

Boundary condition: $\dfrac{10\pi}{n\cdot\Delta t} \le \omega \le \dfrac{\pi}{\Delta t}$

If the auto-correlation function, in its sum definition according to Eq. 2.17, has already been calculated the single function values of $K(k\cdot\Delta t)$ can be applied directly for the determination of the power density spectrum function without a repeated access to the values for the x_{i} of the basic time function. The conversion formula is:

$$S(\omega) = 2\Delta t\sum_{\mathrm{k}=1}^{m} K(k\cdot\Delta t)\cdot\cos(\omega\cdot k\cdot\Delta t) \tag{2.22}$$

Boundary condition: $\dfrac{2\pi}{m\cdot\Delta t} \le \omega \le \dfrac{\pi}{\Delta t}$

Further explications are given in [8,9].

[3] The *spectrograph* of the ZELLWEGER CO., Uster, Switzerland, which is delivered as a module for the *Uster uniformity tester*, has been popular for decades.

2.5 Dynamic Process Analysis and Modelling

2.5.1 Methodical Procedure

Since the most essential terms have been introduced and explained quite independently in the previous sections, now some fundamental rules with respect to the methodical procedure for dealing with process-analytic tasks, which are well-tried, are given. These rules will help to get a general idea of the connections and interactions between those essential terms.

There are two stages in the system analysis, which are classified according to the qualitative representation of their result:

a) the stage of the steady state system analysis aiming on the development of a steady state model of the process, and
b) the stage of the dynamic system analysis aiming on the setting up of a dynamic model of the process, which can include the steady state model as a special case.

According to Table 2.1 (taken from [1]) the mathematical-analytical approach to the modelling of technological flow-processes (which are, in contrast to piece-processes, such processes where the process variables are exposed to continuous changes and not to generally step-like changes) always leads to at least one DEq. or usually to a system of DEqs.

Table 2.1. Fundamental mathematical-analytical approaches to the modelling of technological flow-processes; $x, y, z \ldots$ space coordinates, $t \ldots$ time

	Processes with concentrated parameters	Processes with scattered parameters
Steady state model	Systems of algebraic equations	Systems of ordinary DEqs. with derivatives $\frac{\mathrm{d}}{\mathrm{d}x}$
	Systems of transcendental equations	Systems of partial DEqs. with derivatives $\frac{\partial}{\partial x}, \frac{\partial}{\partial y}, \frac{\partial}{\partial z}$
Dynamic model	Systems of ordinary DEqs. with derivatives $\frac{\mathrm{d}}{\mathrm{d}t}$	Systems of partial DEqs. with derivatives $\frac{\partial}{\partial t}$ and $\frac{\partial}{\partial x}, \frac{\partial}{\partial y}, \frac{\partial}{\partial z}$

Considering the principied type of the chosen methodical approach, the system analysis can be arranged in three classes:

a) the theoretical,
b) the experimental, and
c) the combined theoretical-experimental system analysis.

The latter is based on a continuous interaction between theoretically established relations and experimental investigations, see Fig. 2.12. Practically, this type is the most common one.

The specific steps for the dynamic system analysis or the modelling are:

1. All process and product variables, being involved in the process and probably being connected with each other via cause-effect relations, are to be collected and sorted. For that purpose the establishment of a *cause-effect-scheme*, which inherently goes along with an exact physical-analytical way of thinking and the inclusion of all available a-priori knowledge, is inevitable.
2. If possible, all DEqs. of the system, following from mass-, energy- and momentum balances, have to be set up.
3. The DEqs. coefficients have to be determined theoretically and experimentally. The DEq. is to be solved (step response function and/or frequency-, phase- and amplitude response).
4. If it is impossible to set up a DEq., the cause-effect-scheme has to be split into meaningful smaller cause-effect-blocks. The structure and the parameters of those partial systems must be investigated by actively performed experiments.
5. The model is tested and improved.
6. Mathematical simulations are performed in order to answer the technological questions. The model is applied.

The realisation of the experiments mentioned in 4., aiming on the investigation of the cause-effect relations between the process and product variables, requires the modification of the concerned cause variable under a certain law and the subsequent measurement of the system's response (of the effect variable). The evaluation of the time functions of the varied input variable (test signal) *and* of the output variable allows the determination of the dynamic behaviour of the system. Figure 2.11 (after [4]) summarises the possible test signals for the input variable (cause variable).

The transfer functions are the response time functions for aperiodic determined test signals. From their characteristics the dynamic parameters and the structure of the system can be derived. If periodic determined test signals, for instance sinusoidal signals of tunable frequency, are used, the dynamic transfer function or the complex frequency response of the system can be investigated experimentally. Analogously, the dynamic system characteristics can be determined from them. Correlation and spectral analysis methods belong to the group of methods which employ stochastic signals for the determination of the system parameters and the system's structure. However, their application is coupled with a considerable expenditure. Nevertheless, in several cases it is the only possibility to carry out any dynamic process investigation at all.

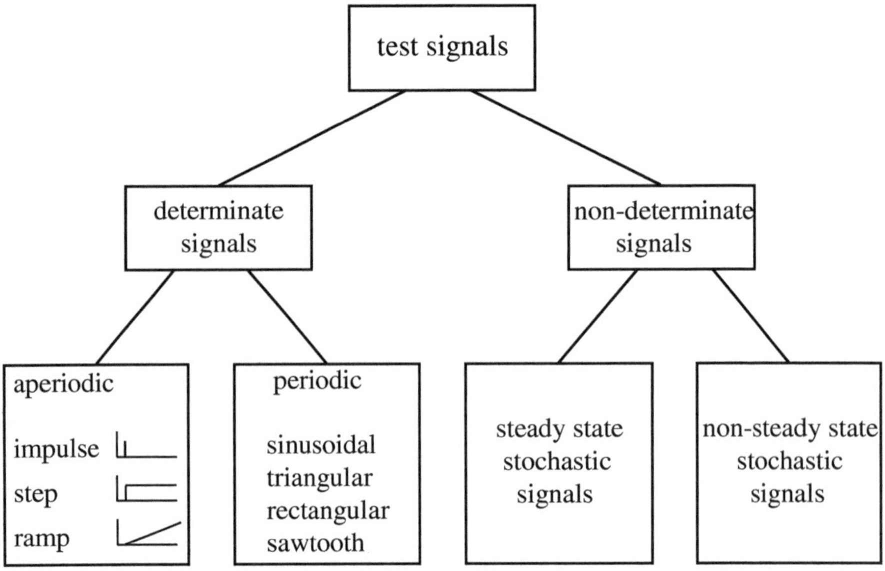

Fig. 2.11. Classification of test signals

Figure 2.12 schematically shows the different methods of dynamic process analysis.

As can be seen here, the problem can be solved theoretically as well as experimentally. That model which starts from the setting up of the DEq. taking all qualitative physical-analytical relationships into consideration and in which only the quantitative fixing of the DEqs. parameters is tested experimentally, must be seen as the most valuable model. All dynamic characteristic functions, for instance the step response function or the complex frequency response, can be calculated if the DEq. of the process or process part behaviour is known.

As soon as the setting up of the DEq. becomes impossible, the questions of interest have to be exclusively solved by experiments. In principle, all experimental methods are equally good, but practically there are big differences with respect to the expense of time and aid and the achievable accuracy. The step or impulse test quickly supplies characteristics in the time range, which can be converted into the qualitatively more valuable characteristic in the frequency range (the complex frequency response or the dynamic transfer function) only with a limited accuracy. Investigations applying a periodic test signal permit a direct access to the complex frequency response – however, with a higher expenditure of time but also with a good accuracy. Those methods which utilise stochastic test signals take a lot of time, as well. Their accuracy decisively depends on the quality of the measuring instruments.

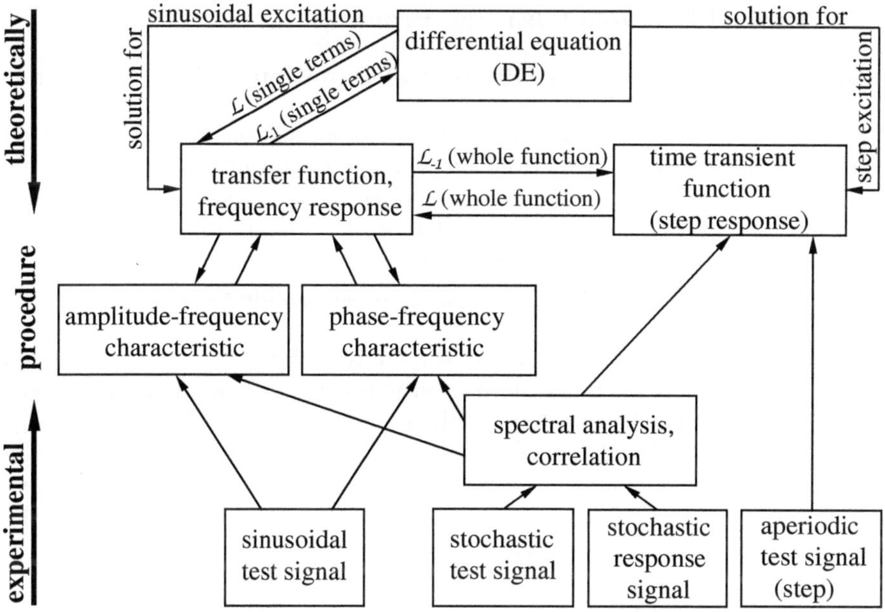

Fig. 2.12. Methods of the dynamic process analysis: L means LAPLACE-transformation from the time to the frequency range, L_{-1} means retransformation of the LAPLACE-transformation from the frequency to the time range

Here, an important problem relating to the purely experimental procedure (often of necessity) shall be mentioned. Of course the experimental procedure leads to a quantitatively correct model of the process under consideration. But such a model cannot contribute to an understanding of the fundamental physical-analytical relations of the process and product variables. The experimentally gained model is usually limited to that process that was employed for its acquisition. It is quite good for the process control but not for a general explication of the connections between several input and output variables. Furthermore, such a model should not be transferred to a similar but larger process (up scaling) without testing its validity again.

Finally it is pointed out that the conversion laws between the dynamic transfer function and the step response, formally given in Fig. 2.12, are theoretically based on the LAPLACE-transformation (abbreviation symbols L and L_{-1}), which will be used in some later examples. Some literature hints will also be given to this later.

2.5.2 Detailed Example (Drawing Process)

Deviation of the Specified Differential Equation and its Solutions

Here, the methods introduced during the previous sections are demonstrated by the setting up, the evaluation and interpretation of a simple dynamic model for the drawing process. For that purpose the following question has to be answered: Which dynamic transfer properties has the process step of drawing relating to changes or fluctuations of the input velocity v_i in its effect on the fineness Tt_o of the yarn leaving the drawing zone?

In this special case the first step of the analytical procedure recommended in Sect. 2.5.1, namely the qualitative coverage of the process and product variables of interest, can be omitted. With some technological basic knowledge the fundamental statistic equation for the output yarn fineness Tt_o, see Eq. 2.7, can be given at once. As already mentioned, the output yarn fineness depends on the input velocity v_i as well as on the output velocity v_o and the fineness Tt_i of the incoming yarn.

The setting up of the DEq. requires the dynamic mass balance equation for the draw field in which the fibre mass is exchanged continuously (the undrawn fibre goes in, the drawn fibre comes out).

As the draw field and, by the way, a lot of other process steps of textile processing processes are systems where a certain mass is stored, the general balance equation (continuity equation) taking the dynamic *processing mode* into consideration can be utilised as the basis for the formulation of the DEq.:

$$\frac{\text{mass inflow}}{\text{time}} = \frac{\text{mass discharge}}{\text{time}} + \text{change of stored mass} \quad (2.23)$$

Specifically for the drawing zone the terms are:

$$\frac{\text{mass inflow}}{\text{time}} = Tt_i \cdot v_i \quad , \quad \frac{\text{mass discharge}}{\text{time}} = Tt_o \cdot v_o$$

The term *change of stored mass* needs, relating to the modelling of the filament drawing process, a short explanation leading to a simplification. The yarn mass stored in the drawing zone is $Tt_o \cdot l$, see Fig.2.1. We assume now that all changes of the fibre fineness take place homogeneously, quasi rubber like, over the whole drawing zone. This assumption is sufficiently exact, if only small changes of process variables around the operating point (<10%) are in the view.

For the situation, that immediately after the inflow of the undrawn fibre the full draw ratio of some 100% would be realised the fibre would reach its output fineness Tt_o shortly after the inflow (neck like deformation). The fibre had to be transported only over the length l of the drawing zone with the output velocity v_o. For the last case (which is not shown here) the model description would be valid for a dead time thread line, which is explained more in detail in Sect. 5.1.3.

The term *change of stored mass* means the mass per time interval which is *additionally* flowing in or being discharged from the drawing zone after a perturbation. Mathematically, this term is the first derivative of the stored mass after the time:

$$\frac{d(Tt_o \cdot l)}{dt} = \frac{dTt_o}{dt} \cdot l = \dot{T}t_o \cdot l \tag{2.24}$$

By inserting this equation in Eq. 2.23 we get the following DEq.:

$$Tt_i \cdot v_i = Tt_o \cdot v_o + \dot{T}t_o \cdot l \tag{2.25}$$

Equation 2.25, which is equal to Eq. 2.8, is the DEq. describing the drawing process under the simplified conditions explained above. The cause variable was v_i and the effect variable was Tt_o. The values of Tt_i, v_o and l should be constant.

Before we continue with the solution of the DEq., an agreement concerning the symbols must be reached. Constants characterising the technological operating point are written with the index m, standing for *mean* or *mean value*. Variables are split into their constant mean value, also marked by the index m, and their fluctuating part, which is symbolised by a Δ in front.

In our example we have to introduce $Tt_o = Tt_{om} + \Delta Tt_o$ and $v_i = v_{im} + \Delta v_i$ for the variables and Tt_{im}, v_{om} and l_m for the constants.

Then, Eq. 2.25 appears as:

$$Tt_{im} \cdot (v_{im} + \Delta v_i) = (Tt_{om} + \Delta Tt_o) \cdot v_{om} + (\dot{T}t_{om} + \Delta \dot{T}t_o) \cdot l_m$$

After multiplication, with $Tt_{im} \cdot v_{im} = Tt_{om} \cdot v_{om}$, which is equivalent to the steady state balance equation, and with the knowledge that the derivative of a constant is equal to zero the DEq. gets the following form:

$$Tt_{im} \cdot \Delta v_i = v_{om} \cdot \Delta Tt_o + l_m \cdot \Delta \dot{T}t_o \tag{2.26}$$

In contrast to Eq. 2.25 this DEq. contains only those variables which have real signal character and not the constant mean values. Consequently, Eq. 2.26 fits better with the definition of the dynamic model, which mainly investigates the fluctuation behaviour of the process and product variables. Thus, Eq. 2.26, represents a view point which is typical for the comparison between actual value and rated value in the automatic control technique. There, only fluctuations around the rated value are also considered [10].

For the calculation of the transfer function the common approach being well-known in the field of DEq. analysis

$$\Delta Tt_o = C_1 \cdot e^{C_2 \cdot t} \tag{2.27}$$

is inserted in the homogeneous DEq. 2.26. Setting the independent cause variable Δv_i equal to zero, we get the so-called **volatile** solution:

$$v_{om} \cdot C_1 \cdot e^{C_2 \cdot t} + l_m \cdot C_1 \cdot C_2 \cdot e^{C_2 \cdot t} = 0 \quad ,$$

which means:

$$C_2 = -\frac{v_{om}}{l_m} \tag{2.28}$$

We insert Eq. 2.28 in Eq. 2.27, extend the approach by the so-called steady state solution and get the modified, complete approach:

$$Tt_o = C_1 \cdot \exp\left(-\frac{v_{om}}{l_m} \cdot t\right) + C_3 \tag{2.29}$$

Aiming on the determination of the coefficients C_1 and C_3 the approach of Eq. 2.29 is inserted into the complete DEq. 2.26:

$$Tt_{im} \cdot \Delta v_i = v_{om}\left[C_1 \cdot \exp\left(-\frac{v_{om}}{l_m} \cdot t\right) + C_3\right] - l_m \cdot \frac{v_{om}}{l_m} \cdot C_1 \cdot \exp\left(-\frac{v_{om}}{l_m} \cdot t\right)$$

For $t \to \infty$ we get:

$$C_3 = \frac{\Delta v_i}{v_{om}} \cdot Tt_{im} \tag{2.30}$$

Inserting Eq. 2.30 into the approach of Eq. 2.29 leads to:

$$\Delta Tt_o = C_1 \cdot \exp\left(-\frac{v_{om}}{l_m} \cdot t\right) + \frac{\Delta v_i}{v_{om}} \cdot Tt_{im} \tag{2.31}$$

For the determination of the constant C_1 we investigate the solution of Eq. 2.31 for the case of $t = 0$, which corresponds to the beginning of the step-like excitation (Δv_i just applied). ΔTt_o is still equal to zero for $t = 0$, that means:

$$0 = C_1 + \frac{\Delta v_i}{v_{om}} \cdot Tt_{im} \quad \Longrightarrow \quad C_1 = -\frac{\Delta v_i}{v_{om}} \cdot Tt_{im} \tag{2.32}$$

Equation 2.32 inserted in Eq. 2.31 results in the complete time transient function or *step response function*:

$$\underline{\Delta Tt_o\overline{|\Delta v_i}} = \Delta v_i \cdot \frac{Tt_{im}}{v_{om}}\left[1 - \exp\left(-\frac{v_{om}}{l_m} \cdot t\right)\right] \tag{2.33}$$

According to Fig. 2.12 the transfer function Eq. 2.33 represents the solution of the DEq. 2.26 for step-like excitations of the system by the independent cause variable. Equation 2.33 is identical with Eq. 2.9 of Sect. 2.3.2 but includes the more detailed symbols for the several mean values and fluctuations. Before the interpretation of Eq. 2.33 it is shown, how the *complex frequency response* can be derived from the DEq. 2.26. As explained in one of the former sections the complex frequency response represents the DEqs. steady state solution for sinusoidal excitations.

For the cause variable we write:

$$\Delta v_{\mathrm{i}} = \widetilde{\Delta v_{\mathrm{i}}} \cdot \mathrm{e}^{\mathrm{j}\omega t} \quad , \tag{2.34}$$

and for the effect variable:

$$\Delta Tt_{\mathrm{o}} = \widetilde{\Delta Tt_{\mathrm{o}}} \cdot \mathrm{e}^{\mathrm{j}(\omega t+\varphi)} \quad , \tag{2.35}$$

where ω is the excitation frequency and φ is the phase shift angle. The Eqs. 2.34 and 2.35 correspond to an oscillating behaviour of the two variables and are written in the complex form, which will directly lead to the questioned complex frequency response (see for instance [10] for further information). When we use Eqs. 2.34 and 2.35 in Eq. 2.26 we gain:

$$Tt_{\mathrm{im}} \cdot \widetilde{\Delta v_{\mathrm{i}}} \cdot \mathrm{e}^{\mathrm{j}\omega t} = v_{\mathrm{om}} \cdot \widetilde{\Delta Tt_{\mathrm{o}}} \cdot \mathrm{e}^{\mathrm{j}(\omega t+\varphi)} + l_{\mathrm{m}} \cdot \widetilde{\Delta Tt_{\mathrm{o}}} \cdot \mathrm{j}\omega \cdot \mathrm{e}^{\mathrm{j}(\omega t+\varphi)}$$

or

$$Tt_{\mathrm{im}} \cdot \widetilde{\Delta v_{\mathrm{i}}} = \widetilde{\Delta Tt_{\mathrm{o}}} \cdot \mathrm{e}^{\mathrm{j}\varphi}(v_{\mathrm{om}} + \mathrm{j}\omega \cdot l_{\mathrm{m}})$$

Calculating the cause/effect-ratio finally leads to:

$$G(\mathrm{j}\omega) = \frac{\widetilde{\Delta Tt_{\mathrm{o}}}}{\widetilde{\Delta v_{\mathrm{i}}}} \cdot \mathrm{e}^{\mathrm{j}\varphi} = \frac{Tt_{\mathrm{im}}}{v_{\mathrm{om}}} \cdot \frac{1}{1 + \mathrm{j}\omega \cdot \frac{l_{\mathrm{m}}}{v_{\mathrm{om}}}} \tag{2.36}$$

Eqation 2.36 represents the *complex frequency response* of the system under investigation and corresponds to Eq. 2.11 in Sect. 2.4.3.
According to Eq. 2.13 the *amplitude frequency response* is equal to the absolute value of the complex frequency response. For its calculation we have to seperate the real and the imaginary part of Eq. 2.36. This can be reached by **extension** of the fraction by the factor $\left(1 - \mathrm{j}\omega \cdot \dfrac{l_{\mathrm{m}}}{v_{\mathrm{om}}}\right)$:

$$G(\mathrm{j}\omega) = \frac{\widetilde{\Delta Tt_{\mathrm{o}}}}{\widetilde{\Delta v_{\mathrm{i}}}} \cdot \mathrm{e}^{\mathrm{j}\varphi} = \frac{Tt_{\mathrm{im}}}{v_{\mathrm{om}}} \cdot \frac{1 - \mathrm{j}\omega \cdot \dfrac{l_{\mathrm{m}}}{v_{\mathrm{om}}}}{\left(1 + \mathrm{j}\omega \cdot \dfrac{l_{\mathrm{m}}}{v_{\mathrm{om}}}\right)\left(1 - \mathrm{j}\omega \cdot \dfrac{l_{\mathrm{m}}}{v_{\mathrm{om}}}\right)}$$

$$G(\mathrm{j}\omega) = \frac{\widetilde{\Delta Tt_\mathrm{o}}}{\widetilde{\Delta v_\mathrm{i}}} \cdot \mathrm{e}^{\mathrm{j}\varphi} = \frac{Tt_\mathrm{im}}{v_\mathrm{om}} \cdot \frac{1 - \mathrm{j}\omega \cdot \dfrac{l_\mathrm{m}}{v_\mathrm{om}}}{1 + \left(\omega \cdot \dfrac{l_\mathrm{m}}{v_\mathrm{om}}\right)^2} \qquad (2.37)$$

The absolute value of Eq. 2.37 leads to the *amplitude frequency response* due to Eq. 2.13:

$$|G(\mathrm{j}\omega)| = \left|\frac{\widetilde{\Delta Tt_\mathrm{o}}}{\widetilde{\Delta v_\mathrm{i}}}\right| = \frac{Tt_\mathrm{im}}{v_\mathrm{om}} \cdot \frac{1}{\sqrt{1 + \left(\omega \cdot \dfrac{l_\mathrm{m}}{v_\mathrm{om}}\right)^2}} \qquad (2.38)$$

Employing Eq. 2.15 the *phase frequency response* is:

$$\varphi(\omega) = \arctan\left[-\frac{\omega \cdot l_\mathrm{m}}{v_\mathrm{om}}\right]. \qquad (2.39)$$

A more elegant and, mainly for the solution of DEqs. of higher order, faster technique for the calculation of the complex frequency response is the LAPLACE-transformation. Cutting out any details, the method is now explained for the (quite simple) DEq. 2.25.

The derivatives with respect to the time $\dfrac{\mathrm{d}^u}{\mathrm{d}t^u}$ are replaced by the so-called LAPLACIAN p^u. Because only first derivatives occur ($\Delta\dot{T}t_\mathrm{o}$) results here $u = 1$:

$$Tt_\mathrm{im} \cdot \Delta v_\mathrm{i} = v_\mathrm{om} \cdot \Delta Tt_\mathrm{o} + p \cdot l_\mathrm{m} \cdot \Delta Tt_\mathrm{o} \quad ,$$

where $p \cdot \Delta Tt_\mathrm{o}$ stands for $\Delta\dot{T}t_\mathrm{o}$ or $\dfrac{d(\Delta Tt_\mathrm{o})}{\mathrm{d}t}$.

It follows:

$$G(p) = \frac{\Delta Tt_\mathrm{o}}{\Delta v_\mathrm{i}} = \frac{Tt_\mathrm{im}}{v_\mathrm{om}} \cdot \frac{1}{1+p} \cdot \frac{l_\mathrm{m}}{v_\mathrm{om}} \qquad (2.40)$$

It can be easily seen that Eq. 2.40, which is called dynamic transfer function $G(p)$ in a narrower sense, proceeds to the complex frequency response $G(\mathrm{j}\omega)$ (Eq. 2.36) when the LAPLACIAN p^u is replaced by the complex frequency $(\mathrm{j}\omega)^u$. The advantage of this direct method, which indirectly includes all boundary conditions, is obvious.

If we look for the step response we will need a transformation from the frequency range back to the time range. For that purpose the retransformation integral of the LAPLACE-transformation must be applied. In our special case that integral is:

$$\underline{\Delta Tt_{\mathrm{o}}|\overline{\Delta v_{\mathrm{i}}}} = \frac{\Delta v_{\mathrm{i}}}{2\pi \mathrm{j}} \int\limits_{-\mathrm{j}\infty}^{+\mathrm{j}\infty} \frac{G(p)}{p} \cdot \mathrm{e}^{pt} \cdot \mathrm{d}p \quad , \tag{2.41}$$

or with Eq. 2.40:

$$\underline{\Delta Tt_{\mathrm{o}}|\overline{\Delta v_{\mathrm{i}}}} = \frac{\Delta v_{\mathrm{i}} \cdot Tt_{\mathrm{im}}}{2\pi \mathrm{j} \cdot v_{\mathrm{om}}} \int\limits_{-\mathrm{j}\infty}^{+\mathrm{j}\infty} \frac{\mathrm{e}^{pt}}{p \cdot \left(1 + p \cdot \dfrac{l_{\mathrm{m}}}{v_{\mathrm{om}}}\right)} \cdot \mathrm{d}p \tag{2.42}$$

The value of the integral Eq. 2.42 can either be extracted from the appropriate literature [11], [12] or solved with the *residue theorem*of the LAPLACE-transformation [13]. The residue theorem states that the value of the integral is equal to the sum of all residues multiplied by the factor $2\pi\mathrm{j}$.

This means for the given example:

$$\underline{\Delta Tt_{\mathrm{o}}|\overline{\Delta v_{\mathrm{i}}}} = \Delta v_{\mathrm{i}} \cdot \frac{Tt_{\mathrm{im}}}{v_{\mathrm{om}}} \cdot \sum_{\nu=1}^{2} \operatorname*{Res}_{p_\nu}[S(p)] \quad , \tag{2.43}$$

where p_ν are the zero values of the integrand of Eq. 2.42. Those are:

$$p_1 = 0 \quad \text{and} \quad p_2 = -\frac{v_{\mathrm{om}}}{l_{\mathrm{m}}}$$

The general equation for the calculation of the residues is:

$$\operatorname*{Res}_{p_\nu}[S(p_\nu)] = \lim_{p \to p_\nu} (p - p_\nu) \cdot S(p)$$

Here $S(p)$ represents the whole integrand of Eq. 2.42. Now, the two residues can be calculated:

$$\operatorname*{Res}_{p_1}[S(p_1)] = \lim_{p \to 0} \frac{(p-0) \cdot \mathrm{e}^{pt}}{p \cdot \left(1 + p \cdot \dfrac{l_{\mathrm{m}}}{v_{\mathrm{om}}}\right)} = 1 \quad , \tag{2.44}$$

$$\operatorname*{Res}_{p_2}[S(p_2)] = \lim_{p \to -\frac{v_{\mathrm{om}}}{l_{\mathrm{m}}}} \frac{\left(p + \dfrac{v_{\mathrm{om}}}{l_{\mathrm{m}}}\right) \cdot \mathrm{e}^{pt}}{p \cdot \left(1 + p \cdot \dfrac{l_{\mathrm{m}}}{v_{\mathrm{om}}}\right)} = -\exp\left(-\frac{v_{\mathrm{om}}}{l_{\mathrm{m}}} \cdot t\right) \tag{2.45}$$

The input of Eqs. 2.44 and 2.45 into the Eq. 2.43 leads to the transfer function

$$\underline{\Delta Tt_{\mathrm{o}}|\overline{\Delta v_{\mathrm{i}}}} = \Delta v_{\mathrm{i}} \cdot \frac{Tt_{\mathrm{im}}}{v_{\mathrm{om}}} \left[1 - \exp\left(-\frac{v_{\mathrm{om}}}{l_{\mathrm{m}}} \cdot t\right)\right] \tag{2.46}$$

Obviously, Eq. 2.46 is the same as Eq. 2.43, which has been acquired by the classical approach with the DEq. 2.25.

Evaluation of Results

We are now ready to deal with the sixth step (application of the model, simulations with respect to technological questions) of rules for the methodical procedure for the dynamic modelling given in Sect. 2.5.1. For that purpose the potential of the mathematical solutions found in the previous section is demonstrated by means of a concrete example. Let us assume a drawing process of a synthetic yarn characterised by the following process and product variables:

mean input yarn fineness	$Tt_{\mathrm{im}} = 30$ tex
mean output yarn fineness	$Tt_{\mathrm{om}} = 10$ tex
($\Rightarrow$ mean draw ratio = 3)	
mean input velocity	$v_{\mathrm{im}} = 300$ m/min
mean output velocity	$v_{\mathrm{om}} = 900$ m/min
assumed step-like or sinusoidal	
shift of the input velocity	Δv_{i} or $\widetilde{\Delta v_{\mathrm{i}}} = 15$ m/min.

Figure 2.13 shows a transfer function calculated with Eq. 2.33 by means of the above values for $l_{\mathrm{m}} = 0.3$ m.

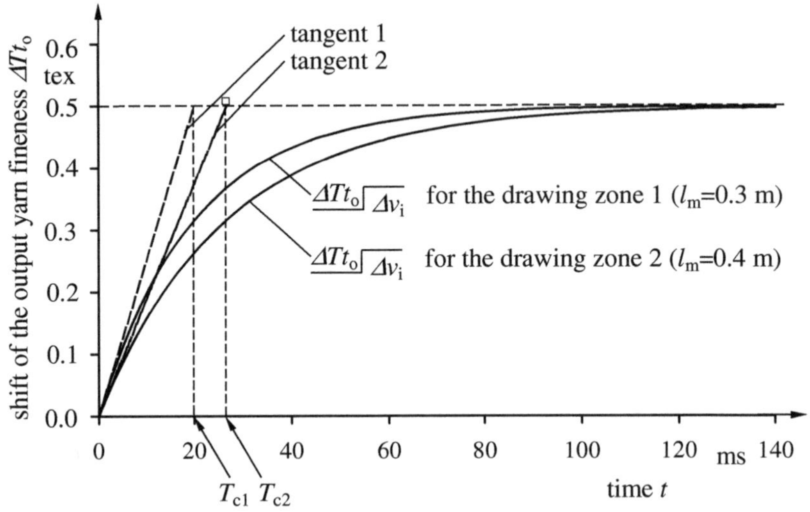

Fig. 2.13. Drawing process; time transient function of the fineness $\Delta Tt_{\mathrm{o}}(t)$ by means of an input velocity step $\Delta v_{\mathrm{i}} = 15$ m/min:
$Tt_{\mathrm{im}} = 30$ tex; $Tt_{\mathrm{om}} = 10$ tex; $v_{\mathrm{im}} = 300$ m/min; $v_{\mathrm{om}} = 900$ m/min

The following information can be extracted from the curves:

a) The change of the output yarn fineness ΔTt_{o} shows an exponential behaviour and reaches a steady state value of 0.5 tex. Consequently, if the disturbance of the input velocity is 15 m/min the outgoing yarn will have a constant fineness of 10.5 tex after a certain transition time.

b) The transition period between leaving the old technological operating point ($Tt_o = 10$ tex) and reaching the new disturbed one ($Tt_o = 10.5$ tex) lasts 100 ms.
c) The proportional change of the fineness after the end of the transition period is equal to the proportional change of the input velocity (5%).

There are a number of further considerations to be made. First, the transition time of 100 ms seems to be quite short. But we have to bear in mind that already 1.5 m of the fibre with fineness values differing from the desired value have left the drawing zone during that period. If the input velocity jumped back from the disturbed value of 315 m/min to the initial value of 300 m/min after those 100 ms we would have to wait for further 100 ms until the output yarn fineness had reached its original value of 10 tex again. Thus, an only 100 ms lasting change of the input velocity to a 5% higher value would have caused a 3 m fibre segment with undefined fineness and – which is even worse for synthetic silks – with different orientation and structure characteristics, which possibly show negative effects on the staining homogeneity. Here, the terms of skitteriness and barre suggest themselves and do not have to be interpreted in detail. As for the rest, the changes of the input velocity cause slippage effects of the fibre on the input godet roll (in diminishing direction), which are often barely recognised, appear for extremely short periods and can be hardly measured.

The parameter in the transfer function which finally determines how fast the effect variable can follow the step-like cause variable, is in our case (and in many other similar cases) the exponent of the exponential function. The reciprocal value of the factor v_{om}/l_m is called the *time constant* T_c of the system:

$$T_c = \frac{l_m}{v_{om}} \tag{2.47}$$

This means for our example:

$$T_c = \frac{0.3 \text{ m}}{900 \text{ m/min}} = 3.33 \cdot 10^{-4} \text{ min} = 0.02 \text{ s} = 20 \text{ ms}.$$

As can be checked easily there is the following correlation between the time since the beginning of a disturbance and the percentage of the alteration range being passed by the effect variable:

$$\begin{aligned} 1 \cdot T_c &\Longleftrightarrow 63\%, \\ 3 \cdot T_c &\Longleftrightarrow 95\%, \\ 5 \cdot T_c &\Longleftrightarrow 99\%. \end{aligned}$$

Therefore it seems to be justified to consider the transition process as finished after a period of $5{\cdot}T_c$ (here: 100 ms). For less exact approximations (error: 5%) even $3{\cdot}T_c$ can be assumed as long enough. Consequently the time constant can be extracted from a purely experimentally acquired transfer function with sufficient exactness (in case of an exponential behaviour, which has to be tested before): One determines the point of the curve, where 63% of the whole alteration range has been passed, and extracts the corresponding time value from the abscissa (Fig. 2.13). Another (less exact) method is the construction of the tangent on the exponential function in the zero value of time. That tangent also meets the parallel to the abscissa which corresponds to the steady state final value of the alteration range of the effect variable at the distance T_c.

Eqation 2.47 shows how the dynamic behaviour of the system (here: the drawing process) can be changed. A faster responding system reacts with a diminishing of the time constant which is equivalent to a shortening of the drawing zone and/or an enlargement of the output velocity, whereas a slower responding system reacts with an enlargement of the time constant which is equivalent to a longer drawing zone and/or a smaller output velocity. If, for a certain reason, the fibre length, being influenced by disturbances, has to be changed, for instance minimised, this will only be possible by an adequate diminishing of l_m, as the disturbance fibre length is proportional to the product $T_c \cdot v_{om}$. The right curve of Fig. 2.13 illustrates how the transition process changes when the drawing zone is enlarged to 0.4 m. For this special case T_c is 26.7 ms and the transition process is practically finished after 133 ms.

Figures 2.14 to 2.16 show the solutions in the frequency range: the frequency response in its transfer locus representation (Fig. 2.14), the amplitude frequency response (Fig. 2.15) and the phase frequency response (Fig. 2.16).
The transfer locus as a summarising representation teaches us that

a) the oscillation $\widetilde{\Delta Tt_o}$, which is caused by sinusoidal exciting oscillation with constant amplitude $\widetilde{\Delta v_i}$ (vector on the real axis), gets smaller for growing excitation frequencies ω and finally vanishes for $\omega \to \infty$,
b) the phase shift angle φ, which is a measure for the delay of the effect oscillation behind the (exciting) cause oscillation, grows from initially zero (for $\omega = 0$, quasi steady state excitation) to a value of $-\frac{\pi}{2}$ (for $\omega \to \infty$).

Amplitude and phase frequency response show those statements separately. The ordinate values of the amplitude frequency response are explicitly given as the amplitude ratio $\widetilde{\Delta Tt_o}/\widetilde{\Delta v_i}$ for our example. The maximum value for this ratio, appearing at $\omega = 0$, is $\frac{0.5}{15}\frac{\text{tex}}{\text{m/min}} = 0.033\frac{\text{tex}}{\text{m/min}}$. This means that at the beginning of the amplitude frequency response, therefore at small

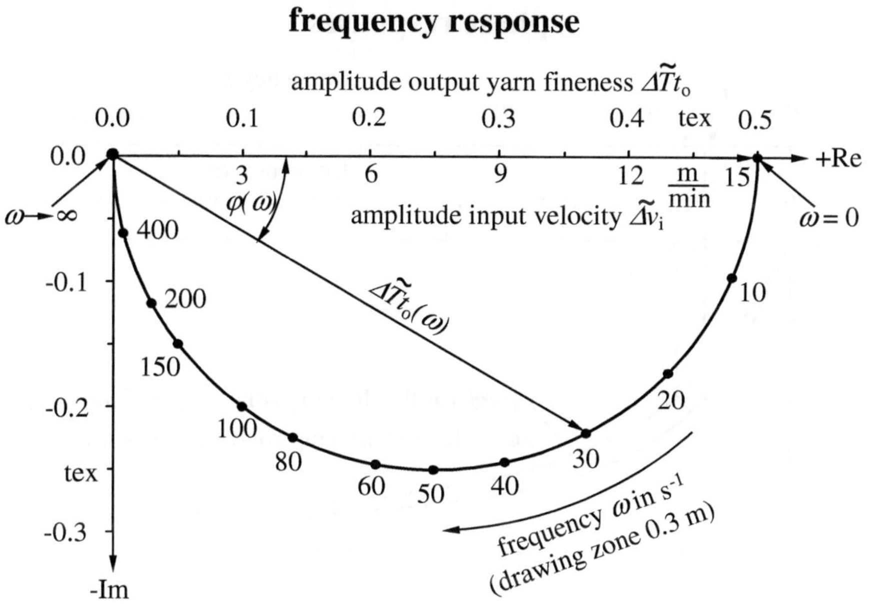

Fig. 2.14. Drawing process; transfer locus of the fineness $\widetilde{\Delta Tt_o}$ by means of periodic input velocity disturbances $\widetilde{\Delta v_i} = 15$ m/min:
$Tt_{im} = 30$ tex; $Tt_{om} = 10$ tex; $v_{im} = 300$ m/min; $v_{om} = 900$ m/min

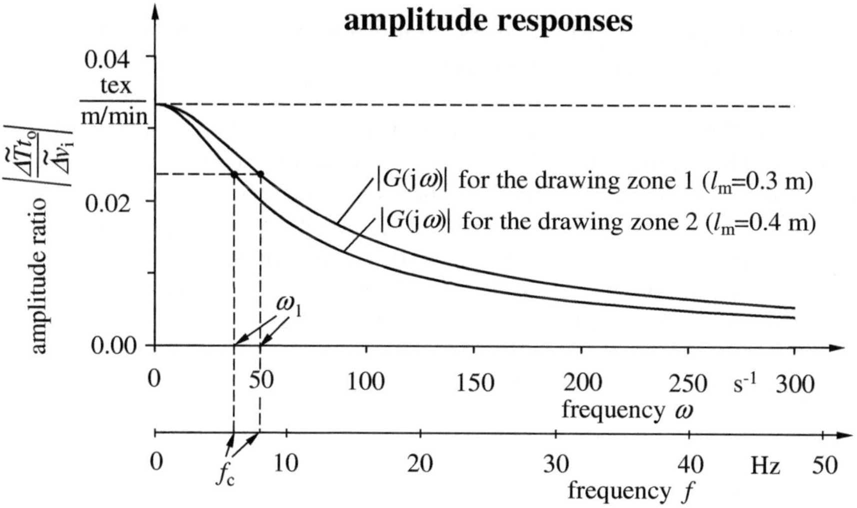

Fig. 2.15. Drawing process; amplitude frequency responses of the fineness $\widetilde{\Delta Tt_o}$ by means of periodic input velocity disturbances $\widetilde{\Delta v_i} = 15$ m/min:
$Tt_{im} = 30$ tex; $Tt_{om} = 10$ tex; $v_{im} = 300$ m/min; $v_{om} = 900$ m/min

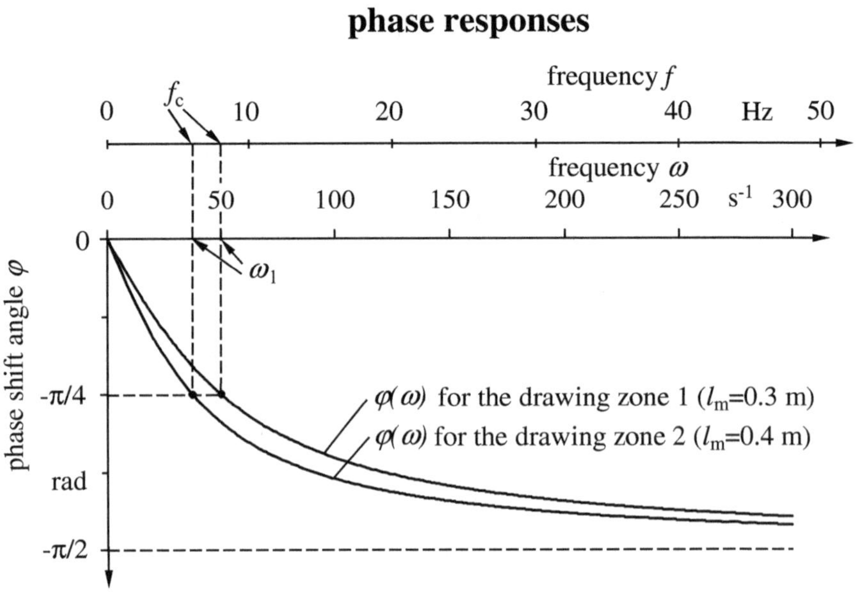

Fig. 2.16. Drawing process; phase frequency responses of the fineness $\widetilde{\Delta Tt_o}$ by means of periodic input velocity disturbances $\widetilde{\Delta v_i} = 15$ m/min: $Tt_{im} = 30$ tex; $Tt_{om} = 10$ tex; $v_{im} = 300$ m/min; $v_{om} = 900$ m/min

disturbance frequencies, a fluctuation amplitude $\widetilde{\Delta v_i}$ =1 m/min results in a fluctuation amplitude $\widetilde{\Delta Tt_o}$ =0.033 tex.

The so-called critical (circuit) frequency ω_c can be taken as a characteristic value for the estimation of the dynamic system properties in the frequency range representation. It is defined as the excitation frequency, at which the effect oscillation amplitude has fallen to $1/\sqrt{2}$ of the value being valid for the case of steady state excitation (for $\omega = 0$). In the transfer locus (Fig. 2.14) the imaginary part of the vector for the effect oscillations is equal to its real part at this point. This is the case for Eq. 2.37 with

$$\omega \cdot \frac{l_m}{v_{om}} = 1 \Longrightarrow \omega = \omega_c = \frac{v_{om}}{l_m}$$

or, for Eq. 2.47 with:

$$\omega_c = \frac{v_{om}}{l_m} = \frac{1}{T_c} \tag{2.48}$$

There, the phase shift angle is $\varphi_l = -\frac{\pi}{4} = -45^\circ$. For our technological example we get: $\omega_c = \frac{900}{0.3}\frac{\text{m/min}}{\text{m}} = 3000\ \text{min}^{-1} \cong 50\ \text{s}^{-1}$. It is clearer to give the values for the circular frequency ω in the more familiar dimension

Hertz corresponding to the number of oscillations per time unit f. The well-known relation $\omega = 2\pi f$ lets Eq. 2.48 appear as

$$f_c = \frac{\omega_c}{2\pi} = \frac{v_{om}}{2\pi \cdot l_m} \tag{2.49}$$

For the example we get $f_c = \frac{50}{2\pi}\text{Hz} = 8$ Hz. The frequency abscissas of Figs. 2.15 and 2.16 contain both values: ω and f. Practically, this means that the drawing zone transfers periodic sinusoidal fluctuations of the input velocity $\widetilde{\Delta v_i} \cdot \sin(\omega t)$ or $\widetilde{\Delta v_i} \cdot \sin(2\pi f \cdot t)$ at frequencies of 0 Hz $\leqq f \leqq$ 8 Hz to the fineness of the outgoing fibre with a transfer factor of at least 0.7 related to the maximum amplitude ratio for $f = 0$ Hz. The input velocity would fluctuate between 285 and 315 m/min. The fineness of the outgoing fibre would fluctuate for a disturbance frequency of

$\approx$ 0 Hz between 9.5 tex and 10.5 tex,
8 Hz between 9.65 tex and 10.35 tex.

If the disturbance frequency exceeds f_c the cause oscillation will be transferred less to the effect variable ΔTt_o. Consequently the drawing zone with its stored fibre mass will dampen the disturbance the better as the disturbance frequency reaches higher values. This behaviour corresponds to the part of the amplitude frequency response which approaches zero for $f > f_c$. As in most practical cases such a system behaviour is formally desired for the following appropriate measure, which directly follows from the explained relations, can be recommended: The *critical frequency* of the technological system for the critical cause-effect relations should be as *small* as possible because then the desired dampening for the dynamic disturbance transfer already begins at lower disturbance frequencies.

After the Eqs. 2.47 or 2.48 this means for the drawing zone: diminishing of v_{om} and/or enlargement of l_m. As the first one lowers the productivity, an enlargement of l_m should be the aim. It is quite clear that other aspects, for instance the technical conditions must be considered as well and consequently compromises have to be made. This point is not further discussed at this place. For comparison, Fig. 2.15 includes an amplitude frequency response which is valid for a drawing process with a drawing zone being enlarged to $l_m = 0.4$ m. As can be calculated with Eq. 2.48 the critical frequency drops to $f_c = 6$ Hz for this case, so an effective dampening of disturbances can be expected already for that frequency.

In a similar way as the time constant of the transfer function does, the phase frequency response $\varphi(\omega)$ (Eq. 2.39) allows statements about the thread length, which leaves the drawing zone before that point of the thread, which corresponds to a disturbance, appears at the end of the drawing zone. This thread length is called **delay thread length** L_d and is calculated as follows:

$$L_{\mathrm{d}} = \frac{|\varphi(\omega)|}{2\pi \cdot \omega} \cdot v_{\mathrm{om}} = \frac{|\varphi(\omega)|}{4\pi^2 \cdot f} \cdot v_{\mathrm{om}} \tag{2.50}$$

For our example we get:

$L_{\mathrm{d}} = 37$ mm for $f_{\mathrm{c}} = 8$ Hz (drawing zone 0.3 m),
$L_{\mathrm{d}} = 50$ mm for $f_{\mathrm{c}} = 6$ Hz (drawing zone 0.4 m).

In Sect. 5.1.5 other problems for which the phase frequency response plays an important role will be discussed.

The next sections cover important aspects of the dynamics of main process steps of fibre/yarn/thread formation and processing technologies. There, the mathematical techniques explained so far will be employed again.

3. Modelling of Steady State Fibre Formation Process in Melt Spinning

3.1 Steady State Single Fibre Formation Process

3.1.1 Definition: What Does Fibre Formation Mean?

Melt spinning is the production of continuous solid filaments from polymeric melt. The fibre formation process includes change in shape, structure and properties of the thermoplastic polymer. The polymer pellets or granules are fed into an extruder where, through heating, their melting temperature is exceeded. The polymeric melt is then transported, under pressure, to the spinneret. Hygroscopic polymers require vacuum drying prior to processing in order to ensure a low water content. The extrusion temperature T_0 is roughly 30-50 K above the melting temperature T_m of the polymer, i. e. T_0=250-270°C for PA 6 or 280-295°C for PET, respectively. A constant mass flow rate of the melt is achieved by a metering pump (the spinning pump) which can be positioned inside the spinning head. Within the spinneret the melt flow is channelled into a number of individual capillary holes, each is responsible for the formation of a single filament. After the melt flow passes through these spinneret orifices into the air, the single filaments cool off, solidify, and are collected into a fibre bundle that is finally wound up. The take-up speed is much higher than the average extrusion velocity at the spinneret exit. The ratio between the spinning velocity v_L (take-up velocity at any distance L) and the (average) extrusion velocity v_0 at the spinneret exit defines the *draw down ratio ddr*:

$$ddr = \frac{v_L}{v_0} . \tag{3.1}$$

Between take-up of the as-spun fibre and the final winding to bobbins an additional drawing procedure (with additional draw ratio DR) may be introduced. The drawing is typically achieved by means of godet pairs. Two or more drawing steps can be performed (Fig. 3.1). Fibre formation in the sense of *spinnability* requires three necessary conditions:

1. Attenuation and acceleration of the filaments, i. e. $ddr > 1$,
2. Stability in time that allows the production of continuous filaments, i. e. no filament breakage,

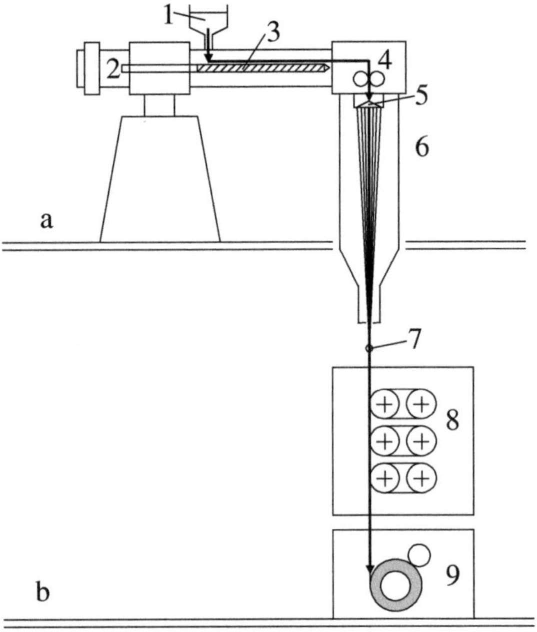

Fig. 3.1. Melt spinning equipment (schematic): a – spinning floor, b – winding floor; 1 – container with polymer pellets, 2 – electrical motor and drive train, 3 – extruder assembly with screw, 4 – spinning head with metering pump (gear pump), 5 – spinneret with capillary holes, 6 – spinning chamber with quenching air, 7 – spin finish applicator, 8 – pairs of godets for online drawing, 9 – (high speed) winder

3. Steady state conditions (stationarity), i. e. uniform filaments without any variation of properties throughout the production time and therefore along the filament length.

The maximum possible draw down ratio for a given polymer is an important characteristic value for the *spinnability* of the polymeric melt. For well-spinnable polymers draw down ratios of several hundred to up to $ddr > 1000$ can be reached in the high speed spinning process. Specifically the high speed fibre spinning shows extremely high deformation and cooling rates. Structural changes and phase transition from liquid to solid state are possibly taking place within *milliseconds*.

3.1.2 Fundamental Balance Equations

At the exit of the capillary holes (at distance $x = 0$) the polymeric melt is extruded with a given constant mass throughput Q and extrusion velocity v_0 at a constant extrusion temperature T_0 (Fig. 3.2). The diameter of the spinneret capillary holes is D_0. The take-up device (godets, winder or air suction device) positioned at any distance L from the spinneret, determines the final take-up velocity v_L of the as-spun filaments. The velocity $v(x)$, the diameter $D(x)$, the temperature $T(x)$, and the filament force $F(x)$ depend upon the axial distance x from the spinneret. Their courses play an essential role in the development of the fibre structure and the resulting textile yarn properties. The theory of steady state fibre formation helps to describe the deformation, cooling, and stress developing processes. It is, in principle, an application of the fundamental physical balance equations [274] of *mass, energy,* and *momentum*, combined with the stress-deformation behaviour of the polymer and the description of its structural changes.

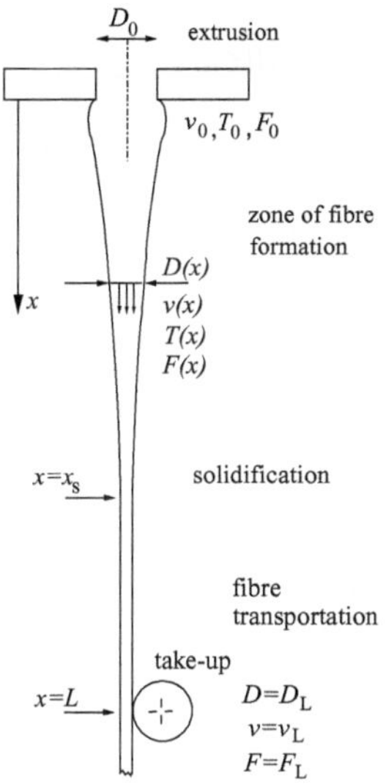

Fig. 3.2. Fibre formation in melt spinning

Mass Balance

The mass balance represents the continuity equation of the melt spinning process. It is largely simplified because no mass exchange takes place between the filament and its environment. The mass conservation formula connects the most important quantities, mass throughput Q, filament cross sectional area $A(x)$ (or filament diameter $D(x)$, respectively), and the (averaged over the cross section) axial filament velocity $v(x)$:

$$Q = \varrho_P(x) \cdot A(x) \cdot v(x) = \text{const.} \tag{3.2}$$

The quantity ϱ_P in Eq. 3.2 denotes the mass density of the polymer. The product $\varrho_P(x) \cdot A(x) = Tt(x)$ is known in textile engineering as fineness (titre). For a circular cross section with diameter $D(x)$ follows

$$Q = \varrho_P \cdot \frac{\pi}{4} D^2 \cdot v = Tt \cdot v \,. \tag{3.3}$$

The fineness Tt quantifies the linear mass density (mass per unit length of the fibre) and has its own special units. The most common are *tex* (resp. *decitex*) and *denier* (den), defined as

$$1\,\mathrm{tex} = \frac{1\,\mathrm{g}}{1000\,\mathrm{m}}\,, \qquad 1\,\mathrm{dtex} = \frac{1\,\mathrm{g}}{10\,000\,\mathrm{m}}\,, \qquad 1\,\mathrm{den} = \frac{1\,\mathrm{g}}{9000\,\mathrm{m}}\,.$$

The fineness of a fibre is called 1 dtex if 10 000 meters of fibre material is equal to 1 gram (resp. the fineness 1 den equals 9 000 meters of 1 gram mass). The relation between *fineness* Tt (in dtex or denier), and *diameter* D (in µm) depends on the mass density ϱ_{P} (in g/cm^3) and is given by

$$Tt = 0.0078\,\varrho_{\mathrm{P}}\,D^2, \quad D = 11.3\sqrt{Tt/\varrho_{\mathrm{P}}} \quad \text{for} \quad Tt \quad \text{in dtex}, \tag{3.4a}$$

$$Tt = 0.0112\,\varrho_{\mathrm{P}}\,D^2, \quad D = 9.44\sqrt{Tt/\varrho_{\mathrm{P}}} \quad \text{for} \quad Tt \quad \text{in denier}. \tag{3.4b}$$

Two examples (with given mass densities) are listed in Table 3.1.

Table 3.1. Equivalence between fineness and diameter

	Fineness Tt	Diameter D
PP – poly(propylene)	1 dtex	12 µm
($\varrho_{\mathrm{P}} = 0.9\,\mathrm{g/cm^3}$)	10 dtex	37 µm
	100 dtex	119 µm
PET – poly(ethylene terephthalate)	1 dtex	10 µm
($\varrho_{\mathrm{P}} = 1.35\,\mathrm{g/cm^3}$)	10 dtex	31 µm
	100 dtex	97 µm

Due of the fundamental character and simplicity of the mass balance Ziabicki [274] distinguishes the principal process variables by means of the continuity equation into *primary*, *secondary*, and *resulting* variables.

Primary variables

- describe the material, its chemical and molecular structure (molecular weight, molecular weight distribution), and the material's characteristics like viscosity, heat capacity, solidification behaviour etc.
- Primary variables are also the parameters which determine the technological conditions of the melt spinning process. These are the mass throughput Q delivered by the spinning pump, the extrusion temperature T_0, the dimensions of the capillary holes (diameter D_0), the length of the spinning line L from spinneret to the take-up device, the take-up velocity v_{L}, and the conditions of the cooling process, its velocity profile v_{air} of quenching air, the temperature profile T_{air}, and the air humidity.

Secondary spinning variables result from the primary variables by application of the continuity equation (3.2). Examples are:

- the (average) extrusion velocity v_0 with

$$v_0 = \frac{Q}{\varrho_P \frac{\pi}{4} D_0^2} ,$$

- the as-spun fineness $Tt_L = Q/v_L$, and the corresponding diameter D_L,
- the draw down ratio *ddr* (3.1) with

$$ddr = \frac{v_L}{v_0} = \frac{v_L}{Q} \varrho_P \frac{\pi}{4} D_0^2 .$$

The formulas show that the secondary variables are combinations of the primary variables, specifically the draw down ratio.

Resulting variables are determined by primary variables as well as the dynamics of the fibre formation process. Some examples are:

- the development of the fibre velocity $v(x)$ along the spinning path,
- the length of the fibre formation zone, i. e. the distance between the spinneret and the solidification point,
- the maximum deformation rate of the polymeric downstream,
- the filament tension force $F(x)$ and the filament stress $\sigma(x)$ at any distance x from spinneret, especially at the solidification point and at the take-up point,
- the filament temperature $T(x)$.

The development of stress and temperature and the structural changes determine the physical and textile properties such as orientation, crystallinity, elongation at break, tenacity and many others.

Modelling the dynamics of fibre formation should lead to a sufficient description of the resulting variables and their correlations to the fibre properties. The main goal is helping to understand the influence of primary variables (material properties and technological process parameters) on the resulting product properties.

Energy Balance

The energy equation describes the development of the filament temperature $T(x)$ from the point of exiting the capillary holes to the points of solidification and take-up.

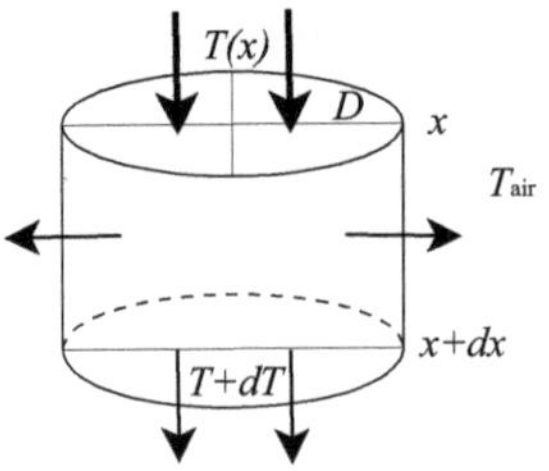

Fig. 3.3. Heat flow balance of a filament volume element (schematic)

Simplified analysis. If we at first assume only heat convection (i. e. no heat conduction and no heat sources within the filament, no heat radiation), then the heat balance of any volume element between x and $x + dx$ is given by the following equation: The difference of heat flow into the unit volume and out of it must be equal to the total heat loss of the unit volume, that is the heat transfer from the surface of the unit volume into the surrounding air (Fig. 3.3).
The heat balance equation can be written as

$$Q \cdot c_\mathrm{p} \cdot T - Q \cdot c_\mathrm{p} \cdot (T + \mathrm{d}T) = \alpha \cdot (T - T_\mathrm{air}) \cdot \pi \cdot D \cdot \mathrm{d}x\ . \tag{3.5}$$

Using the so-called (non-dimensional) NUSSELT number Nu with

$$Nu = \frac{\alpha \cdot D}{\lambda_\mathrm{air}} \tag{3.6}$$

leads to

$$\frac{\mathrm{d}T}{\mathrm{d}x} = -(T - T_\mathrm{air}) \cdot Nu \frac{\pi\, \lambda_\mathrm{air}}{Q \cdot c_\mathrm{p}}\ . \tag{3.7}$$

In Eqs. 3.5–3.7 T_air is the temperature of surrounding air, α is the heat transfer coefficient from the filament surface to the surrounding air, c_p represents the specific heat capacity of the polymer, and λ_air stands for the heat conductivity of air.

The NUSSELT number (or the heat transfer coefficient α) is an essential parameter for calculating the filament temperature profile $T(x)$. A more detailed discussion of heat transfer will follow in Sect. 3.1.3. Here it should be pointed out that the fraction on the right hand side of Eq. 3.7 has the dimension of a reciprocal length:

$$Nu \frac{\pi\, \lambda_\mathrm{air}}{Q \cdot c_\mathrm{p}} = \frac{1}{L_\mathrm{c}}\ . \tag{3.8}$$

Equation (3.7) then becomes

$$\frac{\mathrm{d}T(x)}{\mathrm{d}x} = -\frac{T(x) - T_\mathrm{air}(x)}{L_\mathrm{c}(x)} \quad \text{with} \quad T(0) = T_0\ . \tag{3.9}$$

Example. To get an impression of the cooling of poly(ethylene terephthalate) (PET) filaments the following typical values should be used:

specific heat capacity (PET)	c_p	1500 kJ/kg K
temperature of the melt (PET)	T_0	290 °C
air temperature	T_air	20 °C
heat conductivity of air	λ_air	10^{-5} W/m
NUSSELT number	Nu	≈ 1

Using the parameters above (see also Eqs. 2.3 and 2.6) the cooling length for PET per mass throughput becomes $L_\mathrm{c}/Q \approx 0.2\,\mathrm{m/g \cdot min^{-1}}$. In order to cool off a filament from its initial temperature of 290 °C to a solidification temperature of 70 °C (glass transition temperature of PET) at a throughput of 1 g/min, a cooling length of approximately 0.5 m is needed. As the throughput increases, the required cooling length increases as well, for $Q = 2$ g/min follows $L_\mathrm{c} \approx 1.0$ m, and so on.

Further discussions about energy balance. Heat is transferred not only via convection but also by heat radiation and heat conduction.

Heat radiation strongly depends upon the temperature (power law with T^4 dependence – STEPHAN-BOLTZMANN law). Radiation plays an important role for glass or metal spinning processes where the temperature can reach 1200 °C or more. But in melt spinning of thermoplastics with spinning temperatures of up to 300 °C the contribution of radiation is, consequently, often neglected.

Heat conduction occurs as inner conduction inside the fibres and also as outer conduction in such cases where fibres are brought into contact with other materials of different temperature, like fibre guides, metal plates or godets. The inner conduction is negligible for thin filaments, but needs to be taken into consideration for thick and very thick filaments. The contact of filaments with other objects is sometimes used to force the cooling process. But it is also possible to heat up the fibres in the contact area to initiate phase transitions or to achieve certain drawing effects.

A sensitive contribution to the energy balance of the fibre can arise from the internal heat which is set free when the filament crystallises. Stress-induced crystallisation appears especially in high speed spinning of PET, PA, PP and other crystallising polymers. It is necessary to take the crystallisation heat [59,194–198,200] into consideration for the modelling of such processes. The energy equation (3.7) then has to be expanded with the crystallisation term to

$$\frac{\mathrm{d}T}{\mathrm{d}x} = -(T - T_\mathrm{air}) \cdot Nu \frac{\pi\,\lambda_\mathrm{air}}{Q \cdot c_\mathrm{p}} + \frac{\Delta H}{c_\mathrm{p}} \frac{\mathrm{d}X_\mathrm{c}}{\mathrm{d}x}\,, \tag{3.10}$$

where ΔH is the heat of fusion and X_c is the degree of crystallinity.

The deformation energy can also be an additional source of internal heat. However, in melt spinning the contribution of internal friction processes can usually be neglected.

Momentum Balance

The momentum balance equation can be integrated in order to determine the forces acting on the fibre. In this manner, the forces play an essential role in structure development: the resulting fibre stress σ (that is fibre force F divided by fibre cross-sectional area A) leads to the deformation of the polymer and at least determines the orientational status and the structural arrangement of the polymeric chains.

The analysis of the force contributions reveals the following components which add-up to the total fibre force [274]:

- the initial force at the capillary exit F_0,
- the inertial force F_{inert},
- the gravitational force F_{grav},
- the air drag force F_{drag},
- the surface tension force F_{surf},
- the take-up force F_{L}

Figure 3.4 shows how the forces act on the fibre and how they contribute to the total fibre force.

$$\begin{aligned} F(x) &= F_0 + F_{\text{surf}}(x) + F_{\text{inert}}(x) + F_{\text{drag}}(x) - F_{\text{grav}}(x) && (3.11\text{a}) \\ &= F_{\text{L}} - F_{\text{surf}}(x) - F_{\text{inert}}(x) - F_{\text{drag}}(x) + F_{\text{grav}}(x) && (3.11\text{b}) \\ &= F_{\text{rheo}}(x) \,. && (3.11\text{c}) \end{aligned}$$

Discussion. Some more details should be discussed with respect to each force contribution and its effect on the total force should be discussed by means of simple estimation.

The *surface tension force* F_{surf} should be regarded at first. It is caused by the enlargement of the filament surface per unit volume during stretching and thinning of the fibre:

$$\frac{\mathrm{d}F_{\text{surf}}}{\mathrm{d}x} = \pi \frac{\mathrm{d}}{\mathrm{d}x}\left(\sigma_{\text{surf}} R\right) \tag{3.12}$$

or, after integration with respect to x (assuming constant σ_{surf}) follows

$$F_{\text{surf}}(x) = \pi\, \sigma_{\text{surf}} \big(R_0 - R(x)\big), \tag{3.13}$$

where σ_{surf} is the surface tension (or specific surface energy) of the material. The contribution of the surface tension force to the total force is usually low, except for very low viscous materials. A typical value for poly(ethylene

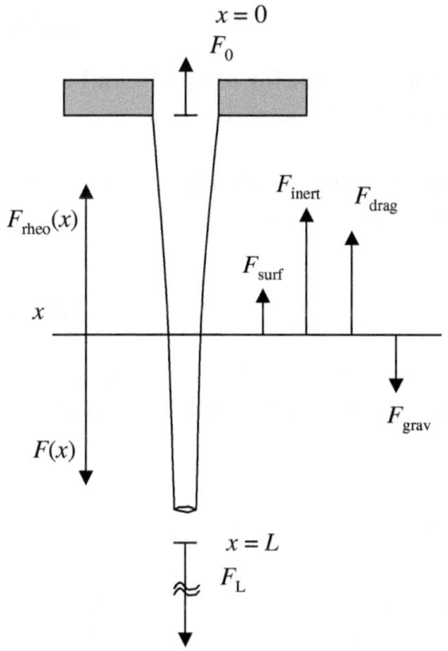

Fig. 3.4. Forces acting on a fibre

terephthalate) (PET) has a magnitude of about $\sigma_{\mathrm{surf}} \approx 0.05$ mN/mm, and with an assumed initial radius of $R_0 = 0.15$ mm one gets a surface tension F_{surf} lower then 0.02 mN.

The *gravitational force* F_{grav} at any distance x is the weight of the filament at this point and can be expressed as

$$F_{\mathrm{grav}}(x) = \int_0^x \varrho_{\mathrm{P}}\, g\, A(x')\, dx' \,, \tag{3.14}$$

where ϱ_{P} is the mass density of the polymer, g is the gravitational acceleration on earth ($g \approx 9.81$ m/s^2), and A denotes the filament cross-sectional area. Assuming a circular PET filament (mass density $\varrho_{\mathrm{P}} = 1.35$ g/cm^3) with an averaged radius of 10 µm and a length of about 1 m, then one can estimate a weight of about 0.01 mN, which is also the order of magnitude of the contribution of F_{grav}.

The *inertial force* F_{inert} is caused by acceleration of the polymeric material from the initial velocity v_0 at the exit point of the spinneret to the velocity $v(x)$ at any distance x, at least to the final take-up velocity v_{L}. The changes in speed directly affect the inertial force:

$$\frac{\mathrm{d}F_{\mathrm{inert}}}{\mathrm{d}x} = Q\,\frac{\mathrm{d}v}{\mathrm{d}x} \tag{3.15}$$

or, after integration with respect to distance x follows

$$F_{\text{inert}}(x) = Q \cdot \big(v(x) - v_0\big). \tag{3.16}$$

The assumption of a filament velocity of 3000 m/min for example and again a radius of 10 μm leads to an estimated value of $F_{\text{inert}} \approx 1$ mN.

The discussion of *air drag force* F_{drag} seems to be a little bit more difficult. The air drag force acting on a filament with circular cross section (radius R) can be written in the form

$$\frac{\mathrm{d}F_{\text{drag}}}{\mathrm{d}x} = 2\,\pi\,R(x)\,\tau_{\text{f}}(x)\,, \tag{3.17}$$

where τ_{f} is the shear stress at the filament surface to the surrounding air. The shear stress is often expressed in terms of the air friction coefficient c_{f}

$$\tau_{\text{f}} = \frac{1}{2}\,\varrho_{\text{air}}\,v^2(x) \cdot c_{\text{f}}\,, \tag{3.18}$$

where ϱ_{air} denotes the air mass density. Unfortunately, the air friction coefficient c_{f} depends on the current state of air flow within the fibre formation region. A short discussion about the interaction between the fibre and its environment, including both heat transfer (Nu-number) and momentum transfer (air friction coefficient c_{f}), is presented in Sect. 3.1.3. Here it is to be remarked that an often used formula to describe the air drag coefficient is that of HAMANA [29]:

$$c_{\text{f}} = 0.37 \cdot Re^{-0.61} \tag{3.19}$$

or to be more general $c_{\text{f}} = a \cdot Re^b$, with the different parameters a and b. Re denotes the non-dimensional REYNOLDS number. The REYNOLDS number itself is given by

$$Re(x) = \frac{\varrho_{\text{air}} \cdot v(x) \cdot 2R(x)}{\eta_{\text{air}}} = \frac{v(x) \cdot 2\,R(x)}{\nu_{\text{air}}}\,, \tag{3.20}$$

where η_{air} is the dynamic viscosity of air, and the fraction $\nu_{\text{air}} = \eta_{\text{air}}/\varrho_{\text{air}}$ is called the kinematic viscosity of air.

Integrating Eq. 3.17 with 3.18 leads to

$$F_{\text{drag}}(x) = 2\,\pi \int_0^x R(x')\frac{\varrho_{\text{air}}}{2}\,v^2(x')\,c_{\text{f}}(x')\,dx' \tag{3.21}$$

and a rough estimation shows that the air drag contribution can reach similar magnitudes compared to the inertial force, especially at high filament velocity v.

3.1.3 Interaction Between Fibre and Environment

The following section deals with heat and momentum transfer from the fibre surface to the environment in more detail. Therefore, the equations for the NUSSELT number Nu and the air friction coefficient c_{f} are presented.

Heat Transfer

The heat transfer from the fibre surface to the ambient medium (usually air) involves the mechanisms *radiation*, *free convection*, and *forced convection* (ZIABICKI [274]). The contributions from radiation and free convection are often neglected or sometimes empirically incorporated into the model by the chosen relationships for the heat transfer coefficient resp. NUSSELT number [25, 26, 38, 59, 62, 63, 68, 196]. However, some more recent investigations also show the influence of radiation [203, 204].

Radiation. Due to radiation, the heat transfer coefficient α_r is strongly dependent upon the temperature T

$$\alpha_\mathrm{r}(T) = \sigma_\mathrm{SB} \cdot \varepsilon_\mathrm{m} \cdot \frac{T^4 - T_\mathrm{air}^4}{T - T_\mathrm{air}}\,, \tag{3.22}$$

where σ_SB is the STEPHAN-BOLTZMANN constant and ε_m is the emissivity of the polymeric material. Replacing the heat transfer coefficient with the non-dimensional NUSSELT number Nu_r (Eq. 3.6) for radiation then follows

$$Nu_\mathrm{r} = \mathrm{const} \cdot D \cdot \frac{T^4 - T_\mathrm{air}^4}{T - T_\mathrm{air}} \tag{3.23}$$

with const $\approx 1.7 \cdot 10^{-6}\,(\mathrm{m} \cdot \mathrm{K}^3)^{-1}$.

Figure 3.5 shows NUSSELT numbers for the radiation heat transfer vs. temperature for different fibre thicknesses. It can be seen that radiation in the temperature range of $T = 50 \ldots 300\,°\mathrm{C}$ for polymer melt spinning has very little effect, only directly below the spinneret ($T = 200 \ldots 300\,°\mathrm{C}$, $D = 300\,\mu\mathrm{m}$) the NUSSELT number can reach values between $0.1 \ldots 0.2$. This is equal $5 \ldots 20\%$ of the convective heat transfer (see below).

Convection. The *free (natural) convective heat transfer* is a typical heat transfer mechanism for stationary systems and is concerned with the movement of the ambient medium (air) caused by thermal expansion and density variation resulting from the related temperature field. The non-dimensional NUSSELT number Nu_n for free (natural) convection can be written as a function of two other non-dimensional numbers, namely the GRASHOF number Gr, and the PRANDTL number Pr [185]

$$Nu_\mathrm{n} = Nu_\mathrm{n}(Gr, Pr)\,, \tag{3.24}$$

where the GRASHOF number is given by $Gr = g\,\beta_\mathrm{air}\,(T - T_\mathrm{air})\,D^3$, with β_air as the thermal expansion coefficient of air, and the PRANDTL number of air is $Pr = \eta_\mathrm{air}\,c_\mathrm{p,air}/\lambda_\mathrm{air}$. Fibre cooling by means of free convection depends on the temperature difference between fibre and ambient medium and becomes an essential factor at low fibre velocities, especially near the spinneret. Near the spinneret the high temperature of the spinning block brings forth additional

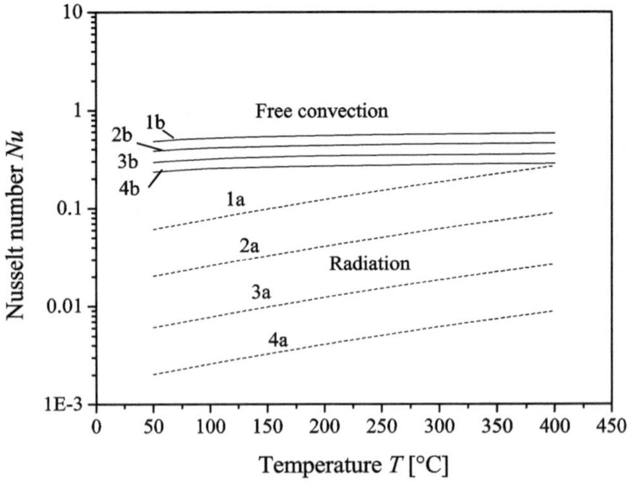

Fig. 3.5. NUSSELT number Nu vs. temperature T for different diameters D, resulting a) from heat radiation (Eq. 3.23), b) from free convection ($Nu = 0.65(Gr \cdot Pr)^{0.07}$, after [186]), diameters: 1 – 300 µm, 2 – 100 µm, 3 – 30 µm, 4 – 10 µm; temperature of ambient air: $T_{\text{air}} = 25\,°\text{C}$

complications regarding the air flow. Careful design of the spinning block is required in order to reduce turbulences and improve the stability of the spinning process.

Resulting NUSSELT numbers for free convection, calculated from an empirical correlation which was determined from analyzing the free convection over thin wires [186] are shown in Fig. 3.5.

Fibre cooling in melt spinning is mainly related to the *forced convective heat transfer*. The filaments move with increasing velocity and can be additionally quenched by cross air flow. The NUSSELT number Nu_{f} for the forced convection can depend upon the exposed length x of the filaments (in parallel flow), but the main contributions are given by the parallel and/or transverse air flow described with the related REYNOLDS numbers:

$$Nu_{\text{f}} = Nu_{\text{f}}(Re_{\parallel}, Re_{\perp}, x, Pr) \; . \tag{3.25}$$

Many authors have developed relations to describe the heat transfer using both boundary layer theory or experimental investigations. Some earlier expressions are summarised by ZIABICKI [184, 274]. The formula developed by KASE and MATSUO is often used [25]:

$$Nu_{\text{f}} = 0.42 Re_{\parallel}^{0.334} \left(1 + \left(\frac{8\, v_{\perp}}{v} \right)^2 \right)^{0.167} , \tag{3.26}$$

which can be rearranged to the more general relationship

$$Nu_{\text{f}} = a \left(Re_{\parallel}^2 + b \cdot Re_{\perp}^2 \right)^c , \tag{3.27}$$

Table 3.2. Relationships for heat transfer (NUSSELT number) in melt spinning

Equation for Nu-number	References and conditions
only parallel flow	
$0.76Re_{\parallel}^{0.38}$	Andrews (1959) [24]
$0.42Re_{\parallel}^{0.334}$	Kase, Matsuo (1965) [25]
$0.10 + 0.15Re_{\parallel}^{0.36}$	Sano (1966) [189]
$0.53Re_{\parallel}^{0.33}$	Copley (1967) [190]
$0.325Re_{\parallel}^{0.3}$	Glicksman (1968) [192]
$0.76Re_{\parallel}^{0.41}$	Conti (1970) [193]
$0.25 + 0.15Re_{\parallel}^{0.36}$	Zieminski (1985) [194]
$0.16Re_{\parallel}^{0.50}$	slow melt spinning of PEEK, Ohkoshi (1993) [202]
$0.42Re_{\parallel}^{0.344}Ra^{0.13}$	melt spinning of PEEK, Golzar (2004) [204]
$3.0Re_{\parallel}^{-0.22}$	melt spinning of PEEK, Golzar (2004) [204]
only transverse flow	
$0.891Re_{\perp}^{0.33}$	$1 < Re_{\perp} < 4$, Hilpert (1933) [205]
$0.821Re_{\perp}^{0.385}$	$4 < Re_{\perp} < 40$, dito
$0.615Re_{\perp}^{0.466}$	$40 < Re_{\perp} < 4000$, dito
$0.32 + 0.67Re_{\perp}^{0.52}$	$0.1 < Re_{\perp} < 10^3$, McAdams (1954) [206]
$0.38Re_{\perp}^{0.6}$	$10^3 < Re_{\perp} < 5 \cdot 10^4$, dito
parallel and transverse flow	
$0.42\left(Re_{\parallel}^2 + 64Re_{\perp}^2\right)^{0.167}$	(*) Kase, Matsuo (1965) [25]
$0.28\left(Re_{\parallel}^2 + 1024Re_{\perp}^2\right)^{0.17}$	(**) Brünig (1999) [207]

here with parameters $a = 0.42$, $b = 64$, $c = 0.167$. $Re_{\parallel}$ and $Re_{\perp}$ are the REYNOLDS numbers related to the parallel and cross air flow, defined as follows

$$Re_{\parallel}(x) = \frac{v_{\parallel}(x) \cdot D(x)}{\nu_{\text{air}}}, \tag{3.28}$$

$$Re_{\perp}(x) = \frac{v_{\perp}(x) \cdot D(x)}{\nu_{\text{air}}}, \tag{3.29}$$

where $v_{\parallel}$ is the axial (difference) velocity between fiber and ambient air, and $v_{\perp}$ is the cross air velocity, $\nu_{\text{air}} = \eta_{\text{air}}/\varrho_{\text{air}}$ is the kinematic viscosity of air, respectively. According to Eq. 3.10 it is possible to recalculate the total NUSSELT number for different spinning conditions, by carefully measuring the fibre temperature $T(x)$ simultaneously with fibre diameter $D(x)$ and/or velocity $v(x)$, if no crystallisation occurs.

$$Nu = -\frac{\mathrm{d}}{\mathrm{d}x} \ln\left(T(x) - T_{\text{air}}\right) \cdot \frac{Q\, c_{\text{p}}}{\pi\, \lambda_{\text{air}}}. \tag{3.30}$$

The experimental investigations can be carried out under different conditions: heated wires or filaments, running in stationary air, or stationary wires or filaments in steady air flow, and so on.

Table 3.2 and Fig. 3.6 show some examples for the relationship between the NUSSELT number and REYNOLDS number.

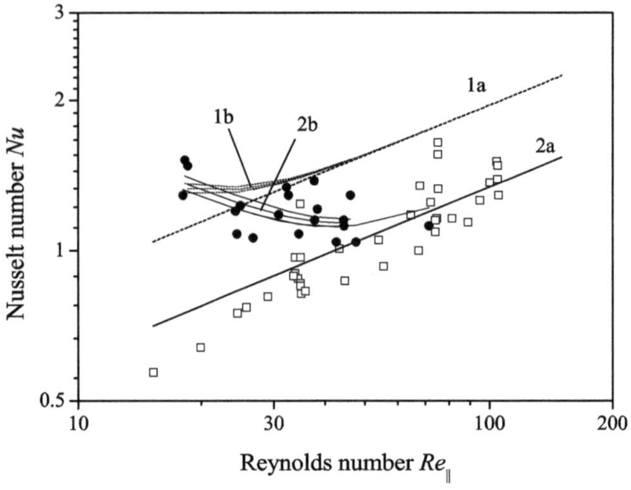

Fig. 3.6. NUSSELT number Nu vs. REYNOLDS number $Re_{\|}$, *symbols*: recalculated from experimental data, *open symbols:* without quenching air ($Re_{\perp} = 0$ (data source: recalculated from BRAGATO, GIANOTTO [187], *closed symbols:* with quenching air ($v_{\text{air}} = 0.4$ m/s, data source: recalculated from HABERKORN et al. [188], 1 – calc. after Eq. (*) in Table 3.2, 2 – calc. after Eq. (**) in Table 3.2, a) without quenching air ($Re_{\perp} = 0$), b) with quenching air ($v_{\text{air}} = 0.4$ m/s, $Re_{\perp} > 0$)

Air Friction

Equation (3.18) describes the friction-caused momentum transfer between the fibre surface and surrounding air. Similar to heat transfer, the air friction coefficient c_{f} was also investigated by several authors [53–55,82,208–211] both experimentally and by using laminar and turbulent boundary layer theory. Most results for the air drag coefficient have led to a relationship in the following manner

$$c_{\text{f}} = a\, Re^{b} \tag{3.31}$$

with non-dimensional REYNOLDS number Re related to the filament diameter and with different parameters a and b. Often used are the numbers given by HAMANA [29] with $a = 0.37$ and $b = -0.61$. A summary of different formulas

Table 3.3. Relationships for momentum transfer (air friction coefficient c_f) in melt spinning: $c_f = a \cdot Re^b$, after Shimizu et al. [59], Table 4

a	a^*	b	References and conditions
4.8	-	-1	Sakiadis (1961), laminar theory
1.8	-	-0.5	Sakiadis (1961), laminar theory [208–210]
5.0	-	-1	Higuchi, Katsu (1960), $0.04 < Re < 0.2$, (a) [72]
1.23	0.4	-0.81	Kase, Matsuo (1967), $3 < Re < 100$, (b) [26]
0.68	0.39...0.41	-0.8	Sano, Orii (1968), $10 < Re < 50$, (d) [73]
0.65	0.5	-0.7	Glicksman (1968), $4 < Re < 100$, (d) [192]
	0.37	-0.61	Hamana (1968) [29]
	1.78		Fukuda (1966), $5 < Re < 20$, (b) [74]
	1.3		Thompson (1953), $20 < Re < 150$, (b) [75]
	0.84		Anderson, Stubbs (1958), $10 < Re < 60$, (b) [76]
	0.77		Shimizu, Okui (1983), $50 < Re < 400$, (b,e) [53]
	0.41		Gould, Smith (1980), $20 < Re < 200$, (b) [211]
	0.56		Selwood (1962), $5 < Re < 54$, (c) [77]
	0.31...0.71		Kwon, Prevorsek (1979), $3 < Re < 30$, (c) [182]
	0.5		Shimizu, Okui (1983), $40 < Re < 250$, (d) [53]
	0.39...0.23		Shimizu, Okui (1983), $25 < Re < 70$, (d) [53]
	0.37		Matsui (1976), (d) [82]

a^* recalculated from measurements under the condition that $b = -0.61$, measurements made (a) on filaments falling in still air, (b) stationary filaments in airstream, (c) moving filaments in still air, (d) filaments spun into still air, (e) filaments spun into airstream

for Eq. 3.31 can be found in literary source [216] and some examples are shown in Table 3.3 [59].

It is surprising that the most published relations do not describe the influence of quenching cross air on the air friction coefficient. But a simple consideration leads to the assumption that with increasing quenching cross air velocity $v_\perp$ the air friction coefficient c_f must also increase. The air flow perpendicular to the fibre axis disturbs the originally axial symmetric boundary layer surrounding the fibre and leads to an additional momentum flux, analogous to the additional heat flux in the case of heat transfer (3.26). Because of the similarity of both transfer effects it was proposed to consider the influence of quenching air by means of both REYNOLDS numbers $Re_\parallel$ and $Re_\perp$ [207]:

$$c_f = c_f(Re_\parallel, Re_\perp) . \tag{3.32}$$

To estimate the parameters in Eq. 3.32 a boundary layer calculation and several fibre spinning experiments with different profiles of quenching air were conducted and evaluated. The investigations led to the following result

$$c_\mathrm{f} = \frac{a}{Re_\parallel}\left(Re_\parallel^2 + b\,Re_\perp^2\right)^c \quad \text{with} \quad a = 0.33,\ b = 4096,\ c = 0.2\ . \qquad (3.33)$$

Figure 3.7 shows the influence of both effects: the air friction coefficient c_f is depicted there dependent upon the parallel REYNOLDS number $Re_\parallel$ and for different cross quenching REYNOLDS numbers $Re_\perp$. Usually in melt spinning the REYNOLDS numbers are in the following ranges: $5 < Re_\parallel < 150$ and $0 < Re_\perp < 5$.

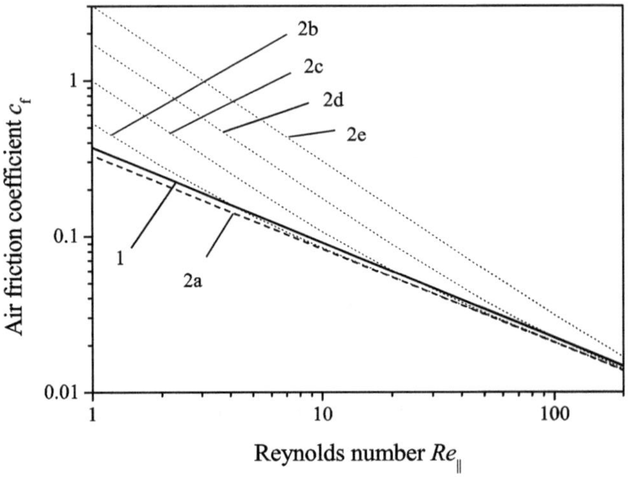

Fig. 3.7. Air friction coefficient c_f vs. REYNOLDS number $Re_\parallel$, 1 – $c_\mathrm{f} = 0.37Re^{-0.61}$ (Hamana [29]), 2 – Eq. 3.33, a) $Re_\perp = 0$, b) $Re_\perp = 0.05$, c) $Re_\perp = 0.25$, d) $Re_\perp = 1.0$, e) $Re_\perp = 4.0$

The influence of the quenching cross air on the air friction coefficient leads to an interesting effect which is unfortunately often neglected in modelling: A strong quenching air profile causes a high cooling rate and therefore a short fibre formation zone with a short fibre length exposed to air friction force. But on the other hand, increasing quenching air flow leads to an increase of the air friction coefficient c_f. Both effects, the shorten fibre formation zone and the higher air friction coefficient are in competition. This means that the total air friction force and at least the resulting fibre stress at the solidification point can decrease with increasing quenching air flow, but it is also possible that they can increase with increasing quenching air flow. How the resulting fibre orientation (and also the resulting fibre properties like the elongation to break) changes with quenching air flow conditions depends on the current spinning conditions.

3.1.4 Response of the Polymer Melt

Shear flow of the polymeric melt takes place inside the capillary of the die. After leaving the die exit, elongational flow takes place within the fibre formation zone until at least the polymer melt stream solidifies to the final fibre. Only the elongational flow after leaving the capillary die exit and no flow behaviour inside the spinneret should be regarded: The elongation $\varepsilon(x)$, depending on distance x from the spinneret, is defined by the logarithmic HENCKY measure

$$\varepsilon(x) = \ln \frac{v(x)}{v_0} , \tag{3.34}$$

where v_0 is the (mean) extrusion velocity at the spinneret exit. Comparing the definition above with the normally used relation for extension (see Fig. 3.8) the equivalence for small elongations becomes obvious: $\varepsilon = \ln(l/l_0) = \ln(1 + \Delta l/l_0) \rightarrow \varepsilon = \Delta l/l_0$ for $\Delta l \ll l_0$.

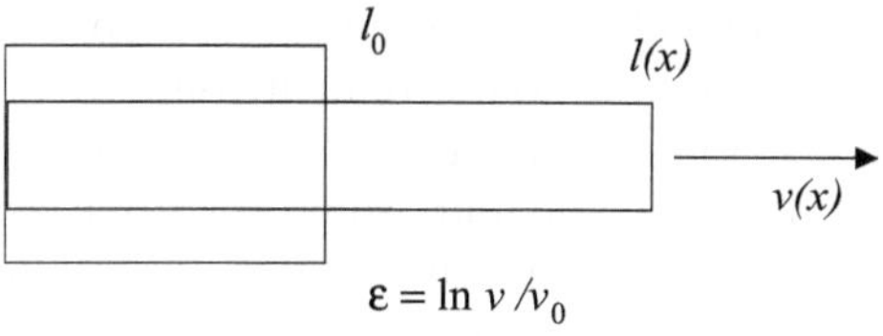

Fig. 3.8. Definition of elongation: HENCKY measure

For modelling the elongational flow, a suitable constitutive equation is necessary. It should describe the polymer response depending on the applied tensile force in a practicable manner, that is the relation between stress σ and strain ε of the fibre from the spinneret to the take-up device sufficient for any treatment in an engineering approach.

The stress-strain behaviour is controlled by the rheological properties of the polymeric material. Therefore, the spinline stress[1] σ is given by the (axial) tensile force $F = F_{\text{rheo}}$ divided by fibre cross section A

$$\sigma(x) = \frac{F_{\text{rheo}}(x)}{A(x)} . \tag{3.35}$$

Several constitutive equations have been developed either on the basis of continuum mechanics or under consideration of the molecular structure of the material by means of statistical methods. For modelling purposes of fibre formation mostly viscous and/or visco-elastic phenomenological relationships [25, 26, 99–102, 155–158, 196] are applied.

[1] In a more detailed theory of flow problems, the stress tensor $\underline{\sigma}$ is related to any kinematic tensor, and tension is described by the normal stress difference $\Delta\sigma = \sigma_{\text{xx}} - \sigma_{\text{rr}}$ of the components of the stress tensor $\underline{\sigma}$

Viscous and Visco-elastic Behaviour

The simplest engineering models for stress-strain behaviour are the basic rheological bodies NEWTON dashpot for purely viscous behaviour and HOOK's spring for purely elastic behaviour. In general, molten polymers are visco-elastic fluids. They show both viscous and elastic behaviour in their response to applied stress (see Fig. 3.9). The relation between viscous and elastic deformation can essentially influence the *spinnability* of a polymer (for further discussion see Sect. 3.3, visco-elasticity also gives reason for the concept of the so-called *apparent elongational viscosity*). The viscous behaviour dominates the elongational flow at comparatively low fibre velocity near the spinneret, but the influence of the elastic part becomes more and more important with increasing velocity (especially near the solidification region). Under certain conditions the *necking effect* can occur, that is the sudden reduction of the filament cross-sectional area within a short range because of a dramatic increase in the filament velocity. The necking region is very sensitive to the relationship of viscous and elastic materials behaviour. After solidication there is only the transportation of the fibre and a (more or less) purely elastic deformation up to the take-up device. Nevertheless it is often possible to get reasonable results for describing and simulating the course of melt spinning of common polymers without taking into account the elastic effects. But on the other hand, including the elastic deformation behaviour, some possibilities for calculating the textile properties are given (e. g. the elongation to break, see Sect. 3.1.6). In the following paragraphs some often used constitutive equations will be presented.

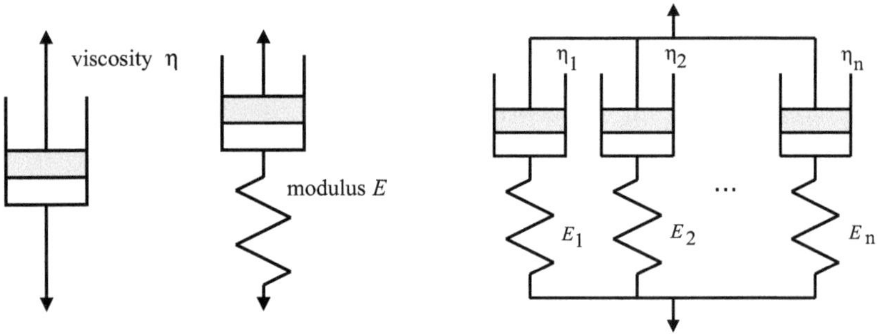

Fig. 3.9. Model of purely viscous (NEWTON) and visco-elastic (MAXWELL) rheological behaviour (*left*); spectral model, series of MAXWELL-bodies (*right*)

Newton model: Purely viscous flow. The mechanical model of the purely viscous elongation is based on the dashpot approach Fig. 3.9 (left side) and the stress-strain-relation is given by the simple constitutive equation

$$\sigma = \eta \cdot \dot{\varepsilon}_{\mathrm{v}} , \tag{3.36}$$

where $\dot{\varepsilon}_{\mathrm{v}} = \mathrm{d}\varepsilon_{\mathrm{v}}/\mathrm{d}t$ is the viscous deformation rate (strain rate) and η is the *elongational viscosity*. For constant applied stress the piston inside the dashpot moves at a constant strain rate. Under the steady state conditions ($\varepsilon = \varepsilon(x(t))$) and with the definition of Eq. 3.34 then Eq. 3.36 becomes

$$\sigma = \eta \cdot \frac{\mathrm{d}v}{\mathrm{d}x} \tag{3.37}$$

because of the equivalence

$$\frac{\mathrm{d}\varepsilon}{\mathrm{d}t} = \frac{\mathrm{d}\varepsilon}{\mathrm{d}x} \cdot \frac{\mathrm{d}x}{\mathrm{d}t} = \frac{1}{v} \cdot \frac{\mathrm{d}v}{\mathrm{d}x} \cdot v = \frac{\mathrm{d}v}{\mathrm{d}x} . \tag{3.38}$$

Example: Constant tensile force. The course of fibre formation can be discussed (after Ziabicki [184]) for the viscous deformation in a simple manner assuming a constant tensile force $F(x) = F_{\mathrm{c}}$. Neglecting all the contributions of inertia, gravitation, air friction, and surface tension to the force balance equation (3.11a) we get

$$\sigma(x) = \frac{F_{\mathrm{c}}}{A(x)} = F_{\mathrm{c}} \cdot \frac{\varrho_{\mathrm{P}} \cdot v(x)}{Q} \tag{3.39a}$$

$$= \eta \cdot \frac{\mathrm{d}v}{\mathrm{d}x} . \tag{3.39b}$$

This leads to the differential equation for the velocity v

$$\eta \cdot \frac{\mathrm{d}v}{\mathrm{d}x} = C \cdot v(x), \qquad C = \varrho_{\mathrm{P}} \cdot \frac{F_{\mathrm{c}}}{Q} \tag{3.40}$$

which can be integrated to

$$v(x) = v_0 \cdot \exp\left(C \cdot x \cdot \Phi(x)\right) , \tag{3.41}$$

where the function $\Phi(x)$ is given by

$$\Phi(x) = \frac{1}{x} \int_0^x \frac{1}{\eta(x')} dx' . \tag{3.42}$$

The constant force F_{c} is related to the boundary conditions

$$v(x=0) = v_0$$
$$v(x=L) = v_{\mathrm{L}}$$

through

$$F_{\mathrm{c}} = \frac{Q}{\varrho_{\mathrm{p}}} \cdot \frac{L}{\Phi(L)} \cdot \ln\left(\frac{v_{\mathrm{L}}}{v_0}\right) . \tag{3.43}$$

The function $\Phi(x)$ describes the so-called *fluidity* of the polymeric material – the mean value of the reciprocal viscosity over certain length. An exponentially increasing elongational flow (Eq. 3.41) results from constant viscosity η (and constant fluidity Φ). But in melt spinning the temperature along the spinning line decreases, thus the viscosity increases with increasing distance, and the slope dv/dx also more and more decreases. Assuming that at the solidification point the viscosity becomes infinitely large, the velocity then reaches its final value v_L. The resulting behaviour is the characteristically "S-shaped" course of velocity vs. distance for purely viscous behaviour. Figure 3.10 shows an example for two different cases of fibre cooling with following conditions:

Fibre temperature T (in °C): $T_1(x) = 20 + 280\exp(-x/0.60)$
$T_2(x) = 20 + 280\exp(-x/0.30)$

Viscosity η (in Pa·s) : $\eta(T) = 0.015\exp(\frac{5200}{T+273})$ for $T \geq 70\,°C$
$\eta(T) = \infty$ for $T < 70\,°C$

Final draw down ratio : $v_L/v_0 = 300$.

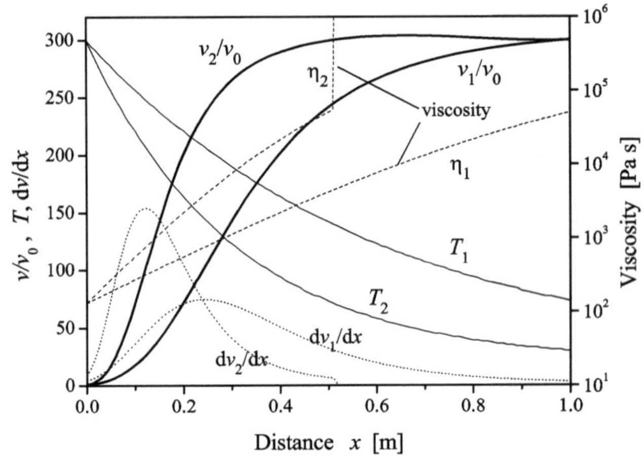

Fig. 3.10. Example of fibre formation with constant tensile force F_0 and two different cooling lengths: 1 – slow cooling, cooling length $L_c = 0.60$ m; 2 – fast cooling, $L_c = 0.30$ m; NEWTONian flow behaviour, draw down ratio $ddr = 300$

Maxwell model: Visco-elastic behaviour. The mechanical model consists of the combination of dashpot and spring in series (Fig. 3.9). The applied force causes the (ideal) spring to an immediate elongational elastic deformation (orientational deformation) ε_o proportional to the applied stress σ. The same stress acts on the dashpot and causes the viscous elongational deformation ε_V. The total deformation ε of the MAXWELL body is now the sum of both parts [280, 281]:

$$\varepsilon = \varepsilon_V + \varepsilon_o \tag{3.44}$$

with

$$\sigma = \eta \cdot \dot{\varepsilon}_{\mathrm{v}} = E \cdot \varepsilon_{\mathrm{o}} , \tag{3.45}$$

where E is the elongational modulus of the spring. Modulus E and viscosity η are the two material properties describing the deformation behaviour of the MAXWELL body. Often the modulus E is used together with the *relaxation time* λ with

$$\lambda = \frac{\eta}{E} . \tag{3.46}$$

The constitutive equation (considering the stationarity behaviour) of the MAXWELL model is given by

$$\sigma + \frac{\eta}{E} \cdot v \cdot \frac{\mathrm{d}\sigma}{\mathrm{d}x} = \eta \cdot \frac{\mathrm{d}v}{\mathrm{d}x} \tag{3.47}$$

or, expressed in terms of the relaxation time λ as

$$\sigma + \lambda \cdot v \cdot \frac{\mathrm{d}\sigma}{\mathrm{d}x} = E \cdot \lambda \cdot \frac{\mathrm{d}v}{\mathrm{d}x} . \tag{3.48}$$

Approaching a spectral series of $i = 1...N$ MAXWELL bodies (Fig. 3.9, right) the corresponding relation is given by

$$\sigma + \sum_i \lambda_i \cdot v \cdot \frac{\mathrm{d}\sigma}{\mathrm{d}x} = \sum_i E_i \cdot \lambda_i \cdot \frac{\mathrm{d}v}{\mathrm{d}x} . \tag{3.49}$$

Generalisations. The constitutive equations above describe a simple one-dimensional model for the extension in spinning-direction. All effects related to the radial stress and deformation components of the uniaxial deformation are neglected. Considering the three-dimensionality of the real deformation process, a more detailed analysis needs to be carried out. This can be done by means of precise formulation with the help of tensor-calculus. Nevertheless, the *thin-filament approximation* in which the radial dependence of axial velocity (and temperature) $\partial v/\partial r = 0$ (and $\partial T/\partial r = 0$) is neglected should further be used. For the incompressible axis symmetrical and uniaxial elongational deformation, the deformation rate tensor $\underline{\dot{\varepsilon}}$ and the stress tensor $\underline{\sigma}$ are given as follows:

$$\underline{\dot{\varepsilon}} = \frac{\mathrm{d}v}{\mathrm{d}x} \begin{pmatrix} 1 & 0 & 0 \\ 0 & -\frac{1}{2} & 0 \\ 0 & 0 & -\frac{1}{2} \end{pmatrix} \quad \text{and} \quad \underline{\sigma} = \begin{pmatrix} \sigma_{\mathrm{xx}} & 0 & 0 \\ 0 & \sigma_{\mathrm{rr}} & 0 \\ 0 & 0 & \sigma_{\mathrm{rr}} \end{pmatrix} . \tag{3.50}$$

A possible generalisation for visco-elastic behaviour using the stress and deformation rate tensors is the *upper-convected* MAXWELL model with relaxation time λ and modulus G:

$$\sigma_{\mathrm{xx}} + \lambda \left(v \frac{\mathrm{d}\sigma_{\mathrm{xx}}}{\mathrm{d}x} - 2 \frac{\mathrm{d}v}{\mathrm{d}x} \sigma_{\mathrm{xx}} \right) = 2 G \lambda \frac{\mathrm{d}v}{\mathrm{d}x} , \tag{3.51}$$

$$\sigma_{\mathrm{rr}} + \lambda \left(v \frac{\mathrm{d}\sigma_{\mathrm{rr}}}{\mathrm{d}x} + \frac{\mathrm{d}v}{\mathrm{d}x} \sigma_{\mathrm{rr}} \right) = - G \lambda \frac{\mathrm{d}v}{\mathrm{d}x} . \tag{3.52}$$

The spinning line tension (3.35) is now related to the normal stress difference $\Delta\sigma = \sigma_{\mathrm{xx}} - \sigma_{\mathrm{rr}}$ with

$$\Delta\sigma = \sigma_{\mathrm{xx}} - \sigma_{\mathrm{rr}} = \frac{F}{A} . \tag{3.53}$$

The PHAN-TIEN-TANNER model [212, 213] is another generalised constitutive equation based on a special type of network theory to describe a nonlinear strain-softening behaviour. It helps to describe the necking effect at higher spinning speeds. The constitutive equations are:

$$K \cdot \sigma_{\mathrm{xx}} + \lambda \left(v \frac{\mathrm{d}\sigma_{\mathrm{xx}}}{\mathrm{d}x} - 2 \frac{\mathrm{d}v}{\mathrm{d}x} \sigma_{\mathrm{xx}} \right) = 2 G \lambda \frac{\mathrm{d}v}{\mathrm{d}x} , \tag{3.54}$$

$$K \cdot \sigma_{\mathrm{rr}} + \lambda \left(v \frac{\mathrm{d}\sigma_{\mathrm{rr}}}{\mathrm{d}x} + \frac{\mathrm{d}v}{\mathrm{d}x} \sigma_{\mathrm{rr}} \right) = - G \lambda \frac{\mathrm{d}v}{\mathrm{d}x} , \tag{3.55}$$

$$K = \exp\left(\frac{\alpha}{G} \cdot \mathrm{tr}\, \underline{\sigma} \right) . \tag{3.56}$$

The parameter α describes the additional strain softening effect. For $\alpha \to 0$ in Eq. 3.56 the upper-convected MAXWELL equation above follows. $\mathrm{tr}\, \underline{\sigma} = \sigma_{\mathrm{xx}} + 2\, \sigma_{\mathrm{rr}}$ denotes the trace of the stress-tensor.

Of course it is also possible to use combinations of different MAXWELL or generalised MAXWELL elements. This leads to a discrete or continuous spectrum with more than one relaxation time and modulus, respectively (Fig. 3.9). Some investigators also describe the viscosity by means of a power law such as

$$\eta(T, \dot{\varepsilon}) = \eta_0(T) \cdot \left(\frac{\mathrm{d}v}{\mathrm{d}x} \right)^{n-1} , \quad n < 1 , \tag{3.57}$$

or a generalised power law equation similar to the CROSS-CARREAU type e. g.

$$\eta(T, \dot{\varepsilon}) = \frac{\eta_0(T)}{1 + a \left(\eta_0(T) \frac{\mathrm{d}v}{\mathrm{d}x} \right)^{b}} , \tag{3.58}$$

respectively, to describe the additional influence of the strain rate $\dot{\varepsilon} = \mathrm{d}v/\mathrm{d}x$ and the deviation of the resulting viscosity from the NEWTONian behaviour with corresponding parameter n, a, b.

Solidification

The glass transition point marks the temperature T_g at which the mobility of the polymer chains prevent any further deformation flow. Some of the typical values are listed in Table 3.4.

Table 3.4. Glass transition temperature T_g of melt spinning polymers

polymer	T_g in °C
PET – poly(ethylene terephthalate)	70
PA 6 – poly(amide) 6	50 . . . 55
PP – poly(propylene)	≈ -20

In reality, the flow deformation often comes to an end *before* the glass transition temperature is reached. The reason for this is that crystallisation can take place within the spinline at certain high levels of orientation and stress. The developing crystalline regions act as additional strain hardening segments. Some researchers (e. g. GEORGE [39]) investigated the solidification temperature T_s where the deformation comes to an end. They found that the solidification temperature depends upon the spinning speed and spinline stress. The solidification temperature T_s showed a step-like behaviour as function of stress σ. The solidification cannot be separated from the appearance of stress-induced crystallisation, especially in the cases of highly crystallisable polymers in melt spinning. Therefore, an empirical relation $T_s = T_s(\sigma)$ between solidification temperature and spinline stress is often used for modelling the fibre formation process .

3.1.5 Structure Development

Structure development is one of the most interesting aspects [175] in the modelling of the melt spinning process because the resulting as-spun fibre structure determines the textile fibre properties like tenacity, residual draw ratio or elongation to break. This is why a practicable and useful simulation should not stop with the calculation of fibre temperature, velocity, or stress, but should also allow for the prediction of the structural parameters and the correlated textile properties. To *quantitatively* describe the fibre formation and resulting fibre structure it is necessary to possess information on how

- the elongational flow influences molecular orientation,
- the orientation influences the crystallisation of the material,
- crystallisation occurs (nucleation and kinetics),
- orientation and crystallisation change the rheological flow properties.

Of course the effect of the heat of crystallisation should also be included. The main parameters for structure development are temperature, stress, and time. Although many researchers have investigated the melt spinning process for a long time (and with investigations still ongoing) the current models do not sufficiently consider these effects. A satisfactory description is still missing specifically for the crystallisation.

Orientation and Birefringence

The polymer deformation, in melt spinning mainly the elongational flow, causes an alignment of the polymer chain molecules. HERRMANN's orientation factor f describes the orientation of the molecular chains with respect to the fibre axis. It can be defined through different methods (e. g. X-ray scattering) as follows [173]:

$$f = \frac{3\left\langle \cos^2\theta \right\rangle - 1}{2}, \tag{3.59}$$

where θ is the angle between the molecular chains and the fibre axis, the brackets denote the mean value. The orientation factor f is equivalent to the second momentum $\langle P_2 \rangle$ of the development of *orientation distribution function* $N(\theta)$ to LEGENDRE polynomials P_n:

$$N(\theta) = a_0 P_0(\cos(\theta)) + a_2 P_2(cos(\theta)) + a_4 P_4(cos(\theta)) + \dots \tag{3.60}$$

due to reasons of symmetry the odd-numbered coefficients vanish, the first three even LEGENDRE polynomials are given as follows:

$$P_0(\cos(\theta)) = 1, \tag{3.61a}$$

$$P_2(\cos(\theta)) = \frac{1}{2}\big(3\cos(\theta) - 1\big), \tag{3.61b}$$

$$P_4(\cos(\theta)) = \frac{1}{8}\big(35\cos^4(\theta) - 30\cos^2(\theta) + 3\big). \tag{3.61c}$$

The orientation distribution function $N(\theta)$ is defined at the sphere (see Fig. 3.11) and of course normalised ($\langle P_0 \rangle = 1$), therefore the second momentum $\langle P_2 \rangle$ gives us the first information about the orientation of the molecular chains:

$$\langle P_0 \rangle = 2\pi \int_0^{\pi} P_0\, N(\theta) \sin(\theta)\, d\theta = 1, \tag{3.62}$$

$$\langle P_2 \rangle = 2\pi \int_0^{\pi} P_2(\theta)\, N(\theta) \sin(\theta)\, d\theta \equiv f. \tag{3.63}$$

The orientation factor vanishes ($f = 0$) if all axes of the polymer chains are randomly distributed. It's maximum is $f = 1$ if all polymer chains align parallel to the fibre axis. The orientation of the molecular chains affects the polarisability and leads to the difference of the refractive index for linear

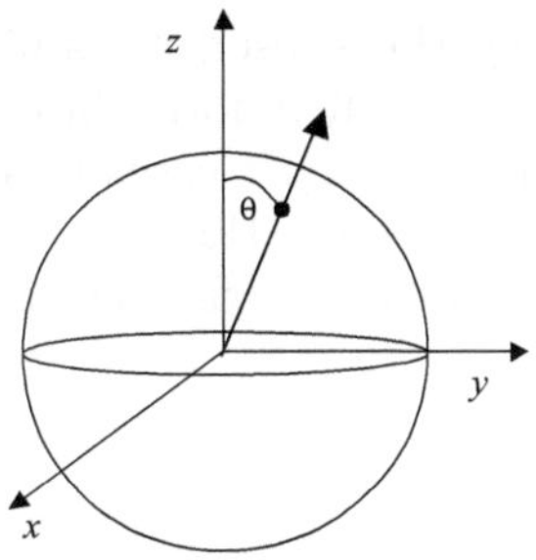

Fig. 3.11. Definition of the orientation distribution function, θ is the angle between fibre axis z and the end-to-end-vector of molecular chains, the distribution of the orientation vectors about the sphere defines $N(\theta)$

polarised light parallel and perpendicular to the fibre axis, called birefringence $\Delta n = n_{\parallel} - n_{\perp}$ (Fig. 3.12). The birefringence is related to the orientation factor as follows

$$\Delta n = f \cdot \Delta n^0 , \tag{3.64}$$

where Δn^0 is the maximum possible birefringence of the material (all molecular chains are aligned parallel to the fibre axis), the so-called *intrinsic birefringence.*

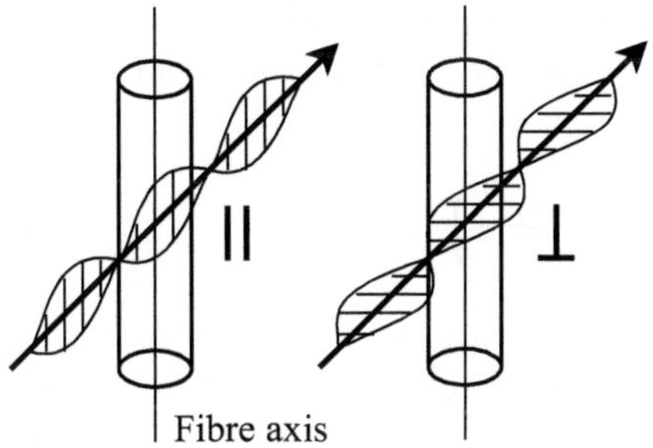

Fig. 3.12. Linear polarised light, parallel and perpendicular to fibre axis

It was found that for noncrystalline amorphous polymers and for moderate stress levels the (amorphous) birefringence Δn_{am} satisfies a simple relation to the applied stress σ, the *stress-optical law* [30]:

$$\Delta n_{\mathrm{am}} = C_{\mathrm{opt}} \cdot \sigma , \tag{3.65}$$

where C_{opt} is the stress-optical coefficient.

The properties of partly crystalline polymers are usually described by a two-phase model consisting of crystalline regions (fractional part X_{c}) and amorphous regions (fractional part $1 - X_{\mathrm{c}}$). Following this two-phase concept, the birefringence of a semicrystalline material can be written as [174]

$$\Delta n = (1 - X_{\mathrm{c}}) \cdot \Delta n_{\mathrm{am}} + X_{\mathrm{c}} \cdot \Delta n_{\mathrm{cr}} \tag{3.66a}$$

$$= (1 - X_{\mathrm{c}}) \cdot f_{\mathrm{am}} \, \Delta n^0_{\mathrm{am}} + X_{\mathrm{c}} \cdot f_{\mathrm{cr}} \, \Delta n^0_{\mathrm{cr}} . \tag{3.66b}$$

The amorphous and crystalline phases are depicted by the subscripts 'am' and 'cr'. An additional component of birefringence, the so-called 'form' birefringence, is caused by the interaction of the two phases. In most cases this form birefringence has only little effect and can be neglected. (However, if rod-like voids are generated in the fibre, sometimes the form birefringence is not negligible).

Crystallisation and Crystallinity

As mentioned above, a satisfactory and easy to handle theory of non-isothermal crystallisation within the spinline is still missing. Nevertheless, the basics of a phenomenological model will be shortly presented in the following section. For more details the interested reader is referred to separate literature [163, 214–240].

The crystallisation can be divided into *nucleation* and *crystal growth*. For higher temperatures (near the melting point) the nucleation is low and the material crystallises slowly, at lower temperatures (near the glass transition) the molecular mobility is low and the crystal growth is slow. The result is a maximum overall crystallisation rate at any intermediate temperature $T_{\max}$. The high cooling rate in melt spinning allows for only a short time interval for the temperature range where crystallisation is possible. Therefore the temperature-dependent crystallisation rate for an isotropic material $K(T, 0)$ is given by the approximation

$$K(T, 0) = K_{\max} \cdot \exp\left(\frac{-4\ln(2)\,(T - T_{\max})^2}{\Delta T_{1/2}^2}\right) , \tag{3.67}$$

where $K_{\max}$ is the crystallisation rate constant at the temperature $T_{\max}$ of maximum crystallisation rate, $\Delta T_{1/2}$ is the half-width of the GAUSSian function. For a non-isotropic material with stress-related orientation $f \propto \sigma$ ZIABICKI [214, 216, 220] proposed a series expansion of the crystallisation rate $K(T, \sigma)$ to include the stress and orientation effects in the form

$$K(T, \sigma) = K(T, 0) \exp\left(C_2\,\sigma^2 + C_3\,\sigma^3 + \ldots\right) , \tag{3.68}$$

where C_i are constants. The linear term in (3.68) vanishes because of thermodynamic symmetry reasons. For moderate stress the crystallisation rate is given as

$$K(T, \sigma) = K_{\max} \cdot \exp\left(\frac{-4\ln(2)\,(T - T_{\max})^2}{\Delta T_{1/2}^2} + C_2\,\sigma^2\right) . \tag{3.69}$$

It can be seen that crystallisation is extremely sensitive to orientation or stress, especially for the high-speed spinning process *stress-induced crystallisation* can be observed. Other approaches [197–200] use the quiescent crystallisation theory from HOFFMAN and LAURITZEN, expanded by the stress-orientation effect. This leads to

$$K(T,\sigma) = K_0 \cdot \exp\left(-\frac{U*}{R(T-T_\infty)} - \frac{C_1}{T \cdot \Delta T + C_2 T^2 \cdot \sigma^2}\right) , \tag{3.70}$$

where U^* is the activation energy for segment motion, R is the gas constant, $\Delta T = T_m - T$ is the supercooling where T_m is the melting temperature, $T_\infty = T_g - 30\,\mathrm{K}$ where T_g is the glass transition temperature, and K_0, C_1, C_2 are material constants.

The development of the crystallinity X_c now is in agreement to the AVRAMI approximation given by the relation

$$\frac{dX_c}{dt} = nK(1-X_c)\left[\ln\left(\frac{1}{1-X_c}\right)\right]^{\frac{n-1}{n}} , \tag{3.71}$$

where n is the AVRAMI exponent, usually an integer value in the range of 1–4. With $n = 1$ at steady state conditions the simpler (and often used) equation follows:

$$\frac{dX_c}{dx} = \frac{K(T,\sigma)}{v} \cdot (1-X_c) . \tag{3.72}$$

Mostly, melt spun fibres are partially crystalline. Assuming the two-phase model of amorphous (density ϱ_{am}) and crystalline (density ϱ_{cr}) regions, the mass density ϱ_p of a partially (X_c) crystalline fibre is given by the rule of mixture:

$$\varrho_p(X_c) = (1-X_c)\,\varrho_{am} + X_c\,\varrho_{cr} . \tag{3.73}$$

Some typical values for fibre densities and also for the densities of the amorphous and crystalline regions are presented in Table 3.5, at a temperature of 20 °C.

Table 3.5. Amorphous, crystalline and typical fibre mass densities at 20 °C

	ϱ_{am} in g/cm^3	ϱ_{cr} in g/cm^3	ϱ-fibre in g/cm^3
PET	1.335	1.455	1.34–1.38
PA 6	1.09	α-modification: 1.23	1.12–1.14
		γ-modification: 1.17	
PP	0.854	0.963	0.90–0.905

Therefore, the precise measurement of the density can be used to determine the crystallinity in the solid state of the as-spun fibre:

$$X_c = \frac{\varrho_p - \varrho_{am}}{\varrho_{cr} - \varrho_{am}} . \tag{3.74}$$

For example, Fig. 3.13 shows the mass density of some spun PET fibres vs. spinning velocity. For a higher take-up velocity, the density increases because stress-induced crystallisation occurs and the crystallinity X_c increases.

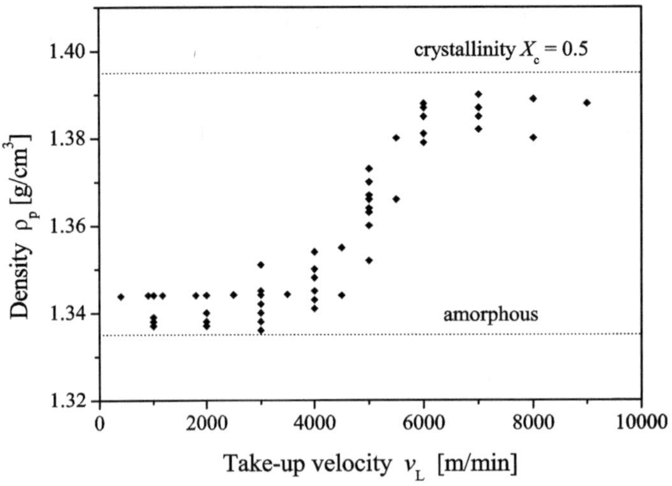

Fig. 3.13. Mass density ρ_p for PET vs. take-up velocity v_L, data points taken/recalculated from [58, 187, 241–246]

3.1.6 Material Properties of Spinnable Polymers

Mass density and specific heat capacity of melt spinnable polymers are not constant but depend on temperature and crystallinity. Elongational viscosity, relaxation time and/or modulus also depend on temperature and structural parameters. Finally the resulting fibre properties like elongation to break or tenacity are correlated with the fibre structure.

Properties of the Melt

Mass density. The mass density ϱ_p of molten polymers can be assumed as linear relation to the temperature T:

$$\varrho_p(T) = \varrho_0 - \varrho_1 \cdot T\,, \quad T \text{ in °C}\,. \tag{3.75}$$

Some often used values for the parameter ϱ_0 and ϱ_1 for modelling the fibre formation are shown in Table 3.6.

Table 3.6. Temperature dependence of mass density (3.75) in melt spinning

	ϱ_0 in g/cm^3	ϱ_1 in (°C)$^{-1}$	temp. range in °C
PET	1.356	0.0005	70–300
PA 6	1.124	0.00056	55–300
PP	0.90	0.0006	25–260

Heat capacity. The specific heat capacity of polymers can be determined by analyzing their (equilibrium) melting behaviour using the established methods of differential scanning calorimetry (DSC). Regarding the thermal scans; the glass transition, the crystallisation, and the melting peaks can be observed. Therefore, the thermal analysis can also be used to calculate the degree of crystallinity of the as-spun fibres. Otherwise the melt spinning process appears in a non-equilibrium state. The process speed and cooling rate are much higher, making it impossible to apply DSC methods for observation of phase transition. The locations and amounts of latent heat which one can attain from the calorimetric scans normally cannot be used for describing the rapid fibre formation process. Regarding the tractability of the fibre formation model, the specific heat capacity c_p is often described by linear relationship to temperature T:

$$c_p(T) = c_{p0} + c_{p1} \cdot T\,, \quad T \text{ in °C}\,. \tag{3.76}$$

Table 3.7 contains some possible values for the parameter c_{p0} and c_{p1} for describing the heat capacity in the modelling of fibre formation.

Table 3.7. Temperature dependence of the specific heat capacity

	c_{p0} in J/(kg·K)	c_{p1} in (°C)$^{-1}$	temp. range in °C
PET	1256	2.51	70–300
PA 6	2180	2.2	> 100
PP	1536	10.1	25–260

Elongational viscosity. The change of elongational viscosity with increasing distance from the spinneret, *essential*, influences the characteristic course of the spinning line (see the example of applying the constant force in Sect. 3.1.4 on page 61). The elongational viscosity η (or the relaxation time λ for the visco-elastic behaviour, respectively) generally depends upon the molecular weight M and the molecular weight distribution of the polymer. The dependence on the molecular weight often can be expressed by the intrinsic viscosity IV, or the relative solution viscosity. Sometimes it is convenient to use the melt flow rate MFR (melt flow index MFI) instead of the relative solution viscosity (e. g. for poly(propylene)). The local elongational NEWTONian viscosity $\eta(x)$ depends on the local temperature $T(x)$, and for the non-NEWTONian case the viscosity additionally depends on the local elongational deformation rate $\dot{\varepsilon}(x)$. Finally, if crystallisation occurs, the effect of the local degree of crystallinity $X_c(x)$ has also to be taken into account.

The NEWTONian viscosity is usually assumed to be approximated by the ARRHENIUS equation with constant activation energy E_a:

Table 3.8. Typical melt viscosities (zero shear viscosities) of spinnable polymers

	PET	PA 6	PP
Molecular weight	19 000	17 000	200 000
Processing temperatures (°C)	285–295	260–270	230–250
Melt viscosity (Pa·s)	120–90	140–120	120–80

$$\eta(T) = \eta_0 \exp(E_a/k\,T)\,, \tag{3.77}$$

where η_0 is a material constant depending on molecular weight or intrinsic viscosity and k is the BOLTZMANN constant. Equation (3.77) works well in the high temperature range above the melting point. The better alternative to the ARRHENIUS formulation within the lower temperature range, as the temperature approaches the glass transition temperature T_g, is the WILLIAMS-LANDEL-FERRY (WLF) equation with T_g and melting temperature T_m as parameters:

$$\eta(T) = \eta_1 \exp\left(\frac{-(T-T_m)}{(51.6+T-T_g)(51.6+T_m-T_g)}\right)\,. \tag{3.78}$$

On the one hand, the WLF equation gives a steep rise in the vicinity of glass transition. However, on the other when solidification is reached the rise of viscosity can be assumed to reach infinity for modelling purposes.

The influence of molecular weight M (expressed by intrinsic viscosity IV or relative solution viscosity) on the elongational viscosity is usually given by the power functions

$$\eta_0 \propto M^b\,, \qquad \eta_0 \propto (IV)^c\,, \tag{3.79}$$

with exponents $b = 3.2\ldots3.4$, and $c = 1.8\ldots2.2$, respectively.

Table 3.8 shows selected parameters for typical melt viscosities of polymers (measured in oscillatory shearing). For modelling purposes, the elongational viscosity is often assumed to be approximately three times the zero shear viscosity. Some examples for the dependency of the elongational viscosity on temperature for PET are shown in Fig. 3.14 .

The dependency of the viscosity on deformation and deformation rate cannot be separated from the general visco-elastic and/or nonlinear rheological behaviour. These problems are also strongly connected with the failure behaviour of the polymer material. A short but only qualitative discussion is given in Sect. 3.3.

Another unsolved problem is the correct description of the effect of crystallinity on viscosity. There is a rapid increase of viscosity if crystallisation occurs. Usually the viscosity $\eta(T, X_c) = \eta(T) f(X_c)$ is expressed as a product of temperature-dependent viscosity and a crystallinity-dependent function $f(X_c)$. There are several proposals [59, 200, 201] how to quantify the influence of crystallinity X_c on viscosity:

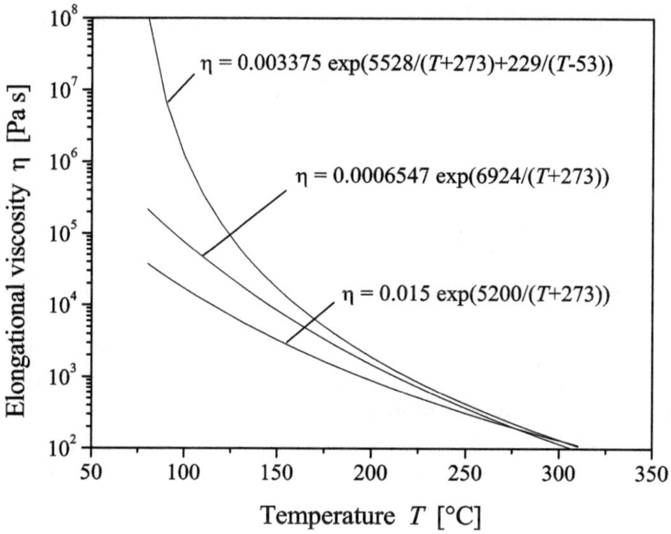

Fig. 3.14. Elongational viscosity η vs. temperature T for PET, ARRHENIUS type equations

$$\eta(T, X_c) = \eta(T) \exp(a\, X_c^b) \;, \tag{3.80}$$

or

$$\eta(T, X_c) = \eta(T) \cdot (1 + 99\, X_c) \;, \tag{3.81}$$

or

$$\eta(T, X_c) = \frac{\eta(T)}{(1 - X_c/X_{crit})^{\alpha}} \;, \tag{3.82}$$

with parameters a, b, α; X_{crit} is a critical value derived from a crosslinking model where $\eta \to \infty$ when $X_c \to X_{crit}$.

Elastic modulus *E*. Little is known about the elastic modulus (or relaxation time) of the melt during elongational deformation. A possible assumption follows the network deformation concept. For example the modulus $E = d\sigma/d\varepsilon_o$ of the GAUSSian entropic network is given by

$$E(\varepsilon_o) = E_0\big(2\,\exp(2\,\varepsilon_o) + \exp(-\varepsilon_o)\big) \;, \tag{3.83}$$

where ε_o is the elongational (orientational) deformation of the network (see also the discussion in the next paragraph), and where the parameter E_0 fits to the properties of the melt. Equation 3.83 implies that the relaxation time $\lambda = \eta/E$ has the same dependence on temperature as viscosity.

Properties of the Solid Fibre - The Network Deformation Concept

Textile properties like elongation to break and/or the tenacity of as-spun fibres are mainly determined by the orientational state reached during melt spinning [280, 281]. The orientation itself depends on the realised spinning stress.

Elongation to break. Regarding the correlation between (amorphous) orientation (birefringence Δn) and elongation to break ε_b leads to the generally accepted assumption that a maximum possible elongation ε_{max} may exist depending on the type of polymer. Extrapolating this relation to a hypothetic value of vanishing orientation with $\Delta n \to 0$ then allows to estimate the maximum possible elongation to break ε_{max}. The same procedure can be applied in principle for the behaviour of elongation to break vs. take-up velocity. Extrapolating $v_L \to 0$ also leads to similar values for the maximum possible elongation to break. Figure 3.15 shows as an example the elongation to break vs. take-up velocity for as-spun PA 6 fibres at various spinning conditions. The elongation ϵ measured in percentages relates to the corresponding draw ratio DR and the logarithmic HENCKY measure ε by[2]

$$DR = 1 + \frac{\epsilon \text{ (in \%)}}{100\%} \quad , \qquad \varepsilon = \ln DR \, . \tag{3.84}$$

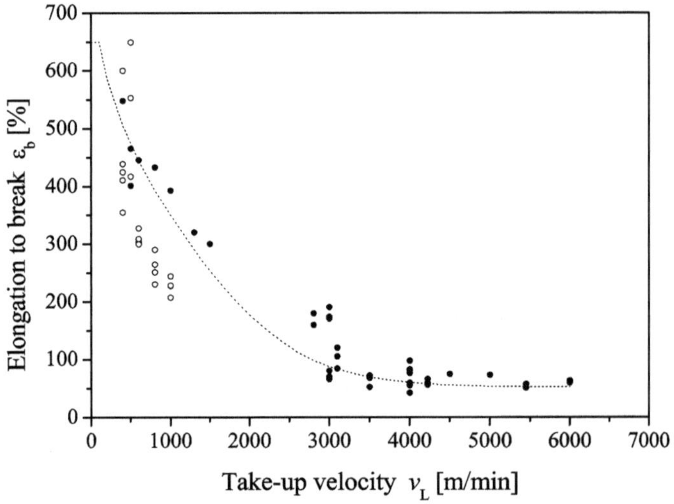

Fig. 3.15. Elongation to break ε_b vs. take-up velocity v_L; PA 6, different spinning conditions, filled symbols (•): molecular weight 16 000 – 18 000, open symbols (○): molecular weight 28 000; data source: Institute of Polymer Research Dresden

[2] The elongation measured in percentage (%) is labeled with ϵ and the corresponding logarithmic HENCKY measure is labeled with ε.

Table 3.9 contains the results for the maximum possible elongation to break ϵ_{max} also expressed as maximum corresponding draw ratio DR_{max} and as logarithmic measure $\varepsilon_{\mathrm{max}}$.

Table 3.9. Maximum possible elongation of spinnable polymers

	PA 6	PET
ϵ_{max} (in %)	650–700	800–900
DR_{max} corresponding draw ratio	7.5–8.0	9.0–10.0
$\varepsilon_{\mathrm{max}}$ (logarithmic HENCKY measure)	2.0–2.1	2.2–2.3

Under the assumptions

- that a maximum possible elongation $\varepsilon_{\mathrm{max}}$ exists and
- that each deformation (or drawing) step i is independent of the step before,

follows for the deformation steps the simple relation (in logarithmic HENCKY measure)

$$\sum_i \varepsilon_i = \varepsilon_{\mathrm{max}} \tag{3.85}$$

or, expressed with the corresponding draw ratios DR_i

$$\prod_i DR_i = DR_{\mathrm{max}} \,. \tag{3.86}$$

Applying this concept of independent deformation steps to a typical melt spinning process, one can easily calculate the residual elongation to break ϵ_{b} of the fibre after spinning and drawing:

$$\epsilon_{\mathrm{b}}\ (\text{in}\%) = \left(\frac{DR_{\mathrm{max}}}{DR_{\mathrm{m}} \cdot DR_{\mathrm{spun}}} - 1\right) \cdot 100\ \% \tag{3.87a}$$

$$= \left(\frac{\exp(\varepsilon_{\mathrm{max}})}{DR_{\mathrm{m}} \cdot \exp(\varepsilon_{\mathrm{spun}})} - 1\right) \cdot 100\ \% \tag{3.87b}$$

or, vice versa the necessary machine draw ratio DR_{m} to any residual elongation to break ϵ_{b} of the final fibre after drawing if the logarithmic elongation measure $\varepsilon_{\mathrm{spun}}$ of the as-spun fibre is known:

$$DR_{\mathrm{m}} = \frac{DR_{\mathrm{max}}/DR_{\mathrm{spun}})}{\epsilon_{\mathrm{b}}/100\% + 1} \tag{3.88a}$$

$$= \exp(\varepsilon_{\mathrm{max}} - \varepsilon_{\mathrm{spun}} - \varepsilon_{\mathrm{b}}) \,. \tag{3.88b}$$

In the equations above $DR_{\mathrm{m}} = DR_1 \cdot DR_2 \cdot \ldots \cdot DR_{\mathrm{N}}$ denotes the *total draw ratio* of the (machine) drawing procedure with N drawing steps; $\varepsilon_{\mathrm{max}}$

is the polymer specific constant describing the maximum possible elongation. The (logarithmic) as-spun elongation $\varepsilon_{\mathrm{spun}}$ reached in the fibre formation process correlates with the elastic elongation and orientational deformation of the visco-elastic rheological model (MAXWELL or PHAN-TIEN-TANNER) as described before: $\varepsilon_{\mathrm{spun}} \approx \varepsilon_{\mathrm{o}}$. Figure 3.16 depicts this concept.

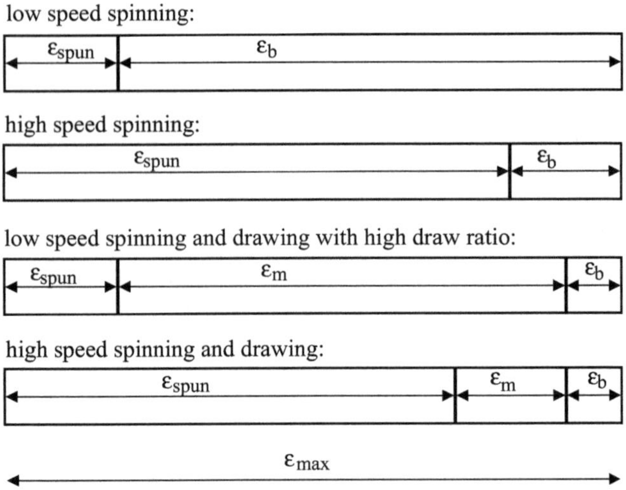

Fig. 3.16. Principle of network deformation concept: independent drawing steps

Tenacity. The tenacity TN (sometimes called strength) is defined as the breaking force F_{b} divided by the initial fineness Tt_0 of the filament at the beginning of the force-elongation experiment $TN = F_{\mathrm{b}}/Tt_0$. Melt spun fibres often show a strain hardening behaviour with increasing elongation. On the other hand, the higher the *pre-orientation* of the fibre (resulting from take-up velocity and spinning stress) the higher the tenacity and the lower the residual elongation to break. For typical melt spinning polymers like PA and PET the product of tenacity and residual elongation to break, the *(true) stress at break,*

$$\sigma_{\mathrm{b}} = TN \cdot \left(\frac{\epsilon_{\mathrm{b}}\ (\text{in}\ \%)}{100\%} + 1 \right) \tag{3.89}$$

is nearly independent of pre-orientation and the resulting elongation and can be taken as a property of the polymer and its molecular weight alone. This assumption in combination with the concept of the independent deformation steps (Eq. 3.87b) allows us to predict the tenacity TN depending upon the machine draw ratio DR_{m} of the spun and drawn fibre if the elongation to break of the fibre is known:

$$TN = \frac{\sigma_b}{\epsilon_b(\text{in }\%)/100\% + 1} \tag{3.90}$$

$$= \sigma_b \cdot \frac{\exp(\varepsilon_{spun})}{\exp(\varepsilon_{max})} \cdot DR_m \; . \tag{3.91}$$

As a consequence, a (true) stress–(true) strain 'master'-curve of the polymer exists to which the stress-strain curves can be shifted [137,280]. The shift of each curve from the origin represents the realised pre-orientation in the spinning process, resp. the spin draw ratio DR_{spun}. Figure 3.17 shows the stress-strain behaviour and Fig. 3.18 shows the resulting master curve after the shifting procedure for low and high speed spinning of PA 6.

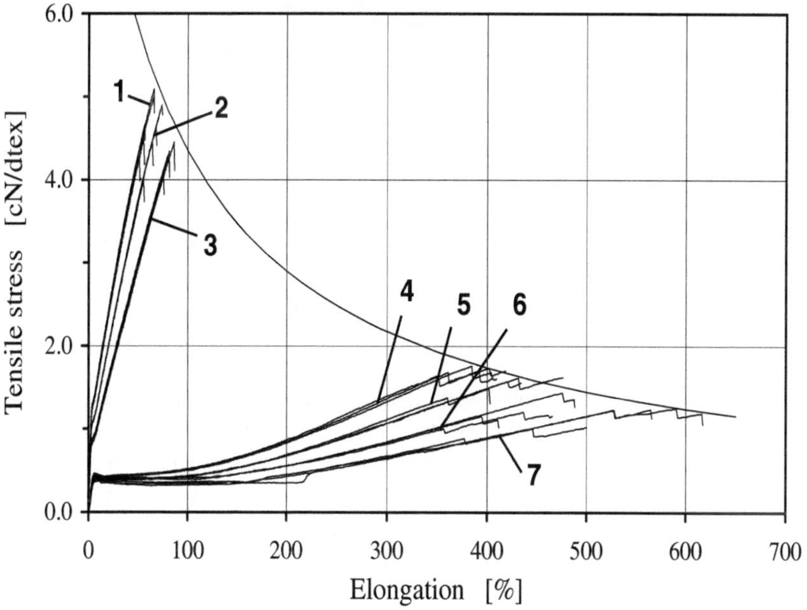

Fig. 3.17. Stress-strain behaviour of as-spun PA 6 fibres; Take-up velocities: 1 – 6000 m/min, 2 – 5000 m/min, 3 – 4000 m/min, 4 – 1000 m/min, 5 – 800 m/min, 6 – 600 m/min, 7 – 400 m/min, data source: Institute of Polymer Research Dresden; hyperbolic envelope: $\frac{8.6\,\text{cN/dtex}}{\epsilon/100\% + 1}$

Remark. The concept of independent deformation steps is quite simple but works well in the manner of an initial approximation. A more exact analysis of properties of spun and drawn fibres shows that ε_{max} is not independent of the drawing procedure. There is a slight increase with increasing draw ratio $\varepsilon_{max} = f(DR_m)$ to a new higher level specifically for higher drawn yarns (FDY). An analogue observation can be made for the maximum possible tenacity TN_{max}.

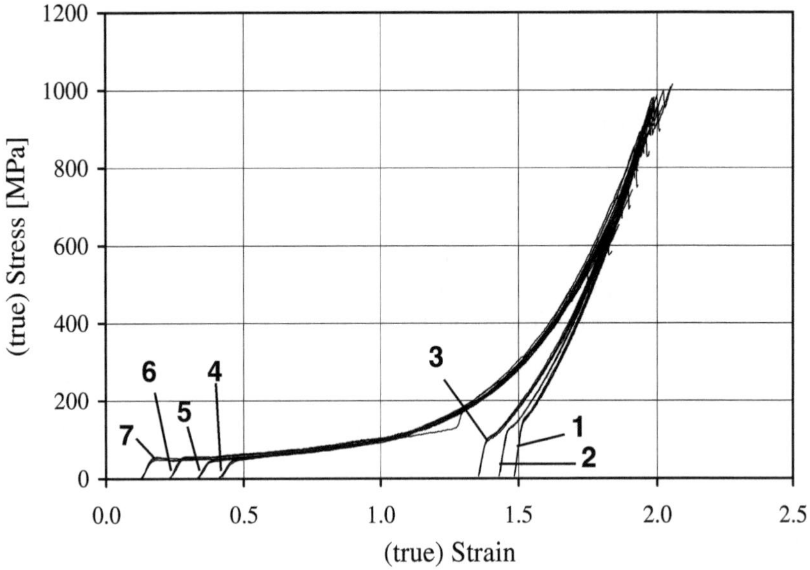

Fig. 3.18. Master curve of PA 6 fibres (true stress - true strain behaviour), shifted curves of Fig. 3.17; Take-up velocities: 1 – 6000 m/min, 2 – 5000 m/min, 3 – 4000 m/min, 4 – 1000 m/min, 5 – 800 m/min, 6 – 600 m/min, 7 – 400 m/min; data source: Institute of Polymer Research Dresden

3.1.7 Practical Modelling: Simulation of Fibre Formation

The complete set of equations which allows the simulation of the fibre formation process differs from case to case, especially with respect to the question of how the authors or investigators describe the deformation behaviour and structure development of the polymer. The available mathematical models show the progress made in the theory of fibre spinning [38,40,59,152–155,157–172,194–197,213,258] but of course they do not satisfy all experimental observations and many of them give no conclusions to the resulting *textile* fibre properties. Generally more agreement exists among the basic balance equations. The following two sections summarise the relations for the fibre formation model and give some examples for the simulation of polyamide 6 melt spinning. The equations used for the model here (specifically the correlations to textile fibre properties) have been developed over many years at the Institute of Polymer Research Dresden as a result of the collaboration with industrial and scientific partners.

Complete Set of Equations

Geometry

$$A(x) = \frac{\pi}{4} D^2(x) \; . \qquad (3.92)$$

Balance equations

$$Q = \varrho_{\mathrm{p}} \cdot v \cdot A \tag{3.93}$$

$$\frac{\mathrm{d}T}{\mathrm{d}x} = -\left(T - T_{\mathrm{air}}\right) \cdot Nu \cdot \frac{\pi\,\lambda_{\mathrm{air}}}{Q\,c_{\mathrm{p}}} + \frac{\Delta H}{c_{\mathrm{p}}}\frac{\mathrm{d}X_{\mathrm{c}}}{\mathrm{d}x}\,, \tag{3.94}$$

$$\frac{\mathrm{d}F}{\mathrm{d}x} = Q\,\frac{\mathrm{d}v}{\mathrm{d}x} + c_{\mathrm{f}}\,\frac{\rho_{\mathrm{air}}}{2}\,v^2\,\pi\,D - \varrho_{\mathrm{p}}\,g\,A\,. \tag{3.95}$$

Remark: The surface tension contribution is neglected.

Rheology: Constitutive equations

Case 1: NEWTON model, purely viscous behaviour

$$\sigma = \eta\,\frac{\mathrm{d}v}{\mathrm{d}x}\,. \tag{3.96}$$

Case 2: MAXWELL model, visco-elastic behaviour

$$\sigma + \lambda\,v\,\frac{\mathrm{d}\sigma}{\mathrm{d}x} = \eta\,\frac{\mathrm{d}v}{\mathrm{d}x}\,. \tag{3.97}$$

Case 3: upper convected MAXWELL model, visco-elastic behaviour

$$\sigma_{\mathrm{xx}} + \lambda\,v\,\frac{\mathrm{d}\sigma_{\mathrm{xx}}}{\mathrm{d}x} = 2\,\lambda\left(G + \sigma_{\mathrm{xx}}\right)\frac{\mathrm{d}v}{\mathrm{d}x}\,, \tag{3.98}$$

$$\sigma_{\mathrm{rr}} + \lambda\,v\,\frac{\mathrm{d}\sigma_{\mathrm{rr}}}{\mathrm{d}x} = -\lambda\left(G + \sigma_{\mathrm{rr}}\right)\frac{\mathrm{d}v}{\mathrm{d}x}\,. \tag{3.99}$$

Case 4: PHAN-TIEN-TANNER model, visco-elastic behaviour with strain softening

$$\kappa \cdot \sigma_{\mathrm{xx}} + \lambda\left(v\,\frac{\mathrm{d}\sigma_{\mathrm{xx}}}{\mathrm{d}x} - 2\frac{\mathrm{d}v}{\mathrm{d}x}\,\sigma_{\mathrm{xx}}\right) = 2\,G\,\lambda\frac{\mathrm{d}v}{\mathrm{d}x}\,, \tag{3.100}$$

$$\kappa \cdot \sigma_{\mathrm{rr}} + \lambda\left(v\,\frac{\mathrm{d}\sigma_{\mathrm{rr}}}{\mathrm{d}x} + \frac{\mathrm{d}v}{\mathrm{d}x}\,\sigma_{\mathrm{rr}}\right) = -G\,\lambda\,\frac{\mathrm{d}v}{\mathrm{d}x}\,, \tag{3.101}$$

$$\kappa = \exp\left(\frac{\alpha}{G}\left(\sigma_{\mathrm{xx}} + 2\,\sigma_{\mathrm{rr}}\right)\right)\,. \tag{3.102}$$

Stress for constitutive equations cases 1 and 2:

$$\sigma = \frac{F}{A}\,, \tag{3.103}$$

for constitutive equations cases 3 and 4:

$$\Delta\sigma = \sigma_{\mathrm{xx}} - \sigma_{\mathrm{rr}} = \frac{F}{A}\,. \tag{3.104}$$

Crystallisation

$$\frac{\mathrm{d}X_{\mathrm{c}}}{\mathrm{d}x} = \frac{K}{v}\left(1 - X_{\mathrm{c}}\right)\,, \tag{3.105}$$

$$K = K_{\max}\,\exp\left(\frac{-4\,(\ln 2)(T - T_{\mathrm{c}})^2}{\Delta T_{1/2}^2} + C \cdot f_{\mathrm{am}}^2\right)\,. \tag{3.106}$$

Birefringence and orientation

$$\Delta n = \Delta n_{\mathrm{am}}(1 - X_{\mathrm{c}}) + \Delta n_{\mathrm{cr}}\, X_{\mathrm{c}} \;, \tag{3.107}$$

$$\Delta n_{\mathrm{am}} = C_{\mathrm{opt}} \cdot \sigma \;, \tag{3.108}$$

$$f_{\mathrm{am}} = \Delta n_{\mathrm{am}} / \Delta n_{\mathrm{am}}^{0} \;. \tag{3.109}$$

Polymer material properties

- mass density $\varrho_{\mathrm{p}} = \varrho_{\mathrm{p}}(T, X_{\mathrm{c}}, ...)$,
- specific heat $c_{\mathrm{p}} = c_{\mathrm{p}}(T, X_{\mathrm{c}}, ...)$,
- elongational viscosity $\eta_{\mathrm{e}} = \eta_{\mathrm{e}}(T, M, \dot{\varepsilon}, X_{\mathrm{c}}, ...)$,
- modulus $E = E(\varepsilon_{\mathrm{o}}, T, X_{\mathrm{c}}, ...)$,
- relaxation time $\lambda(T, \ldots)$,
- solidification temperature $T_{\mathrm{s}} = T_{\mathrm{s}}(\sigma, ...)$.

Transfer equations: heat and momentum

- heat transfer: NUSSELT number $Nu = Nu(Re_{\|}, Re_{\perp})$,
- air friction coefficient $c_{\mathrm{f}} = c_{\mathrm{f}}(Re_{\|}, Re_{\perp})$.

Environment material properties, e. g. for surrounding air

- density of air ϱ_{air} ,
- heat conductivity of air λ_{air} ,
- kinematic viscosity of air ν_{air} .

Structure development and textile fibre properties

- elongation to break ε_{b} ,
- tenacity TN ,
- orientation or birefringence Δn ,
- crystallinity X_{c} ,
- possible draw ratio (to get full drawn yarn, FDY) DR .

Initial and boundary conditions. To solve the coupled system of ordinary differential equations of fibre formation above, the related initial and/or boundary conditions are necessary. Known conditions at the spinneret are

- $T(0) = T_0$, the filament (spinning or extrusion) temperature, and
- $D(0) = D_0$, the filament diameter, resp. the filament velocity (= extrusion velocity) $v(0) = v_0$.

The initial rheological force $F(0) = F_0$ is unknown, therefore the boundary condition for the take-up point L has to be used:

- $v(L) = v_{\mathrm{L}}$, the take-up velocity.

In practice, a shooting method needs to be applied in order to solve the differential equations. The unknown initial rheological force F_0 is estimated first and then varied in an iterative procedure until the velocity v reaches the take-up velocity v_L after solidification within a given tolerance range. The other initial conditions (e. g. the initial crystallisation rate for computing the crystallisation) are usually set to zero at the starting point.

At higher take-up velocities stress-induced crystallisation may occur. The differential equations describing the crystallisation kinetics and the fibre formation procedure become stiffer and for solving them numerically a step width control is necessary. The maximum step width $\Delta x_{\max}$ can be fitted to a given maximum velocity step: $\Delta x_{\max} < \Delta v_{\max} \cdot \left(\frac{dv}{dx}\right)^{-1}$.

Examples of Calculation

The primary interest in simulating the melt spinning process is the possibility to get results on how the large number of process parameters influence the spinning behaviour and consequently the fibre properties. But the simulation results have to be verified by means of experiments in order to assess the quality of the model. These tests must be carried out in a twofold manner. At first, by measuring the physical variables within the fibre formation zone itself, for example by measuring the velocity v vs. distance x from the spinneret. Secondly, by determining the resulting fibre properties after spinning, for example the elongation to break, the birefringence, or the tenacity of the fibres. Many researchers have investigated several aspects to model the fibre formation process. However, their results differ more or less with respect to a satisfactory description of the material behaviour and correlation to the fibre properties. Sometimes the experimental verification seems rather difficult.

The following simulations were carried out exemplary for PA 6 (but would also be possible for other materials) and show some effects of changing the process variables to fibre formation. The computer simulation program and the equations used here for describing the material behaviour and the correlations to fibre properties (mainly for PA 6 and PET) were elaborated and tested over a long time during the scientific cooperation of the authors and their coworkers at the Institute of Polymer Research Dresden, together with several partners from the industry. The strong interaction between experimental and theoretical work is an unalterable requirement for developing and improving any model.

Figure 3.19 shows the fibre velocity v versus distance x from the spinneret exit for a typical melt spinning experiment of PA 6. The points in the figure are determined experimentally by means of laser doppler anemometry (LaserSpeed LSM50, TSI Inc.), the solid lines are calculated using computer simulation for two different rheological models discussed in Sect. 3.1.4. Numerous experiments with different spinning conditions have been conducted in order to improve and verify the model equations. Now the simulation can be used for quick estimation of fibre formation, the simulation needs less than 1 second on a modern personal computer.

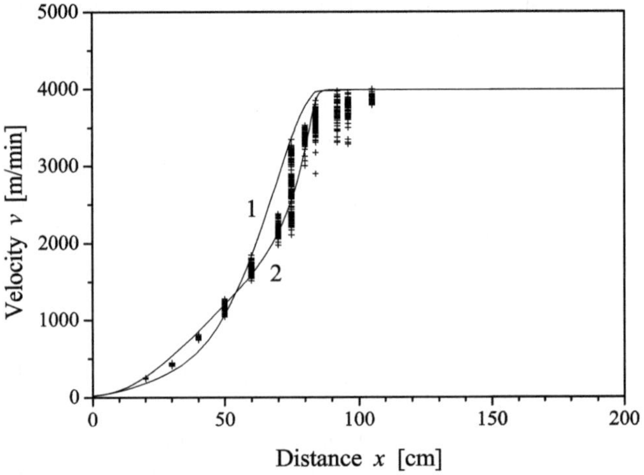

Fig. 3.19. Fibre velocity v vs. distance x from spinneret, *points:* velocity measured by means of laser doppler anemometry, each distance with 100...200 measurements, *solid lines:* calculated, 1 – MAXWELL model, 2 – PHAN-TIEN-TANNER model for rheological behaviour, *Spinning parameter:* PA 6, molecular weight: 17 000, throughput (per hole): 1.5 g/min, diameter of capillary: 0.25 mm, melt temperature: 250 °C, velocity of quenching air: 0.35 m/s, temperature of quenching air: 15 °C, take-up velocity: 4000 m/min; data source: Institute of Polymer Research Dresden

Figure 3.20 shows another comparison between experimentally determined and calculated fibre velocities, respectively. Each point in the diagram represents the mean value of 500...1000 measurements of velocity. The simulation results are in accordance with the experimental data, only for the high take-up speed of 5000 m/min the calculation seems to become a little inaccurate which shows a small gap of about 5 cm for the step (like a jump) to the final velocity. It becomes clear that melt spinning is a highly dynamic process, specifically the course of fibre formation at higher take-up velocities is very sensitive to little changes because of the occurrence of the stress induced crystallisation. *In principle* it seems that the accuracy of simulation which can be reached with current models is in the magnitude of about 10 percent of deviation.

The next simulations carried out for PA 6 and shown in the following diagrams are based on the MAXWELL model in order to describe the rheological behaviour. The molecular (number) weight (Mn) of the polymer is about 17 000. The molecular weight is related to the relative solution viscosity (in H_2SO_4, 96%, 20 °C) of about 2.45.

Figure 3.21 shows the effect of different take-up velocities on the *cooling behaviour* (up to 6000 m/min). It can be observed that there is only weak

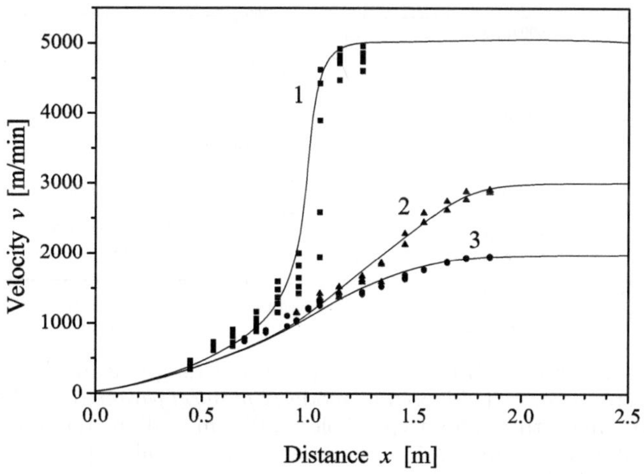

Fig. 3.20. Fibre velocity v vs. distance x from spinneret exit, *symbols:* velocity measured by means of laser doppler anemometry, each point represents the mean value of 500...1000 individual velocity measurements, *solid lines:* calculated with PHAN-TIEN-TANNER model for rheological behaviour, *Spinning parameter:* PA 6, molecular weight: 17 000, throughput (per hole): 2.0 g/min, diameter of capillary holes: 0.30 mm, melt temperature: 265 °C, no quenching air, temperature of environment: 25...28 °C, 1 – take-up velocity: 5000 m/min, 2 – 3000 m/min, 3 – 2000 m/min; data source: Institute of Polymer Research Dresden

coupling between velocity and cooling. On the other hand, the take-up velocity is the most important process parameter to affect the fibre properties. The higher the velocity, the higher the tensile stress and the higher the resulting orientation of the as-spun filaments. The higher the orientation, the lower the elongation to break. In reality, the spinning experiment with 1 g/min throughput and 6000 m/min take-up is not practicable because of the very high increase of spinning stress (see Fig. 3.22).

The two figures (Figs. 3.23 and 3.24) show the graphs of fibre formation for different mass throughputs Q. The mass throughput strongly influences the fibre cooling behaviour and therefore the filament temperature $T(x)$, and also the fineness of the as-spun filaments.

As shown in Fig. 3.24 the variation of the mass throughput Q has only a little effect on the stress at the solidification point. The higher throughput causes an increased length of the fibre formation zone and leads to an increase of the acting fibre force F (inertia and the air friction force increase with increasing mass throughput and distance). However, the larger diameter D, respectively the larger cross-sectional area A, compensate this effect with respect to the stress $\sigma = F/A$. Finally, the variation of troughput leads to nearly similar stress at the solidification point. There is only little changing of the fibre orientation and elongation to break with the variation of mass throughput.

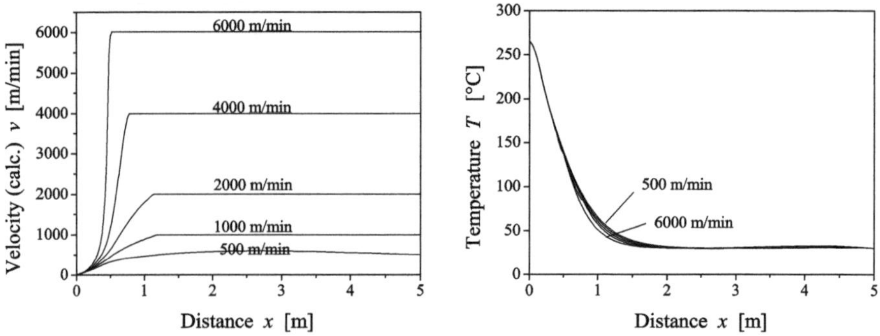

Fig. 3.21. (Calculated) fibre velocity v (*left*) and temperature T (*right*) vs. distance x from spinneret exit, MAXWELL model, *Spinning parameter:* PA 6, molecular weight: 17 000, mass throughput (per hole): 1.0 g/min, diameter of capillary hole: 0.30 mm, melt temperature: 265 °C, no quenching air, take-up velocity indicated: 500, 1000, 2000, 4000 and 6000 m/min

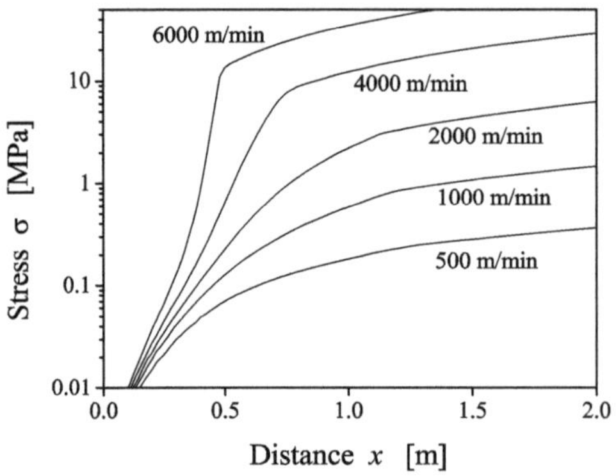

v_L m/min	x_s m	σ_s MPa	ϵ_{spun} %
500	1.20	0.23	416
1000	1.16	0.80	257
2000	1.12	2.99	138
4000	0.76	7.94	79
6000	0.50	13.41	54

Fig. 3.22. *Left:* (calculated) tensile stress $\sigma(x)$ vs. distance x from spinneret, MAXWELL model, *right:* stress σ_s at solidification point x_s and resulting elongation to break ϵ_{spun} of as-spun fibres, *Spinning parameter:* see Fig. 3.21

Combining the changes of both process variables, mass throughput Q and take-up velocity v_L, leads to the diagrams shown in Fig. 3.25. In this figure the graphs for the proportional increase of both $Q \propto v_L$ are depicted, this means *constant* as-spun fineness Tt.

Obviously, the stress at the solidification point increases if the mass throughput and take-up velocity increase in a proportional manner. The resulting fibre orientation also increases and the elongation to break for the as-spun fibres decreases.

Finally, the effect of the cooling conditions will be investigated. Figure 3.26 shows the influence of the environmental air temperature T_{air} and Fig. 3.27 depicts the effect of different velocity profiles v_{air} of quenching air on fibre temperature $T(x)$ and fibre velocity $v(x)$.

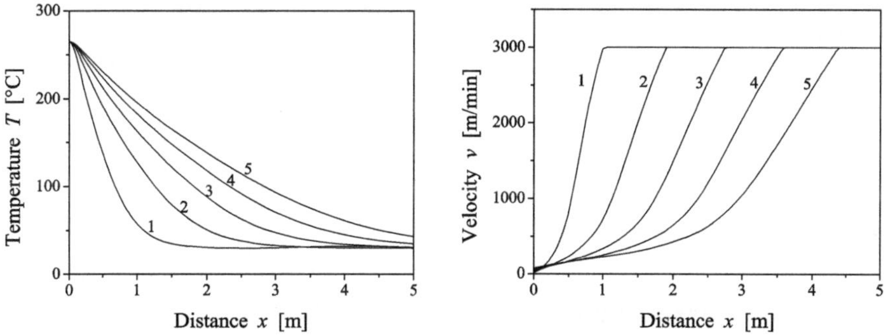

Fig. 3.23. (Calculated) fibre temperature T (*left*) and velocity v (*right*) vs. distance x from spinneret exit, MAXWELL model, *Spinning parameter:* PA 6, molecular weight: 17 000, mass throughput (per hole) indicated: 1...5 g/min (#1...#5), diameter of capillary hole: 0.30 mm, melt temperature: 265 °C, no quenching air, take-up velocity: 3000 m/min

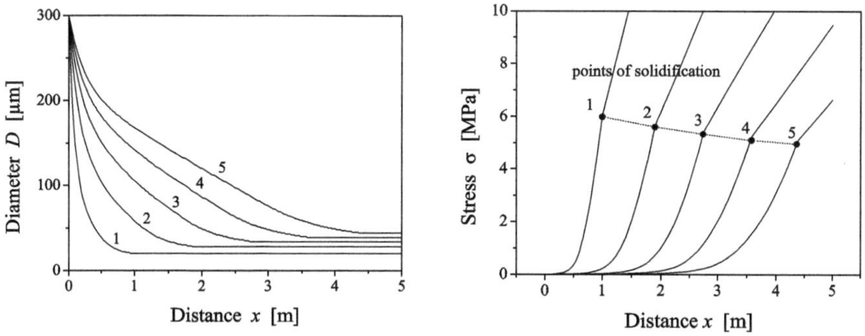

Fig. 3.24. (Calculated) fibre diameter D (*left*) and tensile stress σ (*right*) vs. distance x from spinneret exit, MAXWELL model, *Spinning parameter:* PA 6, molecular weight: 17 000, mass throughput (per hole) indicated: 1...5 g/min (#1...#5), diameter of capillary: 0.30 mm, melt temperature: 265 °C, no quenching air, take-up velocity: 3000 m/min

Table 3.10 summarises the effects of changing certain parameters with respect to fibre orientation, expressed by the elongation to break.

It is interesting to see the model predictions for the practical task of changing the spinning parameter under the conditions of constant fineness *and* constant elongation to break *after* drawing. The following is a typical question that occurs in fibre production: How to change mass throughput, take-up velocity and draw ratio to increase productivity (expressed by the mass throughput) under the condition of unchanging final fibre properties after drawing? An answer can be given with the help of the fibre formation model. The next figures show the main results for the task of spinning a full oriented PA 6 yarn (FDY) with final filament fineness of 4 dtex and final elongation to break of 25% after drawing.

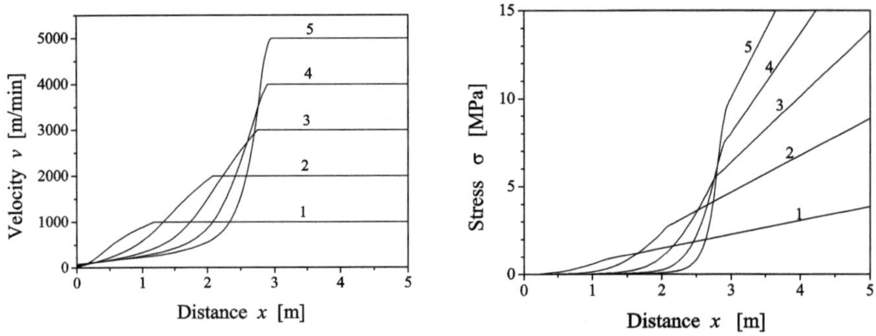

Fig. 3.25. (Calculated) fibre velocity v (*left*) and tensile stress σ (*right*) vs. distance x from spinneret exit, MAXWELL model, *Spinning parameter:* PA 6, molecular weight: 17 000, diameter of capillary holes: 0.30 mm, melt temperature: 265 °C, no quenching air, constant as-spun fineness of $Tt = 10$ dtex: 1 — 1 g/min, 1000 m/min, 2 – 2 g/min, 2000 m/min, 3 – 3 g/min, 3000 m/min, 4 – 4 g/min, 4000 m/min, 5 – 5 g/min, 5000 m/min

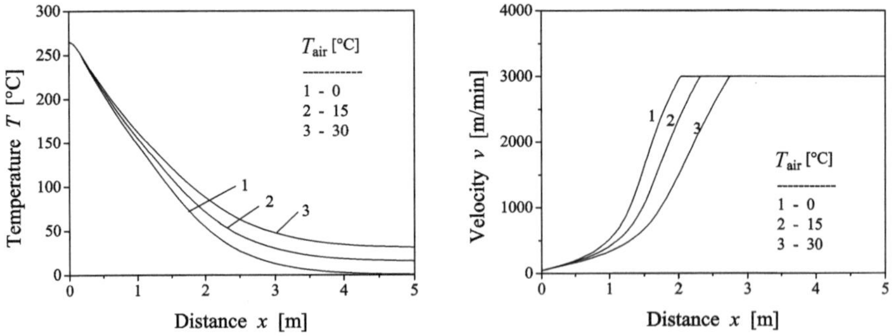

Fig. 3.26. (Calculated) fibre Temperature T (*left*) and velocity v (*right*) vs. distance x from spinneret exit, MAXWELL model, *Spinning parameter:* PA 6, molecular weight: 17 000, diameter of capillary hole: 0.30 mm, melt temperature: 265 °C, mass throughput: 3 g/min, take-up velocity: 3000 m/min, different temperatures of environmental air: 1 – 0 °C, 2 – 15 °C, 3 – 30 °C

Figure 3.28 depicts the simple relations between as-spun fineness Tt and needed draw ratio DR for different mass throughputs Q vs. take-up velocity v_L under the condition that the final filament fineness of (here) $Tt_\mathrm{f} = 4$ dtex should be reached after drawing:

$$Tt = \frac{Q}{v_\mathrm{L}} \quad , \qquad DR = \frac{Tt}{Tt_\mathrm{f}} = \frac{Q}{v_\mathrm{L} \cdot Tt_\mathrm{f}} \; .$$

Each point of these graphs represents the necessary draw ratio DR to reach the final filament fineness of 4 dtex for each combination (Q, v_L). But each combination of the spinning parameters (Q, v_L) is also connected with destinated orientation of the as-spun fibres and therefore the necessary draw ratio to reach the final fineness of 4 dtex leads to very different final elongations

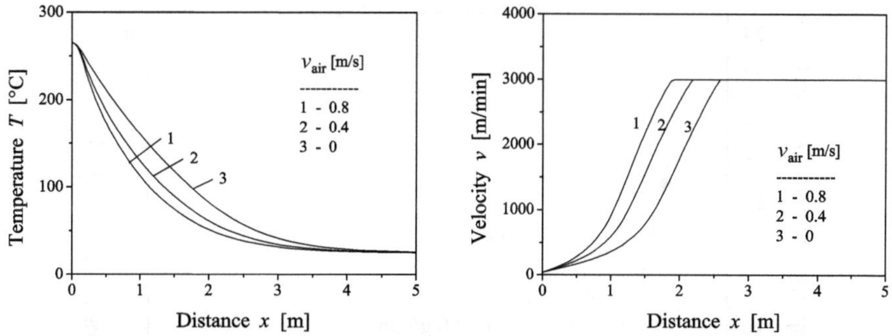

Fig. 3.27. (Calculated) fibre temperature T (*left*) and velocity v (*right*) vs. distance x from spinneret, MAXWELL model, *Spinning parameter:* see Fig. 3.26, different velocity profiles of quenching air: 1 – no quenching air, 2 – $v_{\mathrm{air}} = 0.4$ m/s, 3 – $v_{\mathrm{air}} = 0.8$ m/s, temperature of air: $T_{\mathrm{air}} = 25\,°\mathrm{C}$

to break. Only at certain combinations (in Fig. 3.28 they are marked by the symbol (•)) the elongation to break of 25% can be obtained.

Table 3.10. Variation of technological parameters: effects on orientation resp. elongation to break for as-spun filaments.

Changing	see Fig.	Example	ϵ_b in %
Take-up velocity v_L	3.21	500 m/min	416
($Q = 1$ g/min)		1000 m/min	257
		2000 m/min	138
		4000 m/min	79
Throughput (per hole) Q	3.24	1 g/min	95
($v_L = 3000$ m/min)		2 g/min	98
		3 g/min	101
		4 g/min	102
$Q \propto v_L$	3.25	1 g/min, 1000 m/min	257
($Tt = 10$ dtex)		2 g/min, 2000 m/min	146
		3 g/min, 3000 m/min	101
		4 g/min, 4000 m/min	84
Air temperature T_{air}	3.26	0 °C	109
($v_{\mathrm{air}} = 0$ m/s)		15 °C	105
		30 °C	101
Air velocity v_{air}	3.27	0.0 m/s	102
($T_{\mathrm{air}} = 25\,°\mathrm{C}$)		0.4 m/s	100
		0.8 m/s	97

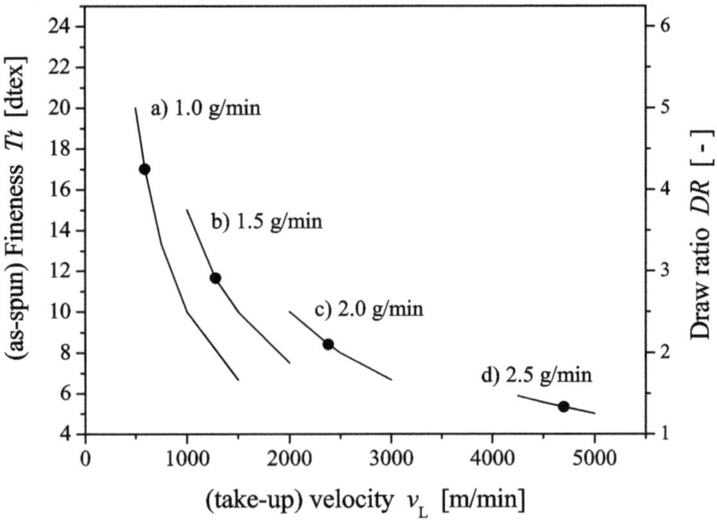

Fig. 3.28. Relation between (as-spun) fineness Tt and needed draw ratio DR vs. take-up velocity v_L to reach final filament fineness of $Tt_f = 4$ dtex after drawing; different mass throughputs Q: a) 1.0 g/min, b) 1.5 g/min, c) 2.0 g/min, d) 2.5 g/min, the symbols (•) mark the combinations where additionally the (calculated) final elongation of 25% after drawing is reached, see also Fig. 3.29, *Spinning parameter:* PA 6, moleculare weight: 17 000, diameter of capillary hole: 0.30 mm, melt temperature: 265 °C, no quenching air

The calculated results for the dependency of elongation to break of the as-spun fibres on take-up velocity v_L for different mass throughputs Q are shown in the upper part of Fig. 3.29. In the lower part the resulting final elongations to break ϵ_b after drawing to the final filament fineness of 4 dtex are depicted. It can be seen that only for certain combinations (Q, v_L) it is possible to reach the required final properties of $Tt_f = 4$ dtex fineness *and* $\epsilon_b = 25\%$ elongation.

In Fig. 3.30 the results of the calculations are summarised and the answer to the question for how to increase productivity, posed at the beginning of this section, is given. The relation between mass throughput and take-up velocity is a nonlinear one: To increase the mass throughput, a super proportional increase of the take-up velocity is necessary if the final fibre properties all remain unchanged.

It is easy now to get the direction of how to adjust the general parameters throughput and take-up velocity for a given task in fibre spinning and also to appreciate the effects of changing spinning parameters on fibre properties.

Therefore the model of fibre formation can be a powerful tool not only in fibre research but also in the production process.

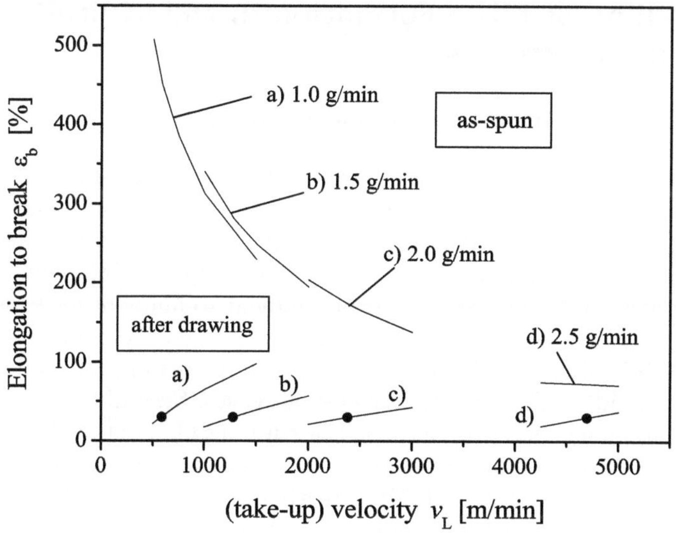

Fig. 3.29. (Calculated) elongation to break vs. take-up velocity v_L *before* drawing (as-spun) and *after* drawing to the final fineness of $Tt_f = 4$ dtex, different mass throughputs Q: a) 1.0 g/min, b) 1.5 g/min, c) 2.0 g/min, d) 2.5 g/min, the points (•) mark the required 25% elongation to break after drawing; *Spinning parameter:* see Fig. 3.28

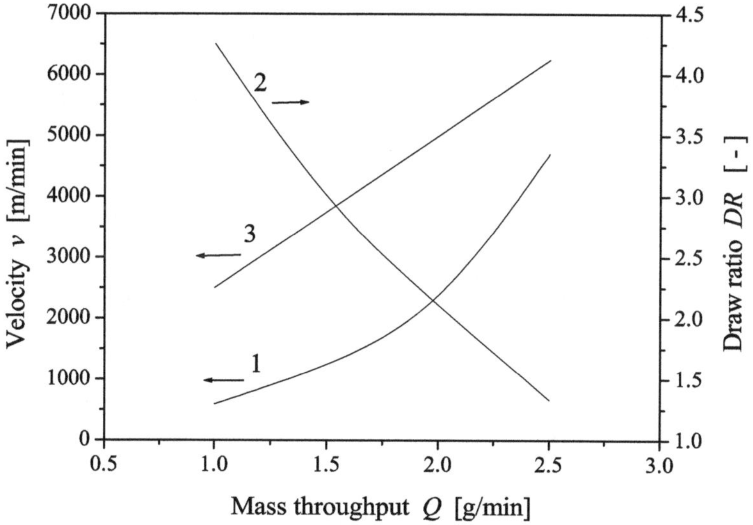

Fig. 3.30. Relationships between (1) take-up velocity v_L, (2) draw ratio DR and (3) wind-up velocity $v_w = DR \cdot v_L$ vs. mass throughput Q (per hole) for FDY spinning process with final filament fineness of $Tt_f = 4$ dtex and (calculated) elongation to break of $\varepsilon_b = 25\%$ after drawing, *Spinning parameter:* see Fig. 3.28

3.2 Modelling of Fibre Formation in the Multifilament Melt Spinning Process

3.2.1 Peculiarities in Multifilament Spinning

In the prior discussion about modelling the stationary single filament process, the environment of the fibre was initially assumed to be the boundary condition with a given temperature $T_{\text{air}}(x)$, resp. velocity profile $v_{\text{air}}(x)$ of the surrounding air. It is important to recognise that in multifilament spinning these boundary conditions can vary from filament to filament location within the filament bundle because of the interaction between the filaments and air. At first, the hot filaments dissipate heat to the quenching air while it flows through the bundle. The air temperature increases. Secondly, the air flow is deflected in filament direction because the running filaments will impart the axial component and reduce the transversal component of air velocity. These effects are illustrated in Fig. 3.31. On the other hand, the cross flow of air will also disturb the axially sucked air flow in the boundary layer of the filaments. For a sufficient number of filaments per square unit the boundary layers for the individual filaments also may overlap. Near the converging point where the filaments are bundled the air flow inside the bundle will be pressed out. As a result, the friction forces for the individual filaments then vanish.

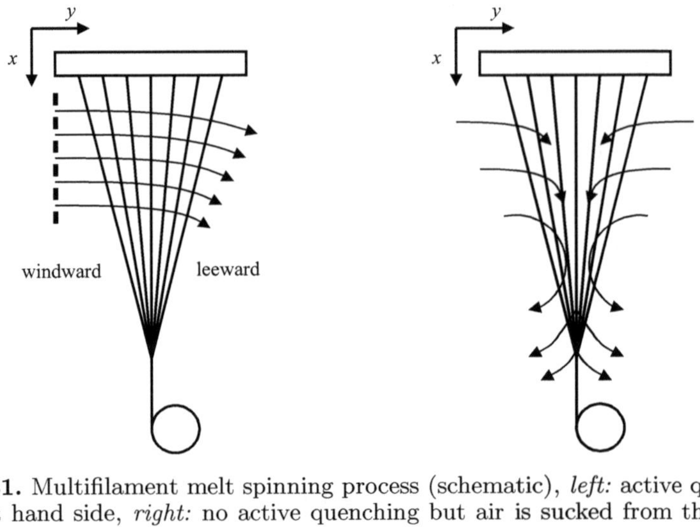

Fig. 3.31. Multifilament melt spinning process (schematic), *left:* active quenching from left hand side, *right:* no active quenching but air is sucked from the bundle itself

All these effects influence the heat and momentum transfer from each filament surface to the environment. Different geometries of the bundle and air flow are possible, too (Fig. 3.32). Therefore, the spinning conditions of the individual filaments within the bundle vary and the spun filaments show differences in structure and properties as a function of their position. These differences lead to non-uniform or scattered properties, respectively, which may influence the further processing in a negative manner.

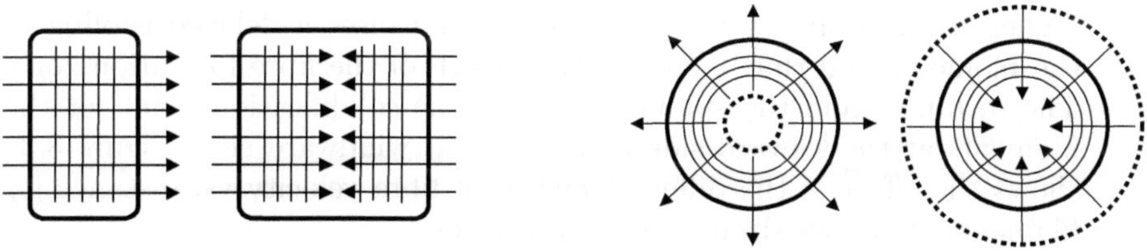

Fig. 3.32. Possible geometries of multifilament spinning. *Left:* one-sided and double-sided cross quenching; *right:* radial blowing from inside to outside and vice versa

Examples. The following two examples give a simple estimation in which a rise in temperature for the quenching cross air flow in a filament bundle can be expected:

1. POY - yarn, PA 6

polymer:	poly(amide) 6
spinneret:	24 holes arranged on two concentric circles,
	diameter of outer circle: $D_1 = 50\,\text{mm}$,
	diameter of inner circle: $D_2 = 40\,\text{mm}$,
	diameter of capillary holes: $D_0 = 0.25$ mm,
melt temperature:	$T_0 = 255\,°\text{C}$,
throughput:	$Q_{\text{PA}} = 24 \times 1.5$ g/min $= 36$ g/min,
take-up velocity:	$v_{\text{L}} = 3000$ m/min...4000 m/min,
quenching:	cross quenching from one side,
	air temperature $T_{\text{air}} = 20\,°\text{C}$,
	air velocity of air $v_{\text{air}} = 0.35$ m/s,
	length of quenching zone $L_{\text{air}} = 1.20$ m .

Assuming that the fibre formation zone ends at the end of the quenching zone (no further deformation occurs) the transported heat $Q_{\text{air}} \cdot c_{\text{air}}$ and the temperature rise ΔT_{air} of air can be estimated as follows

$$Q_{\text{air}} \cdot c_{\text{air}} \cdot \Delta T_{\text{air}} = Q_{\text{PA}} \cdot c_{\text{PA}} \cdot \Delta T_{\text{PA}} \,, \tag{3.110}$$

where c_{air} is the specific heat capacity of air, and the air volume affecting the fibre bundle is approximately given by $Q_{\text{air}} \approx \varrho_{\text{air}} \cdot v_{\text{air}} \cdot L_{\text{air}} \cdot D_1$, while using the diameter D_1 of the outer circle as effective width of quenching air. Using the following parameters

temperature change of polymer:	$\Delta T_{PA} = 200$ K,
specific heat capacity of PA 6:	$c_{PA} = 2.4$ kJ/(kg · K),
specific heat capacity of air:	$c_{air} = 1.0$ kJ/(kg · K),
mass density of air:	$\varrho_{air} = 1.2$ kg/m^3,

leads to the temperature rise of quenching air of $\Delta T_{air,1} \approx 10$ K. The higher temperature at the exit side of air causes a delayed cooling of the filaments and thus increases the length of the fibre formation zone. The distance from the spinneret exit to the solidification point for the filaments at the leeward side is larger as at windward side. Figure 3.33 shows the effect of the different curves of fibre velocity vs. distance for filaments at the windward and leeward side.

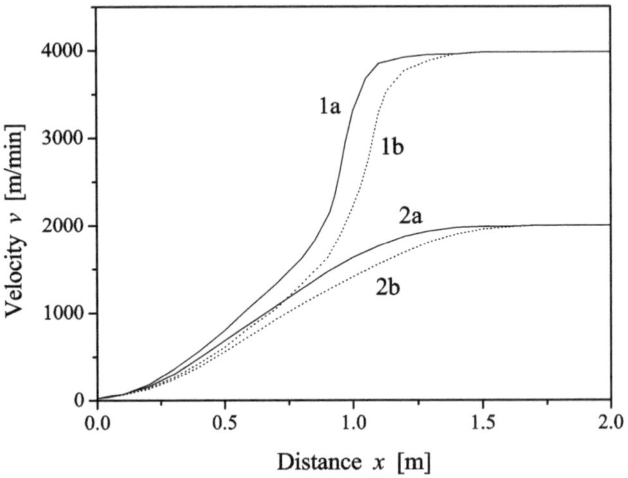

Fig. 3.33. Filament velocity v (measured by means of laser doppler anemometry) vs. distance x from spinneret exit, differences between windward (a) and leeward (b) sides; quenching air velocity: 0.35 m/s, PA 6 melt spinning, throughput: 1.5 g/min, take-up velocity: 1 – 4000 m/min, 2 – 2000 m/min; data source: Institute of Polymer Research Dresden

2. Staple fibre process, PET
   The example concerns staple fibre spinning with the following parameters:

polymer:	poly(ethylene terephthalate), PET
spinneret:	1300 holes, 13 rows with 100 holes in each row, dimension W x B: 26 cm x 6.5 cm, rectangular, diameter of capillary holes: 0.30 mm,
melt temperature:	$T_0 = 290\,^\circ$C,
throughput:	$Q_{PET} = 1300 \times 0.6$ g/min $= 780$ g/min,

take-up velocity:	$v_L = 1200$ m/min,
quenching:	cross quenching from one side,
	air temperature $T_{\text{air}} = 20\,^{\circ}\text{C}$,
	air velocity $v_{\text{air}} = 1.5$ m/s,
	length of quenching zone $L_{\text{air}} = 0.5$ m .

The fibre formation should be finished after a distance of 0.5 m. In analogy to the first example and using the parameters applicable to PET

temperature change of polymer:	$\Delta T_{\text{PET}} = 220$ K,
specific heat capacity PET:	$c_{\text{PET}} = 1.5$ kJ/(kg · K),
effective width of quenching air:	$W = 26$ cm,

then follows the mean temperature rise of quenching air at the exit side of approximately $\Delta T_{\text{air},2} \approx 20$ K

The estimated temperature change of air can be confirmed for the PA 6–POY process experimentally. But, in the case of the second example of the PET–staple fibre process, the experimental observations differ strongly from the estimation: The temperature rise of quenching air at the leeward side varies with distance from the spinneret and reaches *more than* 100 K near the spinneret (see Fig. 3.44 following on page 115). Based on the rather simple heat balance, the approximation carried out is no longer valid because the local effects of air deflection and heat transfer are neglected. A more detailed analysis becomes mandatory.

3.2.2 Models of Interaction Between the Fibre Bundle and the Environment

The temperature and velocity field of air within the fibre bundle and the fibre strands interact and thus influence each other [247, 248]. Therefore, the multifilament process should be treated as a two-phase system [249] where the filaments are embedded in an environmental 'matrix'. At the boundary layers between the two phases the general conditions for heat and momentum transfer have to be satisfied. This method was employed specifically for the wet spinning process by Szaniawski and Zachara [250–254]. But, in case of a common melt spinning process where the 'matrix' is given by air, the properties and the behaviour of the two phases are extremely different. This is why it is possible to regard both phases in a separate manner, and finally to combine a model of multifilament melt spinning from three parts

- the single fibre formation model,
- the model of air velocity and air temperature fields,
- and the model of interaction between both.

Due to the heat and momentum exchange between fibres and air an iterative calculation procedure is necessary. The iteration can be carried out in the following manner (Fig. 3.34):

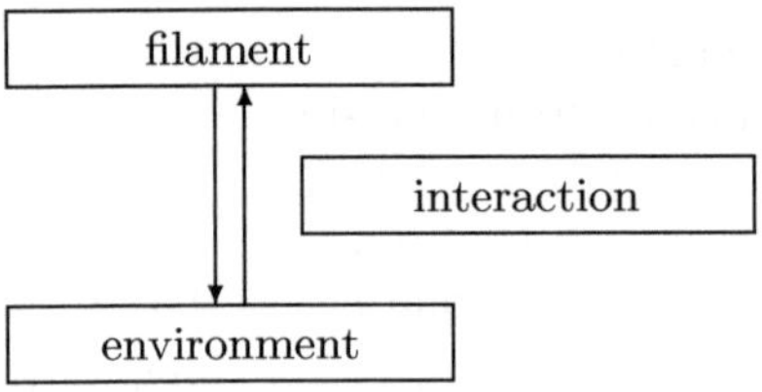

Fig. 3.34. Schematic model for multifilament spinning

1. Calculation of velocity and temperature fields of air on the basis of simple assumptions about fibre formation.
2. Calculation of the fibre formation process for some (not all) single fibres in various positions across the bundle on the basis of the calculated velocity and temperature fields of air.
3. Like step 1, calculation of velocity and temperature fields of air, but now on the basis of the new calculated fibre formation results.

Steps 2 and 3 need to be repeated until sufficient convergence is reached, i. e. no further changes occur. Finally, the calculation of fibre formation for *all fibre positions* should be conducted with the final temperature and velocity fields of air.

Cells method. MATSUO et al. [255], YASUDA et al. [65, 66] and ISHIHARA et al. [256, 257] developed a method to treat the multifilament effects. They divided the fibre bundle into individual cells for each filament row to apply the balance equations of energy and momentum (Fig. 3.35). This method was also used by DUTTA [258, 259] and later similar treatments were developed by means of *FEM* (finite elements method) calculations.

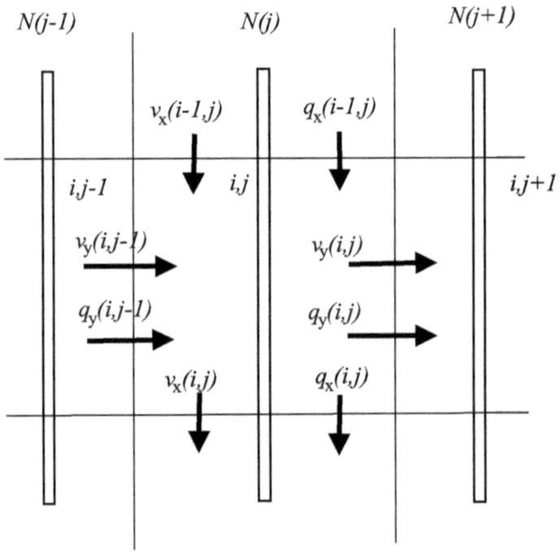

Fig. 3.35. Multifilament bundle divided into individual cells

In Fig. 3.35, $N(j)$ is the spinline of row number j, $v_x(i,j)$ and $v_y(i,j)$ are the components of air velocity in the cell (i,j), and q_x, q_y are the components of heat flow in x and y direction, respectively. The dimensions of each cell are given by $\Delta x \cdot \Delta y \cdot \Delta z$ (dimensions in fibre direction, in blowing air direction, and perpendicular to both). The balance of air flow is composed by four flow elements

$$\big(v_y(i,j-1) - v_y(i,j)\big)\Delta z\,\Delta x + \big(v_x(i-1,j) - v_x(i,j)\big)\Delta y\,\Delta z = 0 \quad (3.111)$$

and the heat flow balance is given by five elements

$$q_y(i,j-1) - q_y(i,j) + q_x(i-1,j) - q_x(i,j) = \big(T(i,j) - T_{air}\big)\,Nu\,\pi\,\lambda_{air}\,. \quad (3.112)$$

The components of heat flow are defined as

$$q_y(i,j) = \varrho_{air}\,c_{air}\,v_y(i,j)\,T_{air}(i,j)\,\Delta z\,\Delta x \quad (3.113)$$

$$q_x(i,j) = \varrho_{air}\,c_{air}\,v_x(i,j)\,T_{air}(i,j)\,\Delta z\,\Delta y\,, \quad (3.114)$$

where $T(i,j)$ and $T_{air}(i,j)$ denote the filament and air temperature in cell number (i,j), respectively, ϱ_{air} and c_{air} are the density and heat capacity of air. The main problem connected with this method is the determination of the exact amount of *air pumping rate* of fibre – this is the difference of the entrained air flows $v_x(i,j) - v_x(i-1,j)$. Some researchers used the estimation of SAKIADIS [208–210] based on the boundary layer theory for continuous cylindrical surfaces. But the cross air flow disturbs the cylindrical symmetry of the boundary layer and a detailed analysis may show that the assumptions made by SAKIADIS are no longer valid in this case. There is no sufficient theory for the flow behavior around a fibre in motion with additional cross air flow. Nevertheless, some calculations using *FEM*-methods more recently have been reported.

3.2.3 Continuum Theory

In order to avoid the difficulties related to the pumping effect of filaments (which are quenched by an additional cross air flow) another model can be used [260, 261]. The basic idea of this model was originally developed by SCHÖNE. The following section describes the method of calculating the velocity and temperature fields of air for symmetric fibre bundles based on the continuum theory of hydrodynamics. The application of this model to the multifilament staple fibre spinning will be demonstrated later.
The main ideas of the continuum method are:

- Momentum balance: the equivalence of both air friction force acting on fibres and acceleration impart of axial velocity component of air.
- Heat balance: the equivalence of both cooling of fibres and heating of quenching air.

- Continuously smoothed sources: the fibres are the sources of momentum and heat imposed to the quenching air. These sources are assumed as continuously distributed density fields in the bundle region.
- Rectangular or axially symmetric geometry is assumed in order to simplify the mathematical treatment and to allow for the use of analytical expressions for the quenching air flow.

Air Velocity

Rectangular geometry. The expressions for components v_x and v_y of the air velocity field $\boldsymbol{v}_{air}$ are given by the NAVIER-STOKES equations and the continuity equation. In cartesian coordinates they are written as follows

$$\varrho_{air}(v_x v_{x,x} + v_y v_{x,y}) = \eta_{air}(v_{x,xx} + v_{x,yy}) + f_x - p_{,x} \; , \tag{3.115}$$

$$\varrho_{air}(v_x v_{y,x} + v_y v_{y,y}) = \eta_{air}(v_{y,xx} + v_{x,yy}) + f_y - p_{,y} \; , \tag{3.116}$$

$$v_{x,x} + v_{y,y} = 0 \; . \tag{3.117}$$

The partial derivatives with respect to the coordinates are symbolised by the subscripts $()_{,x} \equiv \partial()/\partial x$ and $()_{,y} \equiv \partial()/\partial y$, v_x and v_y are the components of air velocity in fibre direction x and in cross flow direction y, see Fig. 3.36, left side. The terms ϱ_{air} and η_{air} represent the density and (dynamic) viscosity of air, p is the pressure, and f_x and f_y are the components of the external force density (see later). The continuity equation (3.117) can be satisfied by introducing the potential flow function $\Psi(x, y)$ with

$$v_x = \Psi_{,y} \; , \qquad v_y = -\Psi_{,x} \; . \tag{3.118}$$

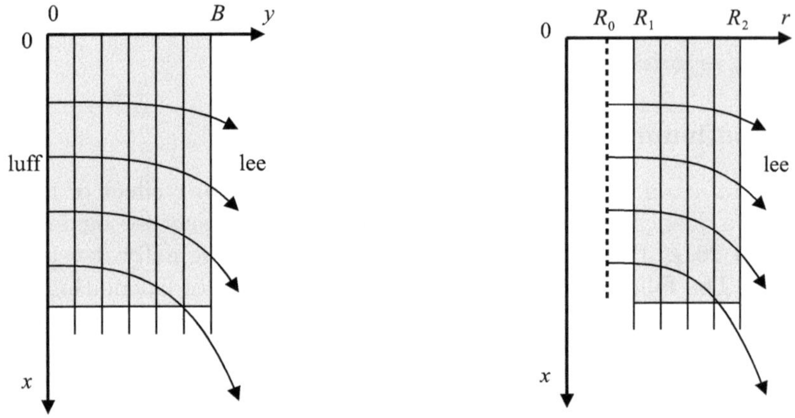

Fig. 3.36. Geometry of multifilament spinning. *Left:* rectangular (cartesian) coordinates, blowing air from the left side, *right:* radial symmetry, blowing air from inner candle (with radius R_0) outside

Discussion I. Some simplifications, which are not so far from real spinning conditions, should be discussed as follows. These assumptions will allow us to use an analytical description of the air flow inside the fibre bundle.

1) At first, we should assume a constant air blowing profile $v_y(x,0) = v_{y0}$, independent of distance x at the windward-side. The interesting region for fibre formation is the one between the spinneret exit at $x = 0$ and the solidification point x_s. In this region the course of fibre velocity v with respect to distance x increases from the (very) low extrusion velocity v_0 to the final take-up velocity v_L in an S-shaped manner. This development of the fibre velocity $v(x)$ should be simply approximated as a linearly with distance x increasing function: $v(x) = x \cdot v_L/x_s$. Furthermore, the relation between the air friction force and the difference of fibre velocity and air velocity $(v - v_x)$ should be assumed as a nearly linearly increasing function $F_{air} \propto x \cdot (v - v_x)$. It is also convenient to approximate the air velocity v_x itself as a function linearly increasing with x. On the other hand, if there are free boundary conditions on the leeward-side, the pressure p also becomes independent of distance x and can be neglected in Eq. 3.115 (This argumentation is only valid in case of the free leeward side, see Fig. 3.32). The advantage of the linearisation is the possibility to uncouple the partial differential equations (3.115) and (3.116), in order to get ordinary differential equations with respect to x and y coordinates. The errors which are made by these linearisations with respect to the real course of fibre formation can be corrected later by means of an iteration procedure in a second step. Using the product

$$\Psi(x,y) = x \cdot g(y) \tag{3.119}$$

now leads to the simplified relations

$$\varrho_{air}\, x \left(g'^2 - g\, g''\right) = \eta_{air}\, x\, g''' + f_x \,, \tag{3.120}$$

$$\varrho_{air}\, g\, g' = -\eta_{air}\, g'' - p' + f_y \,, \tag{3.121}$$

for the reduced potential flow function g, where g is only a function of the cross direction y. The prime ($'$) denotes the derivative with respect to y, $g' \equiv \mathrm{d}g/\mathrm{d}y$.

2) Secondly, we should concentrate on Eq. 3.120. At the moment we are only interested in the velocity field of air and its deflection and acceleration within the filament bundle in the fibre direction. This is why we do not consider the bending of filaments caused by the air stream in cross direction and its backward deceleration effect on the air. Together with the linear increasing force density (that is air friction force per unit volume) in the form

$$f_x(x,y) = x \cdot \varrho_{air}\, k(y) \tag{3.122}$$

we get for Eq. 3.120 the relation

$$g'^2 - g\, g'' = \nu_{air}\, g''' + k \,, \tag{3.123}$$

where $\nu_{\text{air}} = \eta_{\text{air}}/\varrho_{\text{air}} \approx 1.5 \cdot 10^{-5}$ s/m^2 is the kinematic air viscosity. Outside the bundle (region I in Fig. 3.36) the force density term k vanishes and there is a simple analytical solution which satisfies the relation (3.123): $g(y) = c_1 \exp(c_2\, y) + c_3\, \nu_{\text{air}}$, with constants c_{i}. Inside the bundle (region II) the differential equation can be solved numerically. But because of the low air viscosity, the viscous term $\nu_{\text{air}}\, g'''$ becomes insignificant in comparison to the other parts and can thus be neglected. The numerical solutions with and without the viscous part show that it is indeed possible to neglect the viscous part within the bundle, and we then get the most simplified differential equation

$$g'^2 - g\, g'' = k\ , \tag{3.124}$$

which allows us to calculate the air velocity field inside the filament bundle.

Because we are only interested in the air velocity inside the bundle (region II) the coordinate y can now be shifted in such a manner that the left edge of the bundle fits into the origin at $y = 0$, and the right edge of the bundle (its width) is given by $y = B$.

Analytical expressions for the air velocity field. An analytical solution for $g(y)$ in (3.124), resp. for the potential flow function $\Psi(x, y) = x \cdot g(y)$ (3.119) and for the components v_{x} and v_{y} of the air velocity field inside the filament bundle, exists for the special case that the variation of the force density term k (3.122) *across* the fibre bundle in y-direction can be expressed in the manner

$$k(y) = \kappa^2 \exp(2\, \alpha\, y)\ . \tag{3.125}$$

Equation 3.125 includes different possible courses of the force density across the fibre bundle: $\alpha = 0$ describes the constant force density, $\alpha > 0$ the increasing force density, and $\alpha < 0$ is related to a decreasing force density with respect to y.

The analytical solution for the potential flow field in the simplified manner (3.124) is given by an exponential function with complex argument. To achieve physically realistic behaviour, a careful distinction between various cases is required. With the abbreviation

$$\beta = \frac{\alpha \cdot v_{\text{y0}}}{\kappa} \tag{3.126}$$

the results for different cases can be expressed as follows:

Case 1: $1 - \beta^2 > 0$
The solutions for the potential flow function and the velocity components of air in this case are

$$\Psi(x, y) = -x \cdot \frac{v_{\text{y0}}}{\sqrt{1-\beta^2}} \exp(\alpha\, y) \cos\Big(\frac{\kappa\, \sqrt{1-\beta^2}}{v_{\text{y0}}}\, y + c\Big)\ , \tag{3.127}$$

$$v_x(x,y) = x \cdot \kappa \cdot \exp(\alpha\, y) \left[\sin\!\left(\frac{\kappa\sqrt{1-\beta^2}}{v_{y0}}\, y + c \right) - \frac{\beta}{\sqrt{1-\beta^2}} \cos\!\left(\frac{\kappa\sqrt{1-\beta^2}}{v_{y0}}\, y + c \right) \right] , \tag{3.128}$$

$$v_y(x,y) = \frac{v_{y0}}{\sqrt{1-\beta^2}} \cdot \exp(\alpha\, y) \cos\!\left(\frac{\kappa\sqrt{1-\beta^2}}{v_{y0}}\, y + c \right) , \tag{3.129}$$

with

$$c = \arctan\!\left(\frac{\beta}{\sqrt{1-\beta^2}} \right) . \tag{3.130}$$

Case 2: $1-\beta^2 = 0$
The solutions for this (very special) case can be obtained by expanding the functions above to

$$\Psi(x,y) = -x \cdot v_{y0} \cdot (1-\alpha\, y) \exp(\alpha\, y) , \tag{3.131}$$
$$v_x(x,y) = v_{y0} \cdot \alpha \cdot x \cdot y \cdot \exp(\alpha\, y) , \tag{3.132}$$
$$v_y(x,y) = v_{y0} \cdot (1-\alpha\, y) \exp(\alpha\, y) . \tag{3.133}$$

Case 3: $1-\beta^2 < 0$
For this case the square root $\sqrt{1-\beta^2} = \mathrm{i}\sqrt{\beta^2-1}$ becomes imaginary. But using the identities $\cos(\mathrm{i}\,x) = \cosh(x)$, $\sin(\mathrm{i}\,x) = \mathrm{i}\,\sinh(x)$ and $\arctan(\mathrm{i}\,x) = \mathrm{i}\,\mathrm{artanh}(x)$, the solutions can be given again by means of real expressions

$$\Psi(x,y) = -x\, \frac{\alpha}{|\alpha|}\, \frac{v_{y0}}{\sqrt{\beta^2-1}} \exp(\alpha\, y) \sinh\!\left(d - \frac{\kappa\sqrt{\beta^2-1}}{v_{y0}}\, y \right) , \tag{3.134}$$

$$v_x(x,y) = x \cdot \kappa \cdot \frac{\alpha}{|\alpha|} \exp(\alpha\, y) \left[\cosh\!\left(d - \frac{\kappa\sqrt{\beta^2-1}}{v_{y0}}\, y \right) - \frac{\beta}{\sqrt{\beta^2-1}} \sinh\!\left(d - \frac{\kappa\sqrt{\beta^2-1}}{v_{y0}} \right) \right] , \tag{3.135}$$

$$v_y(x,y) = \frac{\alpha}{|\alpha|}\, \frac{v_{y0}}{\sqrt{\beta^2-1}} \exp(\alpha\, y)\, \sinh\!\left(d - \frac{\kappa\sqrt{\beta^2-1}}{v_{y0}}\, y \right) , \tag{3.136}$$

with

$$d = \mathrm{artanh}\!\left(\frac{\sqrt{\beta^2-1}}{\beta} \right) . \tag{3.137}$$

Figure 3.37 depicts the graphs of the velocity component v_y across the bundle in non-dimensional manner (reduced by v_{y0}) for the different cases above.

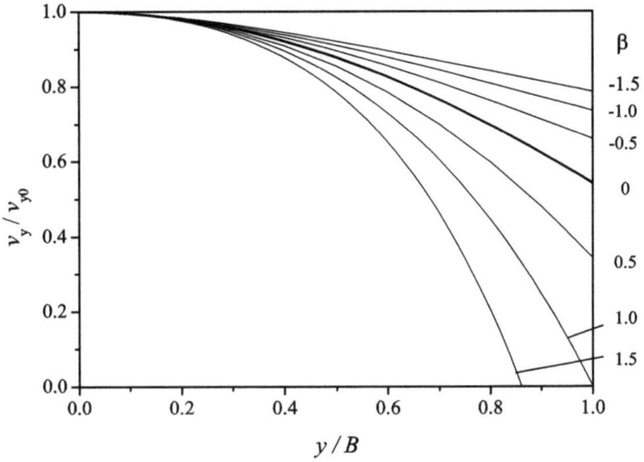

Fig. 3.37. Graphs of (reduced) velocity component v_y/v_{y0} inside the bundle for different cases β, $\kappa B/v_{y0} = 1$, values of β are indicated

Discussion II. Now we will discuss the important case of constant force density $k \neq k(y)$ across the bundle ($\alpha = \beta = 0$, case 1). The expressions for the velocity components v_x and v_y (see Eqs. 3.128 and 3.129) then are reduced to

$$v_x(x, y) = x\,\kappa \sin\left(\frac{\kappa\, y}{v_{y0}}\right) , \tag{3.138}$$

$$v_y(x, y) = v_{y0} \cos\left(\frac{\kappa\, y}{v_{y0}}\right) . \tag{3.139}$$

Comparing the analytical expressions above with the numerical solution of Eq. 3.123 (with consideration of the air viscosity) shows that the approximated analytical solution and the numerical one are in accordance with each other. The air flow velocity inside the bundle is determined mainly by inertness and force density, therefore the viscous term can be neglected without changing the overall flow behaviour.

If there is no active quenching the filament bundle will suck the needed air by itself. This leads to a symmetrical air profile where air enters into the bundle from both sides. The minimum value of air velocity $v_{y0,\text{min}}$ which enters into the bundle can be determined with Eq. 3.138 and the condition

$$v_x(x, 0) = v_x(x, B) = 0 , \tag{3.140}$$

or in equivalence with Eq. 3.139 and the condition for symmetry

$$v_y(x, 0) = -v_y(x, B) = v_{y0,\text{min}} . \tag{3.141}$$

The minimum value of air velocity which results from self-priming and which describes the symmetric flow within the bundle is given by

$$v_{\mathrm{y0,min}} = \frac{\kappa B}{\pi} \,. \tag{3.142}$$

The value of quenching air which is necessary to penetrate the bundle $v_{\mathrm{y0,pen}}$ can be determined by the condition

$$v_{\mathrm{y}}(x, B) > 0 \tag{3.143}$$

and leads to the relation

$$v_{\mathrm{y0,pen}} > \frac{2\,\kappa B}{\pi} = 2 \cdot v_{\mathrm{y0,min}} \,. \tag{3.144}$$

Figure 3.38 qualitatively shows the air flow for the different cases.

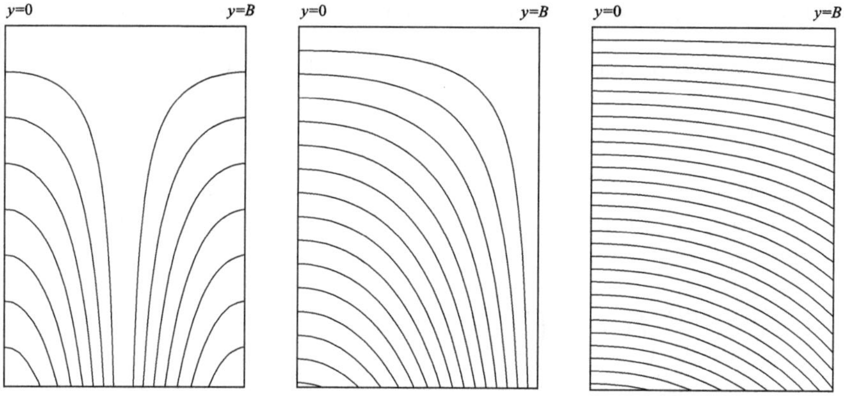

Fig. 3.38. Air flow within the filament bundle. *Left:* symmetric case, no active quenching, the filament bundle is self-priming air, $v_{\mathrm{y0}} = v_{\mathrm{y0,min}}$; *center:* limiting case of small quenching from left side, no penetration, $v_{\mathrm{y0}} = 2\,v_{\mathrm{y0,min}}$; *right:* active quenching from left side, air penetrates the bundle, $v_{\mathrm{y0}} = 4\,v_{\mathrm{y0,min}}$

The assumption of constant force density ($\alpha = \beta = 0$, case 1) across the bundle is applicable for many purposes. An example of this is the staple fibre spinning process with uniform distribution of capillary holes in the spinneret. It will later be investigated in more detail in Sect. 3.2.4. On the other hand, it is also possible to study the effect of air deflection by means of increasing or decreasing the number of capillary holes in the spinneret at each line by assuming a non-constant force density ($\alpha \neq 0$, $k \propto \exp(2\,\alpha\, y)$). The distribution variation of the spinneret capillary holes changes the force density $k(y)$ across the bundle and influences the flow behaviour inside. Figure 3.39 shows the effect of increasing (left graphs) and decreasing (right graphs) force density in comparison to the case of constant force density with respect to distance y.

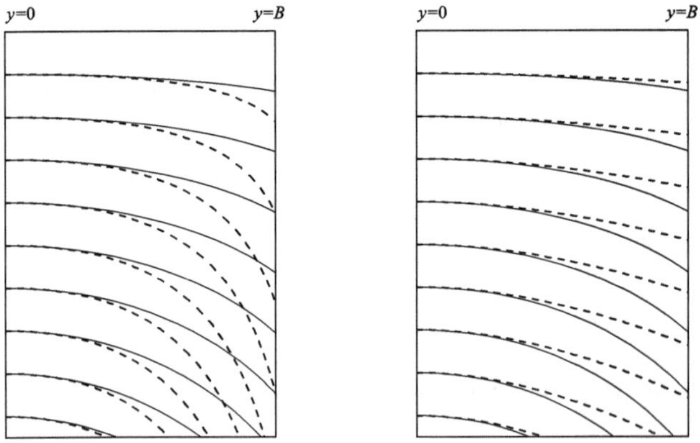

Fig. 3.39. Air flow within the filament bundle. *Left:* Comparison between constant force density (—, $\alpha = 0$) and increasing force density (– – –, $\alpha > 0$); *right:* Comparison between constant force density (—, $\alpha = 0$) and decreasing force density (– – –, $\alpha < 0$)

The minimum necessary air velocity for the self-priming filament bundle is given by Eq. 3.140. In the cases of variable force density $\alpha \neq 0$, it leads to the expression

$$v_{\mathrm{y0,min}} = \frac{\kappa B}{\pi} \sqrt{1 + \frac{\alpha}{|\alpha|} \left(\frac{\alpha B}{2\pi} \right)^2} \, . \tag{3.145}$$

Contrary to the symmetric flow in the case of constant force density (Fig. 3.38, left hand side) the streamlines now become asymmetric.

Iterative improvement of the analytical solution. The Eqs. (3.128) to (3.139) allow fast calculation of the air velocity inside the fibre bundle. However, they are based on the assumptions of constant air blowing velocity v_{y0} and linearly increasing filament velocity $v(x)$. Normally, in real spinning processes both the blowing air velocity and the filament velocity do notexactly follow these conditions. But the effects of the afore mentioned assumptions are mostly insignificant and therefore it is possible to use the developed relations in a modified manner. The external force density $f_{\mathrm{x}}(x, y) = x \, \varrho_{\mathrm{air}} \, k(y)$ and the velocity of blowing air v_{y0} are replaced by their *globally* varying *mean* values (with respect to distance x)

$$k(y) \to k_{\mathrm{m}}(x, y) = \frac{1}{x} \int_0^x k(x', y) \, dx' \, , \tag{3.146}$$

$$v_{\mathrm{y0}} \to v_{\mathrm{ym}}(x) = \frac{1}{x} \int_0^x v_{\mathrm{y0}}(x') \, dx' \, . \tag{3.147}$$

The dependence of the flow function on distance x can now be expressed indirectly through its dependence of k_{m} (or κ_{m}) and v_{ym}, respectively, on distance x. Then, for the case $\alpha = 0$, the Eqs. 3.138 and 3.139 become

$$v_x(x,y) = x\sqrt{k_m(x)}\,\sin\Big(\frac{\sqrt{k_m(x)}}{v_{ym}(x)}\,y\Big)\,, \tag{3.148}$$

$$v_y(x,y) = v_{ym}(x)\,\cos\Big(\frac{\sqrt{k_m(x)}}{v_{ym}(x)}\,y\Big)\,. \tag{3.149}$$

The mean value of the force density $k_m(x,y)$ needs to be calculated by means of iteration of the global momentum balance of air flow inside the fibre bundle with respect to distance x:

$$\sum_{j=1}^{N_r} N(j)\int_0^x c_f(x',y_j)\frac{\varrho_{air}}{2}\big(v(x',y_j) - v_x(x',y_j)\big)^2\,\pi\,D(x',y_j)^2\,dx'$$
$$= W\int_0^B \varrho_{air}\,v_x^2(x,y)\,dy + W\int_0^x \varrho_{air}\,v_x(x',B)\,v_y(x',B)\,dx'\,. \tag{3.150}$$

Here, N_r is the number of filament rows, $N(j)$ is the number of filaments in each row j, y_j is the inter-filament distance at row number j, and B and W are the widths of the fibre bundle in y direction (quenching air direction) and perpendicular direction, respectively. The left-hand side of Eq. 3.150 represents the sum of all air friction forces within the bundle at any distance x. On the right-hand side, the first term describes the momentum in x direction within the bundle, the second term is the expression for the momentum of the air velocity component which has passed the bundle at distance $y = B$. Equation (3.150) is the expression for the global balance between air friction and momentum of the accelerated and deflected air in fibre direction x which is to be satisfied. As an initial value (subscript 0) for the iteration of k_m the result of the linear approximation (v_{y0} = const, $v(x) \propto x$) can be used:

$$k_{m,0}(x) = \frac{2}{B\,W\,\varrho_{air}\,x^2}\sum_j^{N_r} N(j)\,F_{air}(x,y_j)\,. \tag{3.151}$$

Radial symmetry. It is also possible to find similar analytical expressions for the air velocity in the case of radial symmetry and air flow from inside (e. g. from a blowing candle) to the outside (see Fig. 3.36, right-hand side, page 96). Concerning the rectangular cartesian case the procedures and the discussions are analogue to the treatment above. The following paragraph will only summarise the results. The NAVIER-STOKES and the continuity equations for the radial case are

$$\varrho_{air}(v_x\,v_{x,x} + v_r\,v_{x,r}) = \eta_{air}(v_{x,xx} + v_{x,rr}) + f_x - p_{,x}\,, \tag{3.152}$$

$$\varrho_{air}(v_x\,v_{r,x} + v_r\,v_{r,r}) = \eta_{air}(v_{r,xx} + v_{r,rr}) + f_r - p_{,r}\,, \tag{3.153}$$

$$v_{x,x} + \frac{1}{r}\,v_{r,r} + v_{r,r} = 0\,. \tag{3.154}$$

Satisfying the continuity equation and decoupling the NAVIER-STOKES equation is possible by using the radial flow function $\Phi(x,r)$ with

$$\Phi(x,r) = x \cdot h(r) \,, \tag{3.155}$$

where $h(r)$ now is the solution of

$$\varrho_{\mathrm{air}} \cdot x \cdot (h\,h' - r\,h\,h'' + r\,h'^2) = r^3\,f_{\mathrm{x}} \,. \tag{3.156}$$

The prime ($'$) in Eq. 3.156 denotes the derivative with respect to the radial coordinate r. The velocity components are expressed by

$$v_{\mathrm{x}}(x,r) = \frac{1}{r}\,\Phi_{,\mathrm{r}} \,, \tag{3.157}$$

$$v_{\mathrm{r}}(x,r) = -\frac{1}{r}\,\Phi_{,\mathrm{x}} \,. \tag{3.158}$$

With the following approach for the force density

$$f_{\mathrm{x}}(x,r) = x \cdot \kappa^2 \varrho_{\mathrm{air}} \exp\left[2\,\alpha\,(r^2 - R_1^2)\right] \,, \tag{3.159}$$

an analytical solution for Eq. 3.156 can be found. Again, three different cases with respect to

$$\gamma = \frac{2\ \alpha \cdot v_1\,R_1}{\kappa} \tag{3.160}$$

are to be distinguished.

Case 1: $1 - \gamma^2 > 0$
The solutions are

$$\Phi(x,r) = -x \cdot \frac{v_1\,R_1}{\sqrt{1-\gamma^2}} \exp\big[\alpha\,(r^2 - R_1^2)\big] \cos\Big(\frac{\kappa\sqrt{1-\gamma^2}}{2\,v_1\,R_1}(r^2 - R_1^2) + c_1\Big) \,, \tag{3.161}$$

$$v_{\mathrm{x}}(x,r) = x \cdot \kappa \cdot \exp\big[\alpha\,(r^2 - R_1^2)\big] \Bigg[\sin\Big(\frac{\kappa\sqrt{1-\gamma^2}}{2v_1\,R_1}(r^2 - R_1^2) + c_1\Big) - \frac{\gamma}{\sqrt{1-\gamma^2}} \cos\Big(\frac{\kappa\sqrt{1-\gamma^2}}{2v_1\,R_1}(r^2 - R_1^2) + c_1\Big)\Bigg] \,, \tag{3.162}$$

$$v_{\mathrm{r}}(x,r) = \frac{v_1\,R_1}{r\sqrt{1-\gamma^2}} \cdot \exp\left[\alpha\,(r^2 - R_1^2)\right] \cos\Big(\frac{\kappa\sqrt{1-\gamma^2}}{2v_1\,R_1}(r^2 - R_1^2) + c_1\Big) \,, \tag{3.163}$$

with

$$c_1 = \arctan\Big(\frac{\gamma}{\sqrt{1-\gamma^2}}\Big) \,. \tag{3.164}$$

Case 2: $1-\gamma^2=0$

$$\Phi(x,r) = -x\cdot v_1\,R_1\cdot\big[1-\alpha\,(r^2-R_1^2)\big]\exp\big[\alpha\,(r^2-R_1^2)\big]\,, \tag{3.165}$$

$$v_x(x,r) = x\cdot 2\,\alpha^2 v_1\,R_1\big[1-\alpha\,(r^2-R_1^2)\big]\exp\big[\alpha\,(r^2-R_1^2)\big]\,, \tag{3.166}$$

$$v_r(x,r) = \frac{v_1\,R_1}{r}\cdot\big[1-\alpha\,(r^2-R_1^2)\big]\exp\big[\alpha\,(r^2-R_1^2)\big]\,. \tag{3.167}$$

Case 3: $1-\gamma^2<0$

$$\Phi(x,r) = -x\,\frac{\alpha}{|\alpha|}\,\frac{v_1\,R_1}{\sqrt{\gamma^2-1}}\exp\big[\alpha\,(r^2-R_1^2)\big]\sinh\Big(d_1-\frac{\kappa\sqrt{\gamma^2-1}}{2\,v_1\,R_1}(r^2-R_1^2)\Big)\,, \tag{3.168}$$

$$v_x(x,r) = x\cdot\kappa\cdot\frac{\alpha}{|\alpha|}\exp\big[\alpha\,(r^2-R_1^2)\big]\Bigg[\cosh\Big(d_1-\frac{\kappa\sqrt{\gamma^2-1}}{2\,v_1\,R_1}(r^2-R_1^2)\Big) - \frac{\gamma}{\sqrt{\gamma^2-1}}\sinh\Big(d_1-\frac{\kappa\sqrt{\gamma^2-1}}{2\,v_1\,R_1}(r^2-R_1^2)\Big)\Bigg]\,, \tag{3.169}$$

$$v_r(x,r) = \frac{\alpha}{|\alpha|}\,\frac{v_1\,R_1}{r\,\sqrt{\gamma^2-1}}\exp\big[\alpha\,(r^2-R_1^2)\big]\sinh\Big(d_1-\frac{\kappa\sqrt{\gamma^2-1}}{2\,v_1\,R_1}(r^2-R_1^2)\Big)\,, \tag{3.170}$$

with

$$d_1 = \operatorname{artanh}\Big(\frac{\sqrt{\gamma^2-1}}{\gamma}\Big)\,. \tag{3.171}$$

The discussion and application closely follow the analogue cases for the rectangular filament bundle.

Air Temperature

The basis for determination of the air temperature field $T_{\mathrm{air}}(x,y)$ is the energy equation of heat conduction and heat convection inside the fibre bundle. In analogy to the calculation of the air velocity field a two-dimensional steady state flow without free heat convection is assumed. The energy equation in cartesian coordinates is then given by

$$\lambda_{\mathrm{air}}(T_{\mathrm{air,xx}}+T_{\mathrm{air,yy}}) = c_{\mathrm{air}}\,\varrho_{\mathrm{air}}\,(v_x\,T_{\mathrm{air,x}}+v_y\,T_{\mathrm{air,y}}) - q_{\mathrm{fib}}\,, \tag{3.172}$$

where λ_{air}, c_{air}, ϱ_{air} are the heat conductivity, the specific heat capacity, and the density of air, respectively.

The heat source density q_{fib} is equivalent to the heat transfer from fibre surfaces into air. Its mean value (for each fibre row at position y_{j}) is determined similarly to the calculation of force density. The heat quantity per length unit is given by the fibre cooling balance as follows

$$q_{\text{fib}}(x,y) = \frac{1}{B\,W} \sum_{j=1}^{N_{\text{r}}} N(j) \cdot Q \cdot c_{\text{p}} \cdot \frac{\mathrm{d}T(x,y_j)}{\mathrm{d}x} \,, \tag{3.173a}$$

$$\approx \frac{N}{B\,W}\, \pi\, \lambda_{\text{air}}\, Nu(x,y) \cdot \big(T(x,y) - T_{\text{air}}(x,y)\big) \,, \tag{3.173b}$$

where $N(j)$, N_{r} and N are the number of fibres in the jth row, the number of rows, and the total number of filaments within the bundle, respectively. The product $B\,W$ is the cross-sectional area of the rectangular bundle. Instead of the real temperature profile of the air around each fibre, an effective average temperature field T_{air} is now calculated.

In the case of radial symmetry the corresponding relation to Eq. 3.172 for the heat energy is given as

$$\lambda_{\text{air}} \Big(T_{\text{air,xx}} + T_{\text{air,rr}} + \frac{1}{r}\, T_{\text{air,r}}\Big) = c_{\text{air}}\, \varrho_{\text{air}}\, (v_{\text{x}}\, T_{\text{air,x}} + v_{\text{r}}\, T_{\text{air,r}}) - q_{\text{fib}} \,, \tag{3.174}$$

where the heat source density in the radial symmetric case is given as

$$q_{\text{fib}}(x,y) = \frac{N}{R_2^2 - R_1^2}\, \lambda_{\text{air}}\, Nu(x,r) \cdot \big(T(x,r) - T_{\text{air}}(x,r)\big) \,. \tag{3.175}$$

Discussion. There is no analytical solution to the energy equations (3.172) or (3.174) within the filament bundle. The solutions can only be found numerically by transferring the partial differential equations into any adequate difference equations and using for example a five or more-points formula.

The amount of heat transportation by means of thermal heat conduction is insignificant in comparison to heat convection by air:

$$\lambda_{\text{air}}(T_{\text{air,xx}} + T_{\text{air,yy}}) \ll c_{\text{air}}\, \varrho_{\text{air}}(v_{\text{x}}\, T_{\text{air,x}} + v_{\text{y}}\, T_{\text{air,y}}) \,. \tag{3.176}$$

For modelling, this suggests the possibility to simplify the energy equations through neglecting the heat conduction, which means the complete suppression of the left-hand side of Eq. 3.172 or 3.174. However, in such cases where no active quenching occurs or in the range between spinneret and onset of air blowing, the bundle sucks the air by itself (because of the pumping effect – the momentum balance has to be satisfied) and the air flow becomes symmetrical from both sides of the bundle to the center line. The result of these model assumptions is that no filament cooling takes place along the center line if the heat conductivity is not taken into account. Therefore, heat conductivity becomes essential especially for the symmetric flow cases and for the regions without active quenching. Additionally, there are always oscillations of fibres and air, turbulences, and free heat convection in real fibre spinning processes. These effects increase of course the heat transfer described by the heat conduction term $\lambda_{\rm air}$ alone. For a more realistic simulation the influence of heat conduction should be taken into consideration by replacing the heat conductivity $\lambda_{\rm air}$ with an *effectively increased* term $\lambda_{\rm eff}$.

On the other hand, the temperature slope for the air in spinning direction x is small in comparison to the slope in quenching direction y. While assuming $T_{\rm air,xx} \ll T_{\rm air,yy}$ the term $T_{\rm air,xx}$ can be neglected in Eqs. 3.172 and 3.174. Together with $a = \lambda_{\rm air}/(\varrho_{\rm air}\, c_{\rm air})$ and with Eq. 3.173b resp. 3.175 then follows

$$T_{\rm air,yy} = \frac{v_{\rm x}\, T_{\rm air,x} + v_{\rm y}\, T_{\rm air,y}}{a} - \frac{N \cdot \pi Nu\,(T - T_{\rm air})}{B\,W} \tag{3.177}$$

for the rectangular case, resp.

$$T_{\rm air,rr} + \frac{1}{r}\, T_{\rm air,r} = \frac{v_{\rm x}\, T_{\rm air,x} + v_{\rm r}\, T_{\rm air,r}}{a} - \frac{N \cdot Nu\,(T - T_{\rm air})}{R_2^2 - R_1^2} \tag{3.178}$$

for the radial symmetric case.

After the onset of quenching air the influence of heat conductivity within the fibre bundle can be neglected in comparison to the effects of the forced convection. The terms at the left-hand side in Eqs. 3.177 and 3.178 vanish and the equations to calculate the air temperature inside the fibre bundle are simplified to

$$v_{\rm x}\, T_{\rm air,x} + v_{\rm y}\, T_{\rm air,y} = \frac{N \cdot \pi Nu\, a\,(T - T_{\rm air})}{B\,W} \tag{3.179}$$

for the rectangular case, and

$$v_{\rm x}\, T_{\rm air,x} + v_{\rm r}\, T_{\rm air,r} = \frac{N \cdot Nu\, a\,(T - T_{\rm air})}{R_2^2 - R_1^2} \tag{3.180}$$

for the radial symmetric case, respectively.

Some Remarks to Boundary Conditions, Geometry, and Numerical Realisation

For the calculation of the air temperature fields $T_{\rm air}(x, y)$, resp. $T_{\rm air}(x, r)$, it is necessary to possess knowledge about

- the velocity fields of air $\boldsymbol{v}_{\rm air}(x, y)$, resp. $\boldsymbol{v}_{\rm air}(x, r)$ within the bundle
- the fibre temperature profiles $T(x, y_j)$, resp. $T(x, r_j)$ of each row j and the corresponding NUSSELT numbers $Nu(x, y_j)$ resp. $Nu(x, r_j)$ within the bundle, and
- the conditions at the borders of the bundle, the boundary conditions

Air temperature. Knowledge of the air temperature profile at the boundaries of the rectangular region (see Fig. 3.40) is required in order to solve the partial differential equation (3.177) for the temperature field inside the bundle (the radial symmetric case can be treated in an analogous manner):

$$\begin{aligned} &\text{region A}: && T_{\text{air}}(x=0,y) = T_{\text{air,spinneret}}(y) \;, \\ &\text{region B, C}: && T_{\text{air}}(x,y=0) = T_{\text{air,0}}(x) \;, \\ &\text{region D, E}: && T_{\text{air}}(x,y=B) = T_{\text{air,B}}(x) \;. \end{aligned}$$

The air temperature at the spinneret (region A in Fig. 3.40) is assumed to be equal to the spinneret temperature itself, this means that it is also equal to the initial fibre temperature T_0:

$$T_{\text{air}}(x=0,y) = T(0) = T_0 \;.$$

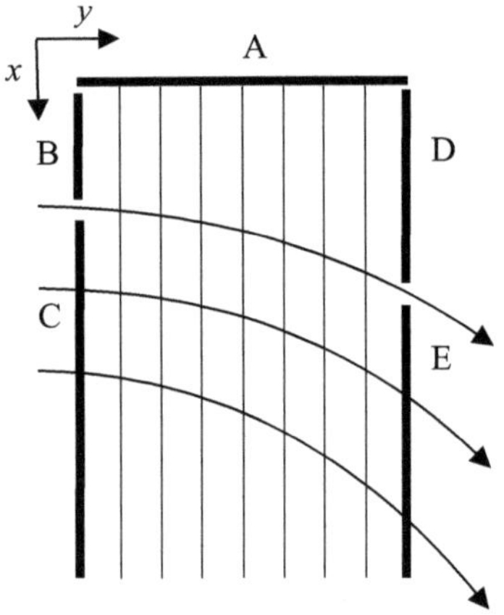

Fig. 3.40. Boundary conditions for rectangular fibre bundle: A – initial fibre temperature, B and D – unknown temperature, C – temperature profile of blowing air, E – temperature profile results from calculation

Near the spinneret and before the onset of quenching, that is the region where the filament bundle sucks air by itself, the temperature profile of the sucked air is normally unknown (region B and D in Fig. 3.40). The air temperature there is influenced by many factors, such as the geometry of the current spinneret design, the realised spinning conditions, the free heat convection, etc. It seems *impossible* to determine the air temperature in this region *a priori*. The only possibility to derive necessary boundary conditions for the calculation is to make an assumption regarding the air temperature based on *temperature measurement* in a real spinning process. The purely theoretical solution of the multifilament melt spinning problem in this region is impossible. This is due to the fact that the temperature profiles at the borders, required as input parameter, depend in a very complex way on the specialities of each respective spinning equipment. The boundary parameter may thus vary from case to case.

After the onset of quenching air (region C in Fig. 3.40) the boundary conditions for the windward side are given through the temperature profile and velocity profile of blowing air which are well-defined parameters in the melt spinning process. The velocity of the blowing air is much higher than the velocity of the self-primed air. Furthermore, the heat transfer becomes much more effective and can be described by means of the forced heat convection *alone*. Neglecting the slope $T_{\mathrm{air,yy}}$ dramatically simplifies the energy equation (3.179). No more information about the temperature profile at the leeward side is now needed (region E in Fig. 3.40).

There is a second difficulty besides the unknown boundary conditions in the region near the spinneret. The heat conductivity of air λ_{air} is too small to show any effect for realistic cooling behaviour along the centre line of the fibre bundle for the case where no active quenching is done by blowing air. The calculated temperature becomes nearly independent from the distance x in the centre. As discussed earlier, this unrealistic result can be improved by using an *effective* value of heat conductivity λ_{eff} (instead of λ_{air}) that takes into account oscillations, turbulences, and free convection effects. The comparison of calculated temperatures with values measured in real fibre spinning trials can be used to fit the effective heat conductivity to the measurements. The results from several experiments suggest a factor between $\lambda_{\mathrm{eff}} \approx (100 \ldots 200)\lambda_{\mathrm{air}}$.

Summarizing the discussion above, it can be said that near the spinneret and for regions without active quenching the air temperature needs to be calculated using the heat energy equation (3.177) (for the rectangular bundle), whereas the heat conductivity of air is taken into consideration and should be replaced with a more realistic effective value λ_{eff}. For the regions with active quenching and much higher air flow velocity the consideration of heat conductivity of air is no longer important and can be neglected. For this case the calculation can be carried out using Eq. 3.179.

Numerical realisation. For the simplest but most interesting case of constant force density across the fibre bundle, the velocity field of air is given by Eqs. 3.148 and 3.149 for the rectangular geometry (and the respective equations for the radial geometry). The improvement made by iterative calculation of the global mean value of force density k_{m} with respect to distance x follows from the global momentum balance equations (3.150) (resp. the radial correspondence). For iteration the use of the *dampened* NEWTON procedure is helpful:

$$k_{\mathrm{m}}^{(i+1)}(x) = k_{\mathrm{m}}^{(i)}(x) - \delta \, \frac{\widetilde{F}\big(k_{\mathrm{m}}^{(i)}(x)\big)}{\widetilde{F}'\big(k_{\mathrm{m}}^{(i)}(x)\big)} \,, \tag{3.181}$$

where $\widetilde{F}$ is given by the air friction momentum balance

$$\widetilde{F} = \sum F_{\mathrm{air},i} - W \int_0^x \varrho_{\mathrm{air}} \, v_{\mathrm{x}} \, v_{\mathrm{y}} \, dx' - W \int_0^B \varrho_{\mathrm{air}} \, v_{\mathrm{x}}^2 \, dy \tag{3.182}$$

for rectangular bundle geometry, and

$$\widetilde{F} = \sum F_{\mathrm{air},i} - 2\,\pi\,R_2 \int_0^x \varrho_{\mathrm{air}}\,v_{\mathrm{x}}\,v_{\mathrm{r}}\,dx' - 2\pi \int_{R_1}^{R_2} \varrho_{\mathrm{air}}\,v_{\mathrm{x}}^2\,r\,dr \tag{3.183}$$

for the radial geometry, respectively. The dampening factor $\delta < 1$ avoids too rapid changes within the iteration process and supresses oscillations. The results of the linearised assumptions can be used as the initial value $k_{\mathrm{m}}^{(0)}$ to start the iteration procedure with

$$k_{\mathrm{m}}^{(0)}(x) = \frac{2}{\varrho_{\mathrm{air}}\,B\,W\,x^2} \sum F_{\mathrm{air},i}(x) \tag{3.184}$$

for the rectangular bundle geometry, and

$$k_{\mathrm{m}}^{(0)}(x) = \frac{2}{\varrho_{\mathrm{air}}\,\pi\,(R_2^2 - R_1^2)\,x^2} \sum F_{\mathrm{air},i}(x) \tag{3.185}$$

for the radial geometry, respectively. For the cases of non-constant force density across the bundle ($\alpha \neq 0$, Eqs. 3.125, 3.159) the convergence region for the iteration procedure is to be carefully observed. The starting values $k_{\mathrm{m}}^{(0)}$ above can only be successfully used if $|2\,\alpha\,B| \ll 1$, or $\left|\gamma\,(R_2^2 - R_1^2)\right| \ll 1$.

The partial derivatives of the energy equations (3.177) and (3.178) can be replaced with difference quotients expressed by any adapted point-formula. If the second derivative $T_{\mathrm{air,yy}}$ (resp. $T_{\mathrm{air,rr}}$) is taken into account for the symmetric case, the result is a (three-)diagonal system of algebraic equations which can be effectively solved with LD-dissection. On the other hand, if the conduction effects can be neglected, a simple step procedure may be carried out. In the latter case the numerical procedure is quite similar to the solution procedures for the earlier described cell-method (see page 94). To accelerate the convergence for the alternating calculations of the multifilament system (fibres – air – fibres – air ...) it is convenient to use only the NUSSELT number and then simultaneously calculate the fibre temperatures again, according to Eq. 3.10.

In principle, the only reason to start the calculation procedure with the model of fibre formation first is its simplicity. At first, the calculation of the fibre formation is done based on any assumption concerning the environmental air. Then the air velocity and air temperature are determined on the basis of the air friction force and the heat exchange from fibres to bundle. After that, fibre formation calculation is carried out again, and so forth until no further significant changes can be observed. Depending upon the starting values and the current spinning parameters (and also on the step width and number of rows) sufficient convergence can be reached after approximately three to five calculation runs. A modern personal computer (1 GHz processor clock rate for example) only needs seconds to complete the calculations.

Geometrical limits. The described method for modelling the multifilament melt spinning process is easy to manage and leads to quick calculations which allow to assess technological parameters with respect to optimised spinning conditions for high quality fibre products. But the model is restricted to the assumption of *free boundary conditions at the leeward side.* The pressure gradient $p_{,\mathrm{x}}$ vanishes only for the free leeward side and it is possible to use the harmonic functions for computing the air velocity components within the bundle. This means that the deflection of air is *only* caused by means of air friction from the filaments alone. There is no influence of any walls of the quenching chamber affecting the leeward side. The quenching air flow originates on the windward side (or from inner side to outer side in radial geometry), and air flow deflection is only caused by the fibres. If there are no fibres then no air flow deflection occurs. On the other hand, if any wall is located at the leeward side then the boundary conditions are not free anymore as in the before mentioned manner, the deflection of air will then be caused by both the filaments *and* the wall. The same effect occurs if quenching air flow is reached from both sides to the centre (or from outer to inner side in case of radial geometry). The effect of the wall can then be replaced with a symmetry line. Furthermore, the pressure gradient $p_{,\mathrm{x}}$ cannot be neglected anymore since the streamlines result from both, air friction from fibres and from the geometrical conditions of the quenching chamber. The further use of the developed model is only possible if some modifications are implemented. Firstly, the streamlines of air have to be determined depending upon the geometry of the quenching chamber. Secondly, the influence of air friction of the fibres needs to be considered. The latter can be achieved, for example, with the described concept of the *varying force density* $k_{\mathrm{m}}(x)$. Figure 3.41 illustrates the basic idea.

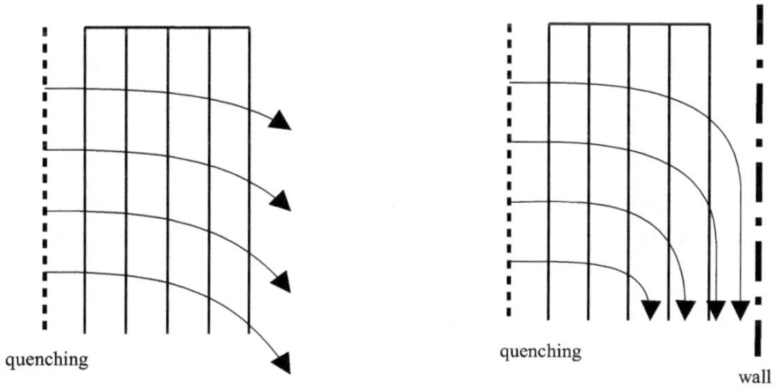

Fig. 3.41. Boundary conditions on leeward side. *Left:* free boundary, air deflection is caused only by fibres; *right:* no free boundary, air deflection is caused by wall *and* fibres, the influence of the wall can be treated by replacing it with a symmetry line

3.2.4 Example 1: Numerical Simulation of Fibre Formation in the Staple Fibre Melt Spinning Process

The PET staple fibre melt spinning with rectangular geometry of the filament bundle, as the introductory discussed on page 92 shall now be treated by means of the outlined model for different quenching conditions. The spinning parameters are as follows

polymer:	poly(ethylene terephthalate) (PET),
spinneret:	1300 holes, 13 rows with 100 holes on each row, dimension W x B: 26 cm x 6.5 cm, rectangular, diameter of capillary holes: 0.30 mm,
melt temperature:	$T_0 = 290\,°\mathrm{C}$,
throughput:	$Q_{\mathrm{PET}} = 1300 \times 0.6$ g/min $= 780$ g/min,
take-up velocity:	$v_{\mathrm{L}} = 1200$ m/min,
quenching:	cross quenching from one side, beginning 5 cm below spinneret, temperature $T_{\mathrm{air}} = 25\,°\mathrm{C}$, velocity (1) $v_{\mathrm{air}} = 1.5$ m/s, length of quenching zone $L_{\mathrm{air}} = 1.0$ m .

The results of the calculation are shown in the following figures. Figure 3.42 depicts the streamlines of air flow within the filament bundle. The deflection of air in fibre direction can be clearly seen on the left hand side of the figure. The air sucking effect from both sides below the spinneret, in the 5 cm region free of quenching air flow, is also shown on the right-hand side of the figure. Next, Fig. 3.43 shows the temperature of air within the bundle (the isotherms); there is a transition from symmetric temperature distribution below the spinneret to asymmetric distribution after the start of quenching at distance $x = 5$ cm.

Figure 3.44 illustrates the experimental verification of the calculated air temperature. The diagram depicts the dependency of air temperature T_{air} on distance x from spinneret at three positions: at windward side, in the center, and at leeward side. The temperature at windward side is a given boundary condition (and thus cannot be used for verification). The temperature inside the bundle seems impossible to be measured, but the temperature at the leeward side, when the air has passed the bundle, can be determined easily.

Unevenness of the Filament Properties

The air flow and air temperature distribution within the filament bundle lead to different environmental conditions for each filament row which influence their cooling and air friction behaviour. It is obvious that any retarded cooling caused by higher air temperature increases the length of the fibre formation zone and affects the resulting filament properties. The longer the fibre formation zone (the distance from the spinneret exit to the solidification

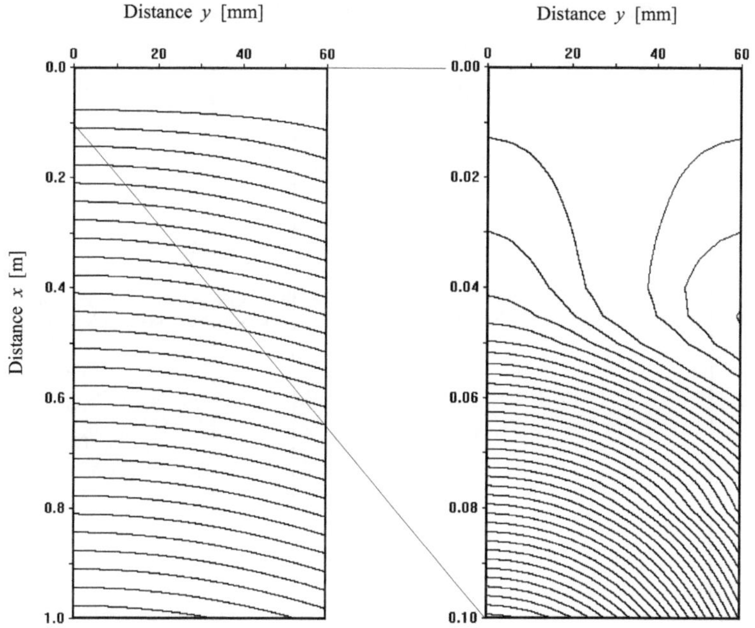

Fig. 3.42. Air flow (streamlines) inside the filament bundle. The magnified region below spinneret ($x = 0 \ldots 0.1$ cm) is presented again at the right part of the figure; spinning parameter: PET staple fibre spinning process, see before

point) the greater the exposition to air friction, and the higher the part of the friction force. On the other hand, the deflection of air into fibre direction (especially at leeward side) reduces the air friction coefficient and therefore can also reduce the friction force. Both effects interact and it depends on the current spinning and geometric conditions if and how the resulting stress at the solidification point changes and thus the connected fibre properties. The following figures show the courses of temperature (Fig. 3.45) and tensile stress (Fig. 3.46) for fibres within the bundle in three selected rows, one at the windward side (row #1), one in the centre of the bundle (row #7), and one at the leeward side (row #13).

Under the prevailing conditions near the spinneret the filaments at windward side and leeward side (the outer sides of the filament bundle) are subjected to higher cooling rates compared to the filaments in the center of the bundle. The symmetric conditions of geometry and air flow in the region near the spinneret also result in symmetric courses of fibre formation. But after the onset of (asymmetric) quenching air the cooling rates change to the (expected) sequence of windward-centre-leeward and the courses of fibre formation also become asymmetric, as expected. The solidification points for the individual filaments are reached at different positions. As discussed in the section about the modelling of fibre formation, the stress at the solidification

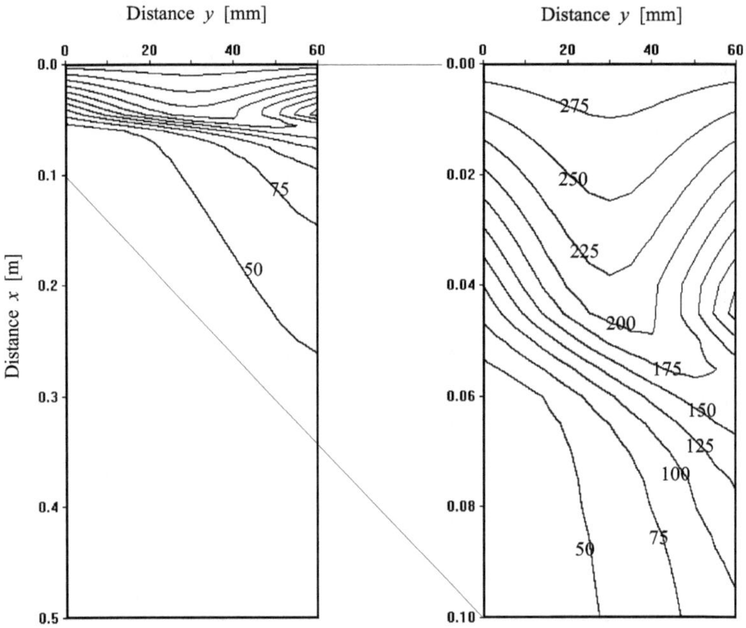

Fig. 3.43. Air temperature (isotherms) inside the filament bundle with quenching air blowing from the left hand side. The magnified region below spinneret ($x = 0 \ldots 0.1$ cm) is presented again at the right part of the figure; temperatures are indicated (°C); spinning parameter: PET staple fibre spinning process, see before

point essentially determines the fibre orientation and its resulting properties. Figure 3.47 shows the stress at the solidification points and the corresponding elongations to break across the fibre bundle. The lowest stress at the solidification point, the lowest orientation and consequently the highest elongation to break (the highest draw ratio) is not attained to the fibres at leeward side of the bundle but inside the bundle at row #9. This (for the moment unexpected) result is a direct consequence of the conditions near the spinneret. The course of graphs in the diagram Fig. 3.47 does not appear very smooth, but the scattering results from the numerical effect of rounding the numbers to limited digits. In principle the numerical accuracy and smoothness of the graphs can be improved.

In order to indicate the tranversal uneveness of the filaments (the variation of the orientation over the bundle cross section) the maximum and minimum elongation to break can be used to define the variation coefficient CV_{ε_b}:

$$CV_{\varepsilon_b} = \frac{\varepsilon_{b,\max} - \varepsilon_{b,\min}}{\varepsilon_{b,\min}} \cdot 100\,\% \,. \tag{3.186}$$

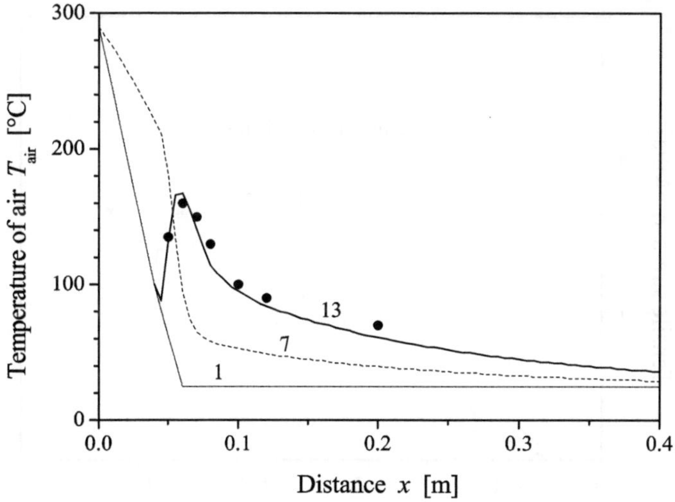

Fig. 3.44. Air temperature inside the filament bundle at three different positions: 1 – windward side (row #1), 7 – center of the bundle (row #7), 13 – leeward side (row #13), symbols: measured temperature at leeward side; data source: Institute of Polymer Research Dresden

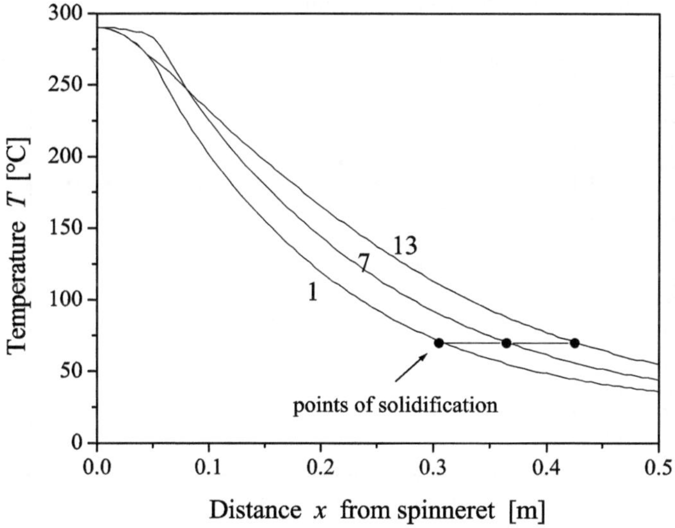

Fig. 3.45. Filament temperature $T(x)$ (calculated) for three different locations: 1 – windward side (row #1), 7 – centre of the bundle (row #7), 13 – leeward side (row #13), the position of solidification points at $T_s = 70\,°C$ is indicated; spinning parameter: PET staple fibre spinning process, see before

With the results shown in Fig. 3.47 the variation coefficient of orientation (resp. elongation to break) is $CV_{\varepsilon_b} = 3.6\,\%$.

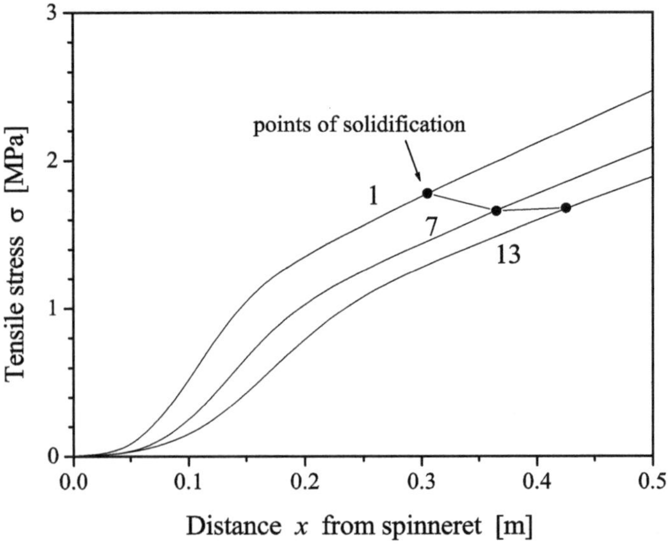

Fig. 3.46. Filament tensile stress $\sigma(x)$ (calculated) for three different locations: 1 – windward side (row #1), 7 – center of the bundle (row #7), 13 – leeward side (row #13), the position of solidification points is indicated; spinning parameter: PET staple fibre spinning process, see before

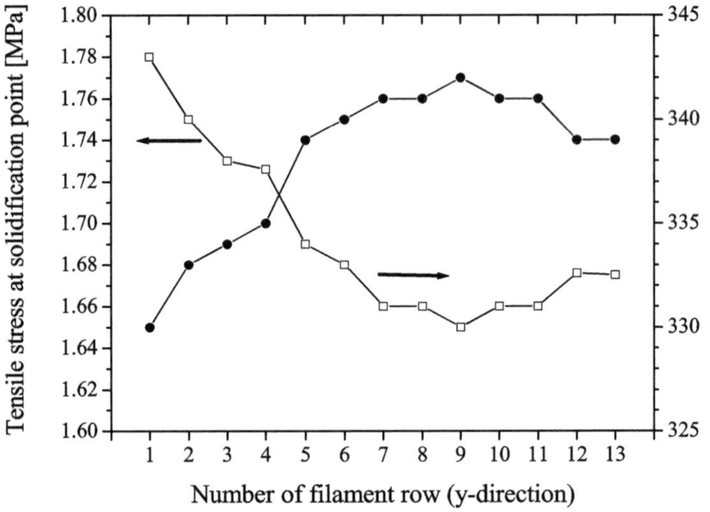

Fig. 3.47. Filament stress $\sigma(x_s)$ at solidification point and corresponding elongation to break (calculated) across the fibre bundle: rows #1 to #13; spinning parameter: PET staple fibre spinning process, see before

The variation coefficient changes by modifying the process parameters, but only small improvements are possible. Neither changes in throughput, extrusion temperature nor take-up velocity can entirely prevent this geometrically associated variance. An example is shown in Fig. 3.48 which depicts the trace of the maximum and minimum elongation to break and the variation coefficient CV_{ε_b} versus the take-up velocity. The throughput in this example was related to the take-up velocity in order to achieve a constant filament fineness.

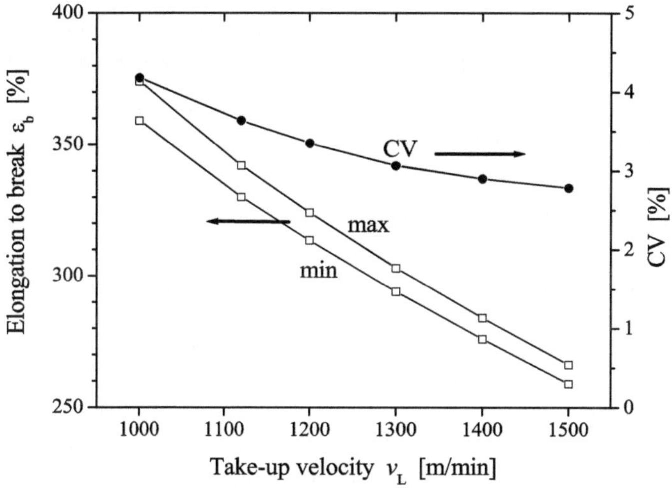

Fig. 3.48. Maximum and minimum elongation to break and corresponding coefficient of variation CV_{ε_b} in dependence on take-up velocity, constant filament fineness

Effect of Quenching Air Profile

The influence of changing the quenching air velocity on both the filament cooling and to the uneveness of filament orientation across the bundle will be briefly discussed in this section. Five different cases are considered, such as no active quenching, and quenching air velocities of 0.5, 1.5, 2.0 and 3.0 m/s. The case of 1.5 m/s air velocity was already mentioned, the figures of the previous section can be directly compared to the following ones. The air blowing starts again at distance $x = 5$ cm below the spinneret, the different profiles of blowing air velocity and temperature are shown in Fig. 3.49.

The resulting streamlines and the air temperatures within the filament bundle for the three cases (#A, #B, #D) are shown in Fig. 3.50.

It is obvious that the different velocity profiles affect the fibre properties. Table 3.11 summarises the maximum and minimum elongations to break together with the corresponding CV-value for each case. Additionally, results of the calculations with $v_{\text{air}} = 1.5$ m/s (#C) and $v_{\text{air}} = 3.0$ m/s (#E) are

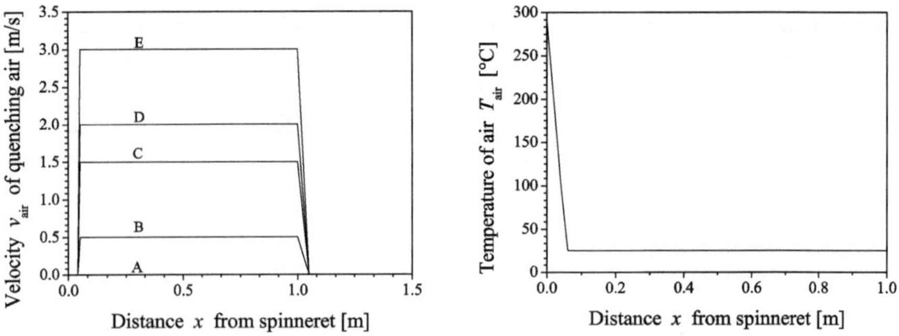

Fig. 3.49. Velocity profiles (*left*) and temperature profile (*right*) of quenching air, as used for calculation example

shown. It becomes clear that increasing the quenching air velocity increases the eveness but also makes the process unstable if a critical value of air velocity is reached. On the contrary, extremly high air velocities do not result in a further decrease of uneveness. One reason for that is that the initial rheological force F_0 plays a more important role for fibre formation at low take-up velocities than inertial or air friction force. The initial force F_0 is essentially determined from the conditions near the spinneret. Therefore it should be possible to maximise the eveness by means of an adapted mass throughput or by adapted temperature regime, respectively.

Table 3.11. Maximum and minimum elongation to break for different quenching air velocity profiles

	Air velocity	ε_{max} in %	ε_{min} in %	CV_{ε_b} in %
A:	no active quenching	397	363	9.4
B:	$v_{air} = 0.5$ m/s	361	334	8.1
C:	$v_{air} = 1.5$ m/s	342	330	3.6
D:	$v_{air} = 2.0$ m/s	338	325	4.0
E:	$v_{air} = 3.0$ m/s	332	320	3.8

In general, stress at the solidification point becomes less dependent upon spinning conditions at high take-up velocities. This is because the inertia force then dominates the tensile stress. But at lower take-up velocities (which are commonly used in the staple fibre spinning process and in the example above) the stress at the solidification point can be influenced by spinning conditions near the spinneret.

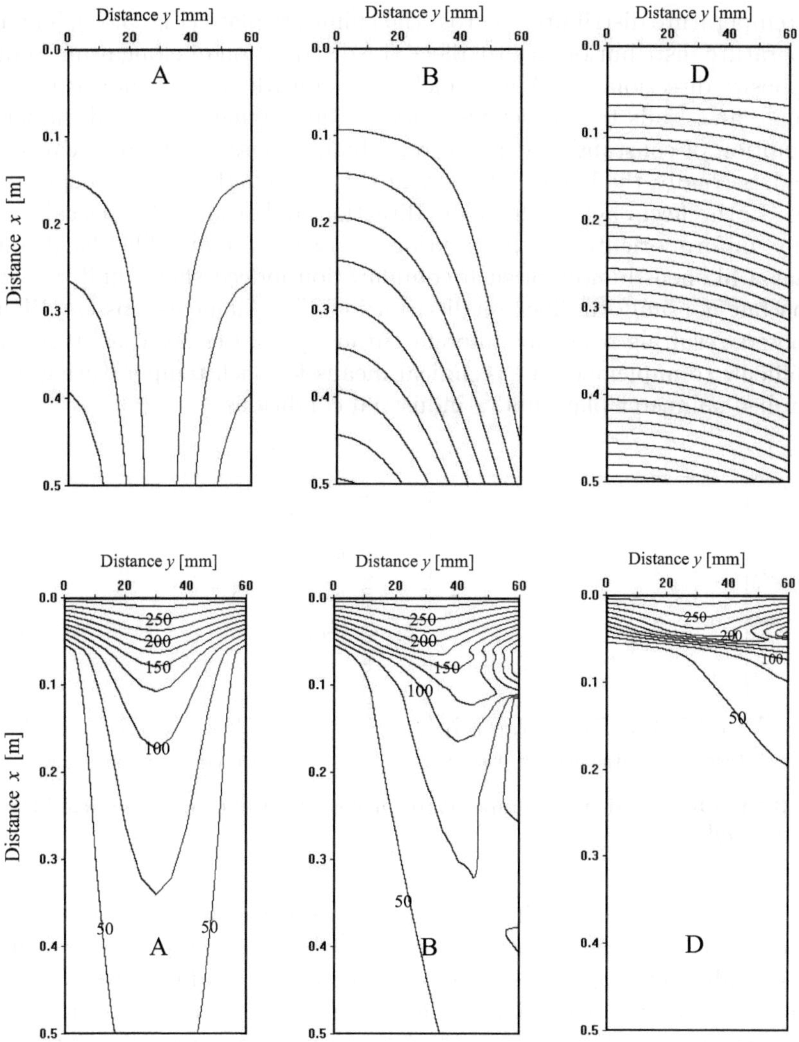

Fig. 3.50. Streamlines (*upper row*) and temperature (*lower row*) of air inside the filament bundle, *A:* no active quenching, *B:* $v_{\text{air}} = 0.5$ m/s, *D:* $v_{\text{air}} = 2.0$ m/s, the temperature of the isotherms is indicated

Effects of Adapted Temperature Distribution and Mass Throughput

The nearly parabolic characteristic of the elongation to break behaviour which is shown in Fig. 3.47 can be corrected by a gradual adjustment of the mass throughput and/or melt temperature distribution for the filaments on each individual row of the spinneret. One possibility is to create a specific temperature distribution across the spinneret plate (Fig. 3.51). The melt temperature distribution has to affect the distribution of elongation to break in opposite direction. The higher melt temperature at the windward side for example first leads to a lower viscosity of the polymer melt and, under the assumption of constant melt pressure, a higher mass throughput can be expected. Secondly, the higher temperature also affects the cooling length of the fibres. Both effects act in the same direction and result (theoretically) in a totally uniform orientation distribution across the bundle. The disadvantage is that a filament bundle of such configuration indeed shows uniform orientation but non-uniform fineness distribution. The filaments possess different diameters resulting from the different throughputs. Besides that, it seems to be difficult to implement the technical means for such temperature gradient controlled spinneret plate under industrial conditions.

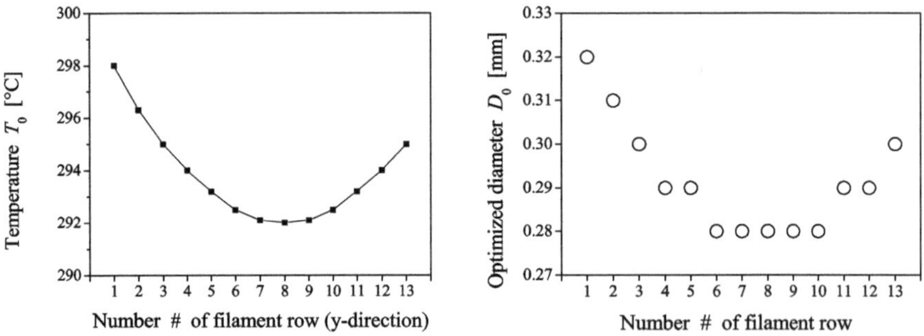

Fig. 3.51. Optimised melt temperature profile (*left*) and optimised capillary diameters (*right*)

Another possibility to influence the mass throughput in the described manner is to adjust the diameters of capillary holes in the spinneret. A similar idea was already proposed earlier, with diameters gradually reducing from windward to leeward side. The model calculation now allows to predict the exact stepwise adaption of the hole diameters of each row. It indicates the adaption of larger diameters at the windward side followed by the leeward side and holes of smaller diameters in the centre. Figure 3.51 provides a proposal for five stages of spinneret hole diameters. The corresponding Table 3.12 shows that, by adapting these five grades, the initial variation coefficient of the elongation to break of nearly 4% drops to a quarter, and, adapting only three grades still results in a drop of variation to half of the initial value. The disadvantage of uneven fineness of course remains.

In principle it should be possible to combine the adapted melt temperature profile and the adapted capillary hole diameter distribution to minimise *both* effects, that of uneven orientation *and* uneven fineness. But in practice the realisation seems difficult and may only be achieved under very special conditions.

Table 3.12. Optimised spinneret hole diameters

Number of stages	Diameter in [mm]	Fineness in [dtex]	CV_{ε_b} in [%]
1	0.30	18	4
3	$0.29 \cdots 0.31$	$16 \cdots 19$	2
5	$0.28 \cdots 0.32$	$14 \cdots 21$	1

Effect of Retarded Cooling

A more effective and also easy to realise approach which does not change the individual filament mass throughputs is the carefully designed asymmetrical cooling program for the filament bundle. Since the cooling rate for the filament rows in the centre of the tow near the spinneret cannot be increased and is always lower than that at the outer sides, the conclusion is to delay the cooling rate of the filament rows at the windward and leeward side by means of adapted booster heating. The lateral quenching exemplary adopted here results in non-uniform cooling down of the windward side, the centre, and the leeward side, respectively. A compensation of the non-uniform cooling can be attained via asymmetrically designed booster heating near the spinneret. The model calculation now allows to find the optimum temperature profile at windward and leeward sides that provides (theoretically) for reduction to zero of the CV value of the orientation. In practice the aim is to minimise the variation of the orientation and the resulting uneveness of elongation to break in order to achieve high quality yarns and to realise stable process conditions and high productivity.

Figure 3.52 shows an optimised velocity and temperature profile of quenching air, and Fig. 3.53 depicts the effect of the optimised profile to the air temperature within the bundle. The air quenching starts 10 cm below the spinneret. The region between spinneret at $x = 0$ and beginning of air blowing is heated to different temperatures for the windward and the leeward side, respectively.

Figure 3.53 can be directly compared with Fig. 3.43. The next diagram (Fig. 3.54) shows the courses of fibre temperatures for three different positions inside the bundle. A comparison with Fig. 3.45 illustrates the retarded cooling. The higher temperature near the spinneret mainly influences the initial rheological forces F_0 of the fibres and therefore also the total forces at each solidification distance. The result of booster heating is the nearly uniform distribution of stress at solidification points versus the filament rows and at last nearly uniform filament orientations. Figure 3.55 shows the calculated elongations to break across the fibre bundle in comparison with the original (unoptimised) variant. The coefficient of variation for the elongation to break CV_{ε_b} changes from initially 3.6% to 0.8%.

As the mean orientation of the bundle is simultaneously reduced, a higher draw ratio in the following drawing stages becomes possible and opens up the

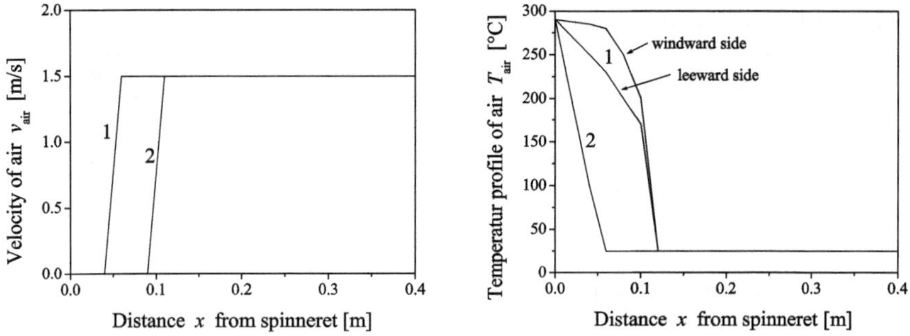

Fig. 3.52. Optimised profiles (#1) of air. *Left:* Velocity profile at windward side; *right:* Air temperature profiles at windward and leeward sides, the profiles of the example before (#2, see Fig. 3.49-C) is shown for comparison

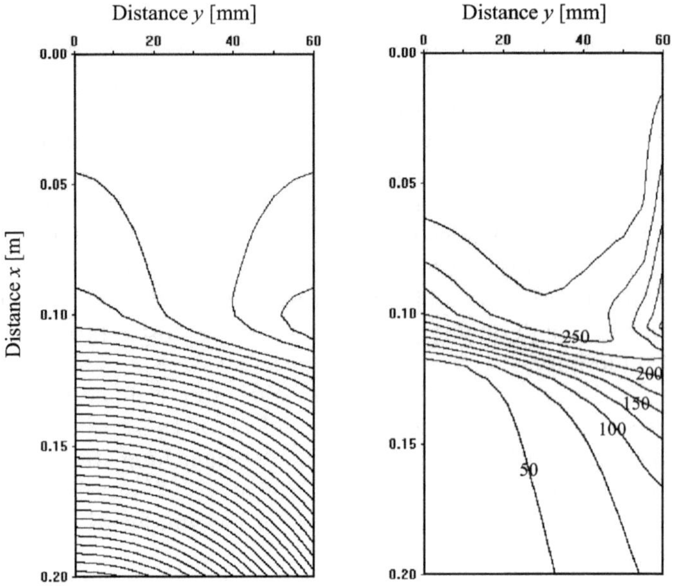

Fig. 3.53. Streamlines (*left*) and temperatures (*right*) of air inside the filament bundle for optimised air profiles, the temperature of the isotherms is indicated. Only the region 20 cm below spinneret is shown, please compare to Figs. 3.43 and 3.42

additional opportunity of increasing the mass throughput (increased productivity) with equal target fibre fineness. For the discussed example this would lead to an increased productivity of about 4...5%. A further advantage of the method of retarded cooling is its adabtability and flexibility. The method allows for easy and quick adaptions to changes in production technology [262].

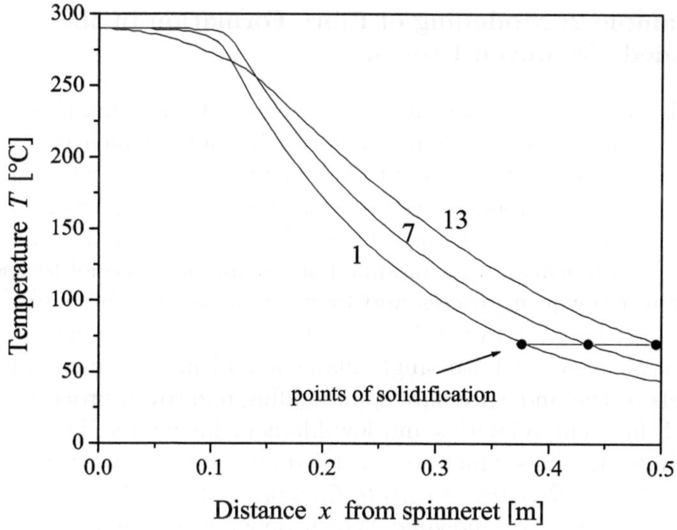

Fig. 3.54. Filament temperature $T(x)$ (calculated) for three different locations: 1 – windward side (row #1), 7 – center of the bundle (row #7), 13 – leeward side (row #13), the positions of solidification distances are indicated; optimised air profile

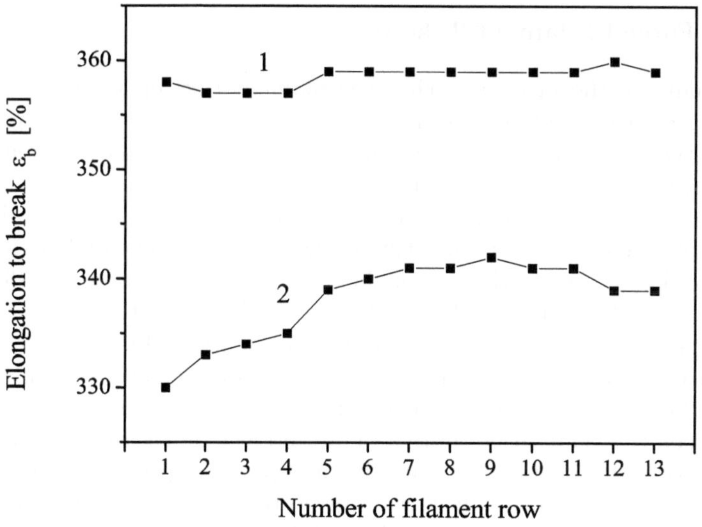

Fig. 3.55. Elongation to break (calculated) across the fibre bundle (row #1 to #13) for the optimised air profile, $CV_{\varepsilon_b} = 0.8\%$ (1), results of the origin variant (Fig. 3.47) as comparison, $CV_{\varepsilon_b} = 3.6\%$ (2)

3.2.5 Example 2: Modelling of Fibre Formation in the Spunbonded Nonwoven Process

The spunbonded nonwoven process represents a typical multifilament spinning process because of its high number of filaments which can be up to a multiple of thousands. Take-up of the filaments is usually achieved by means of air friction, either in an excess-pressure or in an underpressure process, or sometimes a combination of both. HAJJI, MISRA, SPRUIELL et al. [263, 264] discussed the application of a modified single filament model to the *Reicofil* spunbonding nonwoven process and found good agreement between experimental and predicted data for their investigation. The following section deals with the application of both single filament and multifilament fibre formation models to the underpressure spunbonding nonwoven process in order to attain high filament velocities and low filament finenesses. The discussion is based on investigations which were carried out by the authors together with the *Saxon Textile Research Institute Chemnitz* [265]. The questions for the investigations were the following: Which dependencies exist for the underpressure process between the filament velocity and the filament fineness on the one hand and the spinning and take-up condition on the other? What are the best energetic conditions for the air suction device? How is the take-up to be designed to enable high filament velocities? Some answers can be given with the help of the applied fibre formation model.

Friction Forced Filament Take-up

The specialty of the process is that the filament take-up is not realised by means of godets or a winder like in the conventional yarn spinning processes but via drag of an air stream in spinning direction (Fig. 3.56). Contrary to the fibre spinning process where the final fibre velocity is fixed, the take-up velocity resulting from air drag force is not known from the beginning and thus cannot be treated as an initial spinning parameter. The higher the velocity of the axial air stream the higher the friction and drag force transferred to the fibre – thus the higher the fibre velocity. Furthermore, the higher the fibre velocity the lower its diameter (at constant mass throughput) and the lower the resulting friction and driving of the fibre. Consequently, the intentional effect will be reduced as the resulting air drag itself depends upon the difference between the velocities of fibre and air and also on the fibre fineness. The final fibre velocity, its fineness and all corresponding properties result from the balance between the air drag transferred to the fibre and the inherent fibre force contributions like the rheological, inertial, and gravitational forces, respectively.

When trying to solve the differential equations of the fibre formation model it becomes obvious that the initial value F_0 for the force at the spinneret is unknown. This causes no difficulties in the case of conventional take-up by means of godets or a winder because the initial rheological force in the

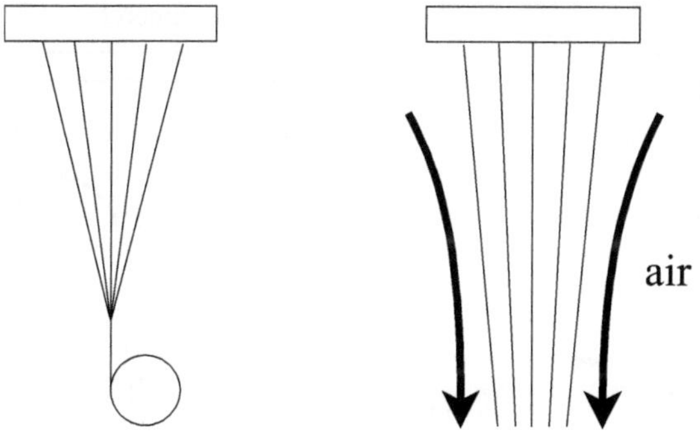

Fig. 3.56. Comparison between conventional yarn melt spinning process (*left*) and nonwoven process with air drag take-up (*right*)

model can be determined via an iteration procedure, that is, until the final take-up velocity is reached within an adequate numerical tolerance. However, the situation regarding the take-up by means of air drag is quite different because the final take-up velocity results from an equilibrium of forces and therefore is unknown from the beginning.

The simplest idea to solve this problem is to assume the initial rheological force is equal to zero $F_0 = F(0) = 0$ [263, 264]. But the problem can be more accurately solved in general by means of an additional iteration if the rheological force is known (or can be estimated) at *any* distance x of the fibre path. Especially the usage of a take-up channel (see below) opens up the possibility to get a much better assumption for the missing boundary value.

Figure 3.57 illustrates the basic idea of modelling the effect of a take-up channel. The possible technical realisation of an underpressure nonwoven equipment is also depicted. The positions of entrance and exit of the take-up channel (measured from the spinneret at $x = 0$) are x_1 and x_2, respectively.

The tensile force F for a single filament at entrance x_1 of the take-up channel is given by (see Eq. 3.11a on page 50 – the force balance)

$$F(x_1) = F_0 + Q \cdot (v(x_1) - v_0) + \int_0^{x_1} \frac{\varrho_{\mathrm{air}}}{2} c_{\mathrm{f}}\, \tilde{v}^2\, \pi\, D\, dx - \int_0^{x_1} \varrho_{\mathrm{p}}\, g\, \frac{\pi}{4}\, D^2\, dx\,, \tag{3.187}$$

where $\tilde{v}$ is the difference between the velocities of fibre and surrounding air, respectively, $\tilde{v} = |v - v_{\mathrm{air}}|$. The tensile force must be equal to the force applied inside the take-up channel:

$$F(x_1) = F_e + \int_{x_1}^{x_2} \frac{\varrho_{\mathrm{air}}}{2} c_{\mathrm{f}}\, \tilde{v}^2\, \pi\, D\, dx - \int_{x_1}^{x_2} \varrho_{\mathrm{p}}\, g\, \frac{\pi}{4}\, D^2\, dx\,. \tag{3.188}$$

If the fibre is already solidified at the entrance $x = x_1$ of the channel, which means the fibre velocity and diameter are fixed and the air velocity inside

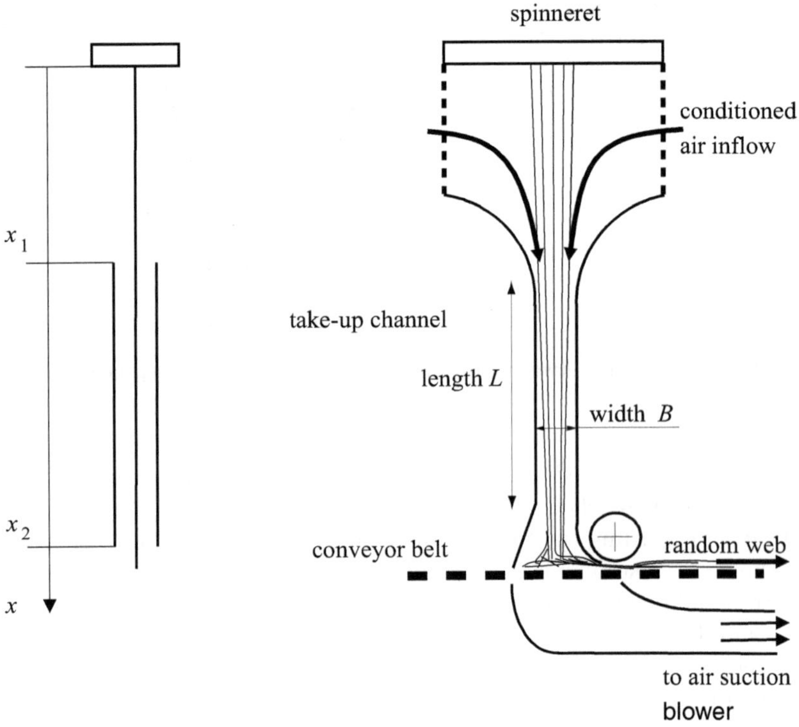

Fig. 3.57. Principle of the take-up channel (*left*) and possible realisation of an underpressure spunbonding nonwoven process (*right*)

the channel can also assumed to be constant, then (3.188) is simplified to

$$F(x_1) = F_e + L\,\frac{\varrho_{air}}{2}\,c_f\,\tilde{v}^2\,\pi\,D - L\,\varrho_p\,g\,\frac{\pi}{4}\,D^2\,. \tag{3.189}$$

The term $F_e = F(x_2)$ is the tensile force at the exit of the channel and $L = x_2 - x_1$ is the channel length. Both equations can be combined together for all distances x. If careful attention is payed to the signs of air friction force contribution inside and outside of the take-up channel follows

$$F(x) = F_0 + Q\cdot(v(x) - v_0) + \Theta \int_0^x \frac{\varrho_{air}}{2}\,c_f\,\tilde{v}^2\,\pi\,D\,dx - \int_0^x \varrho_p\,g\,\frac{\pi}{4}\,D^2\,dx\,, \tag{3.190}$$

with

$$\Theta = \begin{cases} 1 & \text{for} \quad v(x) > v_{air} \\ -1 & \text{for} \quad v(x) < v_{air} \end{cases}\,. \tag{3.191}$$

The initial tensile force F_0 correlates with the initial fibre velocity gradient dv/dx at $x = 0$ and therefore determines the final velocity after integration.

The only reasonable assumption for the equilibrium of forces at any distance is that the tensile force vanishes at the exit of the take-up channel

$$F(x_2) = 0 \, . \tag{3.192}$$

This condition replaces the boundary condition of a fixed take-up velocity. With the air profile $v_{\text{air}}(x)$ given and the equations of fibre modelling, it is now possible to calculate the final fibre velocity and all connected characteristics of fibre formation.

High Filament Velocities – Realised by Means of an Underpressure Spunbonding Nonwoven Process

Subject of the investigations described below is the design of an underpressure spunbonding nonwoven equipment that realises high filament velocities up to 3000 m/min and filament finenesses of 1...2 dtex [265]. Boundary conditions are: the maximum total height of the equipment of less than 3 m and the limited power of the air suction blower. With help from the fibre formation model it was possible to accompany the design process and to give useful hints for the construction.

Inside the upper part of the underpressure spunbonding nonwoven equipment, the air entrance chamber, the air flow behavior should be a symmetrical (Fig. 3.57). Assuming a channel width of $B = 20$ mm, height of entrance chamber of $H = 1$ m, and mean value of the air velocity inside the channel of $v_{\text{channel}} = 3000...6000$ m/min, then follows for the air entrance velocity v_{entr}:

$$v_{\text{entr}} = \frac{B}{H} \cdot v_{\text{channel}} = 1...2 \text{ m/s} \, .$$

The amount of entrance air velocity is comparable to the quenching air velocity for the staple fibre multifilament spinning example treated in the section before. Therefore the model of symmetrical flow behaviour can be used to calculate the multifilament effects on individual fibre formation, for example to estimate the differences in cooling behaviour of filaments located at the inner and outer side of the bundle, respectively. Figure 3.58 elucidates the application of the model for melt spinning of polypropylene.

For this process it must be assured that the filament temperatures inside the take-up channel are lower than any given critical temperature in that manner, and that no sticking occurs while the filaments touch each other or the walls of the channel. The solidification which represents the end of the fibre formation zone is not fixed because it depends on the spinning conditions, especially on the mass throughput, the melt temperature, and the kind of polymer material (with its property of heat capacity). Additionally, the cooling rate is retarded in the centre of the filament bundle. Therefore each set of spinning conditions determines a minimum distance x_{min} where the temperatures of all filaments is lower than the critical temperature:

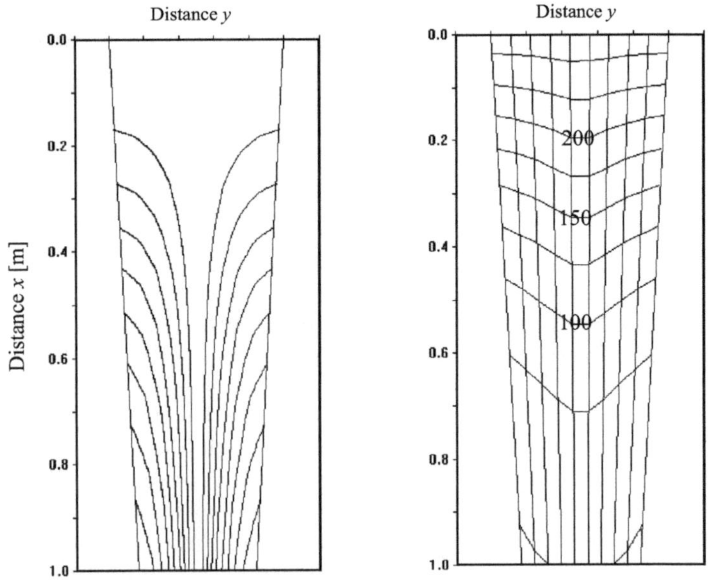

Fig. 3.58. Example of multifilament model application to the underpressure spun-bonded nonwoven process. *Left:* streamlines of air, *right:* filament temperatures (isotherms), temperatures in °C indicated

$$T(x_{\min}) < T_{\text{crit}} \,. \tag{3.193}$$

The next figure (Fig. 3.59) shows the distances where the polypropylene filaments reach the temperature of $T_{\text{crit}} = 100\,^{\circ}\text{C}$ as example.

It can be seen that for the indicated spinning conditions these distances are always lower than 1 m; therefore the minimum distance of $x_{\min} \approx 1$ m according to the mass throughput of $Q = 1$ g/min is a sufficient distance between the spinneret and the entrance of the take-up channel. The calculation also confirms the assumption that the final filament velocity depends nearly linearly on the length L of the take-up channel. The calculation also provides relationships to describe the dependence of final velocity on air velocity inside the channel (see Figs. 3.60, 3.61).

At last the theoretical investigations allow to estimate the pairs of air velocity v_{channel} within the channel and corresponding length L of the channel to reach a destined filament velocity (Fig. 3.62). From the figures it can be seen that the conditions to reach 3000 m/min filament velocity are approximately the following: channel length ≈ 1 m, driving air velocity ≈ 100 m/s.

The task of reaching a defined air velocity inside the take-up channel of a given length and width is due to the power and energetic conditions of the air suction blower and depends upon pressure losses of the channel together with other parts of the spinning device. The longer the channel, the

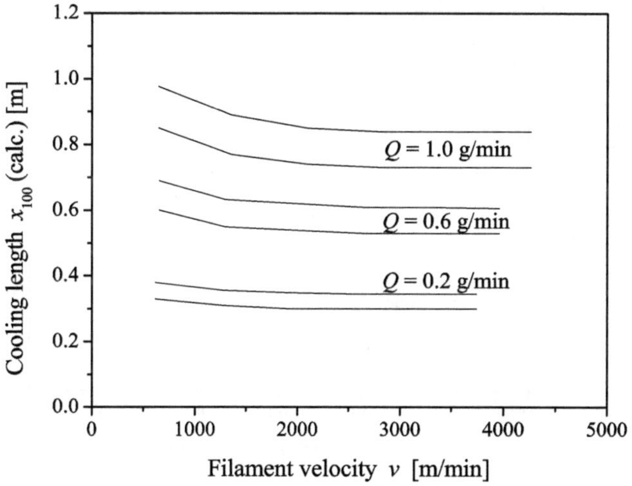

Fig. 3.59. Cooling length x_{100} (calc.) vs. final filament velocity v (upper and lower limit for filaments at center and boundaries of the bundle); polymer: polypropylene (PP), melt temperature: 260 °C, air temperature: 25 °C, three different mass throughputs (per hole) are indicated

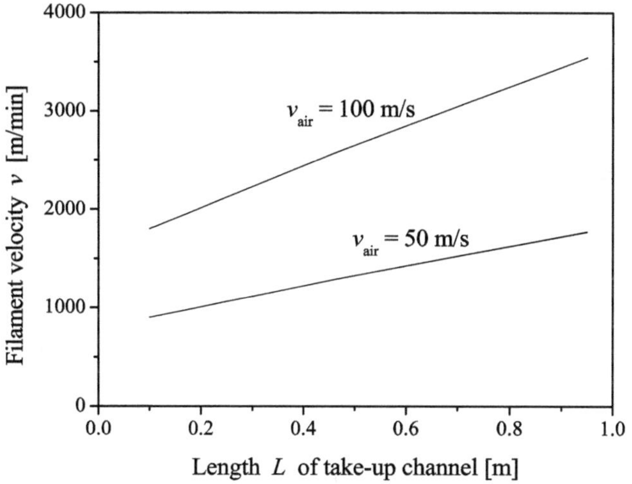

Fig. 3.60. Final filament velocity v (calc.) vs. length L of the take-up channel, two different velocities of air inside channel (indicated), other spinning conditions: see Fig. 3.59

higher the pressure losses and the lower the resulting air velocity inside, but the higher the exposed length of the filaments to air friction force. On the other hand, there is an optimum channel width B for a given length L to achieve maximum possible air velocity. The smaller the channel width from this optimum value the higher the pressure losses, and the higher the channel

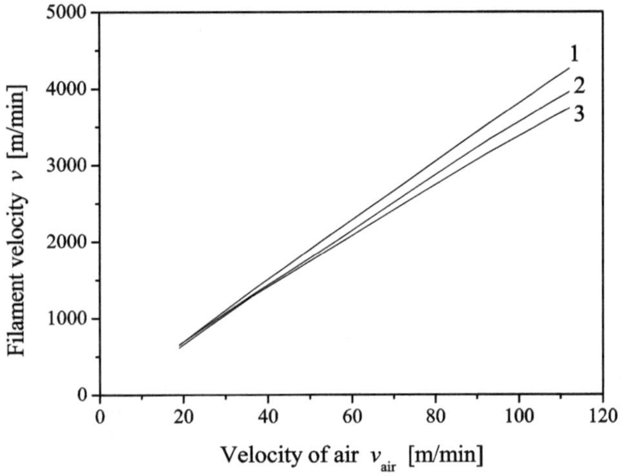

Fig. 3.61. Final filament velocity v (calc.) vs. air velocity $v_{channel}$ inside take-up channel, three different mass throughputs (per hole) indicated: 1 – 1 g/min, 2 – 0.6 g/min, 2 – 0.2 g/min, other spinning conditions: see Fig. 3.59

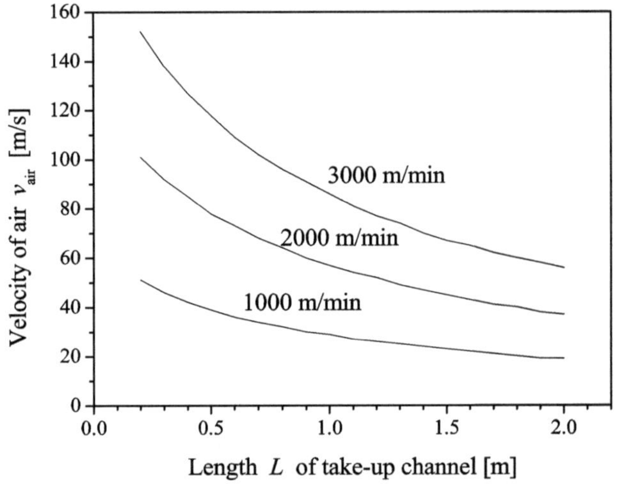

Fig. 3.62. Combinations for length L of take-up channel and air velocity $v_{channel}$ to destined final filament velocity v (indicated), mass throughput (per hole) 0.6 g/min, other spinning conditions: see Fig. 3.59

width from the optimum, the higher the cross section area and the lower the resulting air velocity. For the investigated equipment [265] the optimum channel width was $B = 20$ mm. At last Fig. 3.63 shows the finally reached filament velocities and finenesses depending upon the air velocity inside the channel for the investigated equipment.

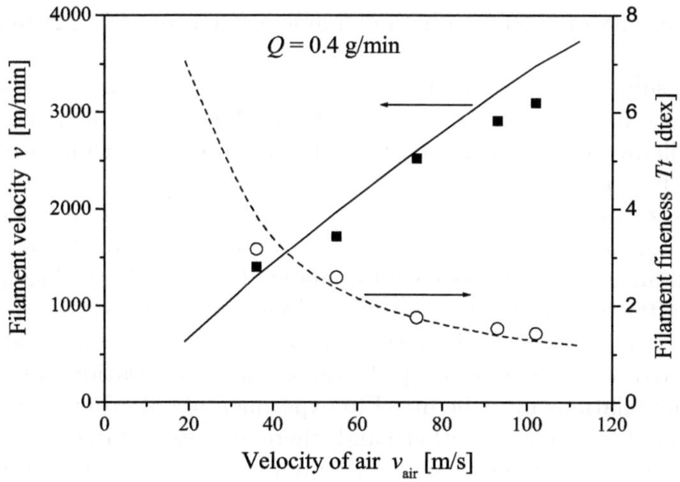

Fig. 3.63. Final filament fineness Tt and final filament velocity v in dependence on the velocity of air inside take-up channel v_{channel}, mass throughput (per hole): 0.4 g/min, *lines*: calculated, *symbols*: measured values

3.2.6 Summary

Process modelling and simulation seems to be an effective way in order to solve engineering problems. Computer simulations can indeed be a helpful tool for understanding the principle behaviour of the considered process. But one has to keep in mind that information generated by the computer program is always based on any physical and/or mathematical model of the real process, and this describes the essential aspects of the treated process only in a relatively qualified manner. The simulations can never replace the experimental work and the empirically collected experiences, but they can powerfully support the laboratory and industrial research in order to avoid time-consuming and/or expensive investigations regarding process modification, extension, or optimisation, respectively. Especially the last mentioned task of optimisation is typical for engineering procedures in melt spinning: the prediction of resulting fibre properties after changing the process parameters, geometrical conditions, the used material, or to optimise the process with respect to productivity and quality. The example treated above shows how the computer simulation was used to support the design of a special melt spinning device for underpressure spunbonded nonwoven equipment. Here, the theory of fibre formation together with the *engineering* modelling of the entire process allowed for the determination of major dependencies and relations regarding material properties, process conditions and characteristics of fibre formation. This knowledge, combined with additional empirical results enabled the successful design of the device which provided for the desired filament velocity of 3000 m/min, and filament finenesses of 1...2 dtex, respectively.

The results of any simulation are always to verify on the basis of experimental data. It is possible to measure on-line the filament velocity, the filament diameter, the temperature, and orientation (birefringence) depending upon distance from the spinneret and to correlate the process conditions with the resulting fibre properties. A serious model should be based on such experimental background. Secondly, such a model should also be tractable and for engineering purposes easy to handle. The models of fibre formation, available after more than 40 years of development, do not satisfy all these requirements. Some important effects cannot be described well so far but only in an approximative manner (e. g. the theory of crystallisation in fibre formation). Some relations must be fitted to experimental data due to the lack of a satisfactory theory. On the other hand, there are also reliable empirical data missing for some polymer types, especially for the non-isothermal behaviour under high stresses and high deformation rates. Therefore, the model of fibre formation consists of both well investigated and exact relations but also of approximations valid only for special conditions or within certain ranges. For any successful engineering the need for the combination of *both* model simulation *and* empirical experience is obvious.

3.3 Limits of Fibre Formation in Melt Spinning and Spinnability

3.3.1 Maximum and Minimum Fineness

The limits of the melt spinning process can be estimated by some simple physically based considerations [266, 267]. At first, we want to look at the maximum possible fineness Tt_{max}. The fineness of the as-spun filament is given by the quotient $Tt = Q/v$. Additionally, for the final drawn and full oriented fibre the necessary draw ratio DR to reach final fineness Tt_{f} has to be considered: $Tt_{\text{f}} = Q/(v \cdot DR)$. This means that in order to reach the maximum possible fineness, the highest possible mass throughput Q and the minimum possible take-up velocity v has to be used. The draw ratio itself depends upon the spinning conditions, mainly on the take-up velocity. One has to regard that cooling down of the polymer stream strongly relates to the mass throughput (see Sect. 3.1, Fig. 3.23 on page 85) to realise the maximum fineness. The higher the mass throughput per hole the longer the distance to the solidification point. A simple demand for stable melt spinning and avoidance of fibre breaks means that the fibres need to be cooled down below the solidification temperature before their first contact with the preparation applicator or any other fibre guidance element, respectively. The minimum length L_{min} of the spinning tube has to exceed the solidification distance: $L_{\text{min}} > L_{\text{s}}$. After solidification in general no further deformation of the fibres occurs. But sometimes it is required that a certain minimum fibre temperature has to be reached in order to avoid sticking effects between the filaments and the guidance elements or between the filaments themselves. In Fig. 3.64 on the left-hand side the distance L_{50} is required to reach the temperature $T = 50°C$ and the solidification distance L_s for spinning of PA 6 at three different mass throughputs depending upon the take-up velocity is shown. For higher take-up velocities ($v \geq 3500$ m/min) where stress induced spinline crystallisation takes place the solidification occurs above the glass transition temperature of $T_g \approx 50°C$. But for maximum fibre fineness the lower take-up velocities ($v \leq 3000$ m/min) should be preferred so that the fibre temperature at the solidification point equals the glass transition temperature of $T_g = 50°C$ for PA 6. The figure on the right-hand side depicts the calculated cooling distance $L_{50} \approx L_s$ vs. mass throughput Q for different take-up velocities of up to 3000 m/min. The upper curve in the picture represents the relationship $L_{50}(Q)$ for the minimal possible take-up velocity v_{min} which itself is determined by the minimal possible take-up distance (this is the solidification distance) and the resulting gravitational force. The calculation based on the minimal possible take-up velocity v_{min} was carried out in such way that the resulting spinline stress shows a nearly constant or an only very small increasing behaviour vs. the spinning distance x, that means that the minimum take-up velocity has to be greater at least than the velocity of freely falling fibre. At the solidification point the spinline stress for all variants reaches an extremly low value: $\sigma_s = 10 \ldots 15$ kPa. The processing for PA 6 then results in fibres with no (or very low) orientation and possible draw ratios of about $DR \approx 6$.

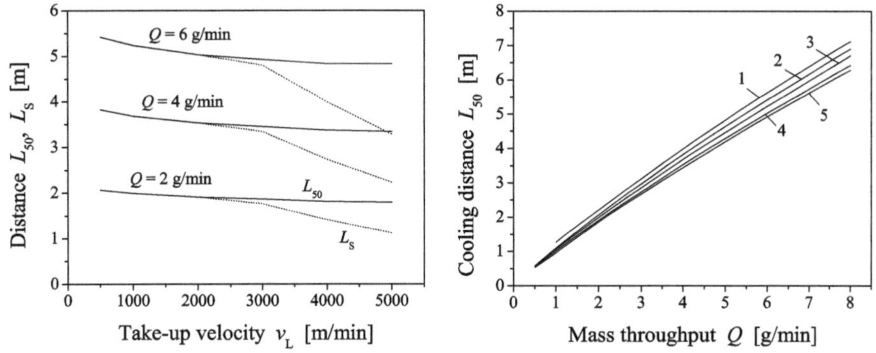

Fig. 3.64. *Left:* (calculated) distance L_{50} required to reach fibre temperature $T = 50°C$ (*solid lines*) and solidification distance L_s (*dotted lines*) vs. take-up velocity v, mass throughput (per hole) indicated; *right:* cooling distance L_{50} vs. mass throughput Q, different take-up velocities: 1 – minimum possible (150 ... 300 m/min), 2 – 500 m/min, 3 – 1000 m/min, 4 – 2000 m/min, 5 – 3000 m/min, *Spinning parameter:* PA 6, molecular weight: 17 000, melt temperature: 265°C, no additional air quenching

Figure 3.65 shows on the left-hand side the thermal limitations for regarding the maximum possible fineness (*after drawing* to final elongation to break of 25%) determined in terms of necessary cooling length vs. take-up velocity. The figure on the right-hand side shows the maximum possible fineness Tt_{max} depending upon the throughput Q after drawing with draw ratio $DR = 6$, applying the minimal possible take-up velocity v_{min} and under the condition of an adapted optimum cooling distance.

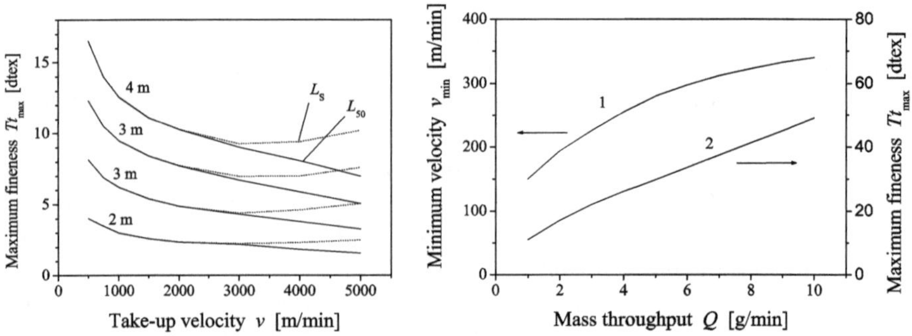

Fig. 3.65. *Left:* limitation curves for maximum fineness Tt_{max} (after drawing) vs. take-up velocity v, cooling length L_{50} is indicated; *right:* (calculated) maximum possible fineness Tt_{max} (1) (after drawing) and minimum possible take-up velocity v_{min} (2) vs. mass throughput Q (per hole), adapted spinning length; *Spinning parameter:* PA 6, molecular weight: 17 000, melt temperature: 265°C, no additional air quenching

The maximum possible finenesses presented here are theoretical values, of course, and estimated without any additional air quenching. Inclusion of quenching should further stabilise the (low tension) spinning process. Resultingly, it can be stated that regarding the filament fineness the upper limit for conventional spinning equipment with cooling lengths of $L = 4 \ldots 6$ m for PA is in the range of $30 \ldots 40$ dtex after drawing. For PET the required cooling lengths are somewhat smaller, and the maximum possible fineness slightly increases. In order to achieve any further increase of filament fineness the conventional cooling by means of air is to be replaced by more effective cooling methods, e. g. cooling and solidification in a water quench.

Remark. It must be mentioned that if the melt is extruded at very high throughputs (high flow rates) the melt stream becomes distorted. These distortions are known as *melt fracture* [268] where the extrudate shows instabilities in form of spiral or gross flow behaviour or surface roughness and the so-called sharkskin effect. These phenomena have been studied by many investigators and are still up to now not fully understood.

The minimum possible fineness after drawing should be reached with minimum throughput Q, maximum possible take-up velocity v and (if possible) high draw ratio DR:

$$Tt_{\mathrm{DR}} = \frac{Q}{v \cdot DR} \,. \tag{3.194}$$

The fibre orientation and therefore the resulting possible draw ratio are determined by the stress at the solidification point and depend upon the take-up velocity (see Sect. 3.1, Fig. 3.22 on page 84). The higher the take-up velocity the higher the tensile stress and also the degree of orientation and the lower the residual draw ratio. Figure 3.66 (*left-hand side*) shows the resulting fineness before and after the drawing (to final elongation of 25%) of PA 6 fibres in a spinning process with increasing take-up velocity and constant mass throughput. The advantage of getting minimum fineness after drawing is not very effective due to the fact that, with increasing take-up velocity the orientation increases and thus the draw ratio decreases. On the other hand, the reduction of throughput directly decreases the fineness and only slightly increases the fibre orientation resp. reduces the residual draw ratio. However, the effect of increasing orientation in the case of decreasing throughput is smaller than in the case of increasing velocity. To get minimal fineness, the reduction of throughput is much more effective than increasing the spinning speed (Fig. 3.66, *right-hand side*).

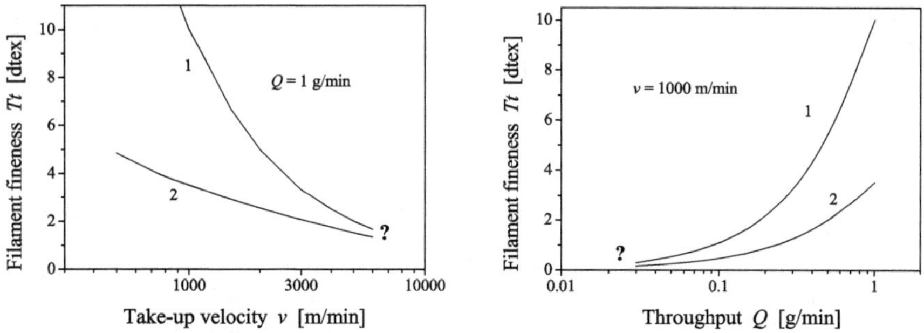

Fig. 3.66. (Calculated) filament fineness before drawing (1) and after drawing (2) of PA 6 vs. take-up velocity v (*left*) resp. vs. mass troughput Q (*right*), constant throughput and take-up velocity, respectively; the limits (marked with symbol '?') are unknown and depend strongly on the material and the processing parameter

Increasing the take-up velocity and decreasing the mass throughput rapidly border on the physical limits of the process. The higher the take-up speed the higher the resulting spinline stress and the extensional deformation rate of the material. This may cause filament breakages due to brittle cohesive fracture or visco-elastic ductile failure. In the next section some aspects of the failure behaviour will be discussed.

An estimation of the limiting conditions with respect to fibre stress in the spinning line can be made as follows: After solidification the tensile stress must not exceed the critical stress level that may cause secondary deformation. Figure 3.67 shows on the left-hand side the fibre stress σ_L at certain distances L depending upon take-up velocity v_L. The example is calculated for single PA 6 filaments with mass throughput $Q = 1$ g/min. The distance to solidification for low take-up velocities when temperatures reach 50°C is about $L_s \approx L_{50} \approx 1$ m. For higher take-up velocities the distance L_s becomes smaller, however, the cooling distance L_{50} remains nearly constant. The filament stress strongly increases with take-up velocity, at longer distances L it is mainly caused by air friction. To avoid unwanted secondary deformation, the critical stress for PA 6 is assumed to be $\sigma_{crit} \approx 50$ MPa. The condition $\sigma_L < \sigma_{crit}$ leads to the picture at the right-hand side of Fig. 3.67, which depicts the maximum possible take-up velocity and the resulting minimum fineness for the given mass throughput depending upon the take-up distance L before and after drawing to 25% of final elongation. The shorter the take-up distance L the higher the stress limited maximum possible take-up velocity and the lower the resulting minimum fineness.

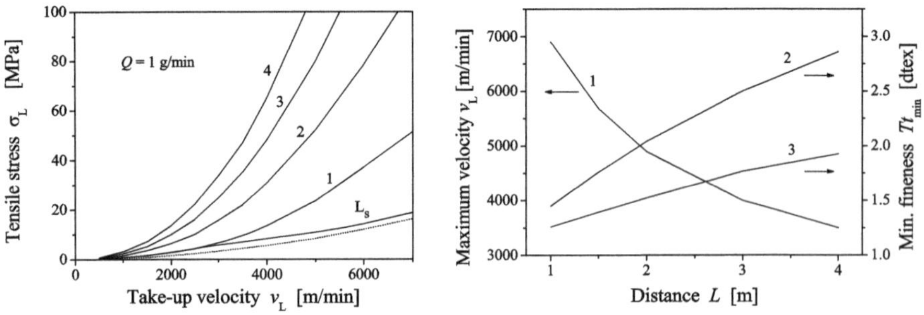

Fig. 3.67. *Left:* (calculated) filament stress σ_L vs. take-up velocity for different distances L, distances are indicated: $L = 1$ m (1), 2 m (2), 3 m (3), 4 m (4), and solidification distance L_s, *dotted line*: calculation without air friction; *right:* maximum possible take-up velocity (1) for the condition $\sigma_L \leq 50$ MPa and minimum fineness Tt_{min} before (2) and after drawing (3) to 25% final elongation; PA 6, mass throughput $Q = 1$ g/min

For the example above, with constant throughput of 1 g/min, it is not possible to reach a final fineness of 1 dtex by only increasing the spinning speed. For the production of fine filaments the reduction of throughput appears at the same time necessary. To get an impression of this second variant in the next diagram (Fig. 3.68) the resulting tensile stress at distance $L = 1$ m with respect to the accessible fineness is compared if (1) the throughput is kept constant and the take-up velocity increases (this is the example above) and if (2) the take-up velocity is kept constant and the throughput decreases, respectively. It can be clearly observed that reducing the throughput leads to much lower stress at any certain distance and opens up the way to spinning fine and ultra fine fibres.

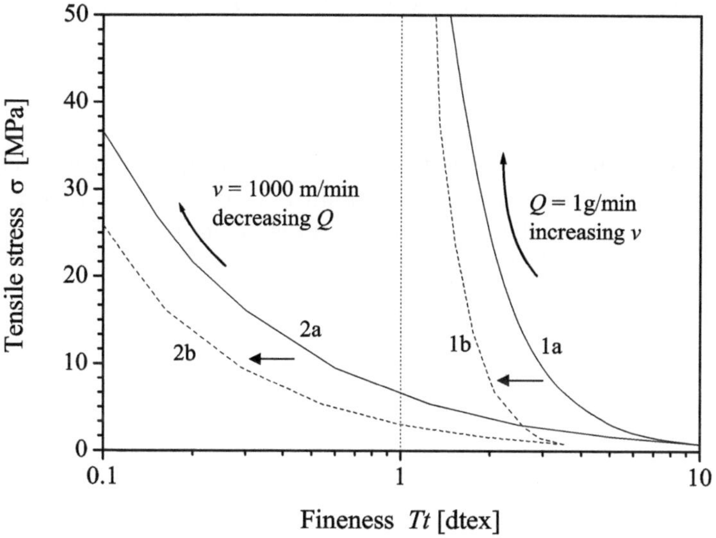

Fig. 3.68. (Calculated) filament tensile stress σ_L at distance $L = 1$ m vs. filament fineness Tt, (1) – constant throughput $Q = 1$ g/min, increasing take-up velocity, (2) – constant velocity $v = 1000$ m/min, decreasing throughput, (a) – fineness before drawing (as spun filaments, *solid lines*), (b) – fineness after drawing to 25% elongation (*dotted lines*)

Remark. Of course minimum air friction (and minimum stress level) results at solidification distance L_s. In practice, the take-up distance should be as long as necessary and as short as possible. To reduce air friction, it is effective to place the guide, where the single filaments are bundled together, as close to the spinneret as possible. Additionally, a further reduction of fibre stress seems possible via complete avoidance of air friction. The friction component disappears if air and fibres have an equal speed, realised by means of finely adjusted air stream *in the fibre direction* through using a special shaped (e. g. conical) spinning tube. In this case the tensile stress is mainly determined by inertia: $\sigma \approx \varrho \cdot v_L^2$ (see dotted line in Fig. 3.67) and nearly independent of throughput and the current take-up distance. For very low mass throughput the cooling distance becomes very low and also the resulting stress contribution from inertia. Now, the surface tension and the contribution from initial rheological force F_0 at the spinneret becomes more and more important. F_0 is determined from the mean fluidity of the material in the deformation region and can be influenced within certain limits by means of a post heating zone below the spinneret in order to control the cooling and also the deformation behaviour. Finally, the limit for possible minimum fineness of continously spun and drawn filaments could be $Tt_{\min} \approx 0.1$ dtex.

On the other hand, the unlimited reduction of throughput leads sooner or later to capillary break, the filaments divide into single drops. Under limitation of very thin filaments the surface tension is contributed substantially at the spinneret exit. The following consideration supports this assumption. Melt spinning of filaments is also a generation of filament surface: The surface generation rate $\dot{A}_F$ for a filament is given by

$$\dot{A}_F = \pi \cdot D_F \cdot v_F = 2\sqrt{\pi \cdot v_F \cdot \dot{Q}_V}\ , \tag{3.195}$$

where D_F is the final filament diameter, v_F is the final velocity, and $\dot{Q}_V$ is the volumetric rate of throughput. Assuming N single drops with radius R_d, the production rate of the drops surface $\dot{A}_d$ is given by

$$N \cdot \dot{A}_d = \frac{\dot{Q}_V \cdot 4\pi R_d^2}{4/3\,\pi R_d^3} = \frac{3}{R_d} \cdot \dot{Q}_V\ , \tag{3.196}$$

and the ratio of surface generation between drops and filaments is given by the relationship

$$\frac{N\dot{A}_d}{\dot{A}_F} = \frac{3}{2R_d}\sqrt{\frac{\dot{Q}_V}{\pi v_F}}\ . \tag{3.197}$$

At decreasing throughput (volume rate $\dot{Q}_V$) the relation (3.197) becomes at a certain level < 1 and the division into single drops is the energetically favorable case. Of course the relationships above only provide for a qualitative discussion and are not applicable to any quantitative calculation. For the more detailed analysis ZIABICKI showed that for capillary break the ratio between surface tension and viscosity of the melt plays an important role [274].

3.3.2 Visco-elastic Failure and Nonlinear Effects

Increasing spinning speed and/or decreasing throughput leads to filament breakage. Failure behaviour in spinning polymer melts includes:

- capillary break,
- cohesive, brittle fracture,
- ductile failure.

For low take-up velocities and low deformation rates the NEWTONian viscous constitutive equation may sufficiently describe the rheological flow and deformation behaviour of the molten polymer. But even at higher deformation rates the visco-elastic polymers show transient effects which characterise the time dependency between polymer relaxation and deformation, described by any relaxation time spectrum. To discuss the effect of visco-elastic failure the simple MAXWELL model (3.47) should be used. It is clear that the MAXWELL model, and only one single relaxation time λ, cannot completely characterise the rheological behaviour of the fiber forming polymers in melt spinning but the effect can be made quantitatively visible.

Equation (3.47) on page 63 can be written as

$$\eta \cdot \frac{\mathrm{d}v}{\mathrm{d}x} = \sigma + \lambda\, v \, \frac{\mathrm{d}\sigma}{\mathrm{d}x}, \qquad \lambda = \frac{\eta}{E}, \tag{3.198}$$

and together with $\sigma = F/A = \varrho_\mathrm{p}\, v\, F/Q$ then follows

$$\frac{\mathrm{d}v}{\mathrm{d}x} = \frac{1}{\eta} \cdot \frac{\sigma\left(1 + \lambda\, v \frac{1}{F}\frac{\mathrm{d}F}{\mathrm{d}x}\right)}{1 - \sigma/E}\,. \tag{3.199}$$

It can be seen from Eq. 3.199 that the deformation rate $\mathrm{d}v/\mathrm{d}x$ becomes infinitely large if the spinning stress σ or/and the compliance $1/E$ reach some critical value. Assuming constant tensile force $F(x) = F_\mathrm{c} = C \cdot Q/\varrho_\mathrm{p}$ as in the example on page 61 the deformation rate is simplified to

$$\frac{\mathrm{d}v}{\mathrm{d}x} = \frac{1}{\eta} \cdot \frac{\sigma}{1 - \sigma/E} = \frac{1}{\eta} \cdot \frac{C \cdot v}{1 - C \cdot v/E}\,. \tag{3.200}$$

The ratio between stress and deformation rate defines the *apparent elongational viscosity*

$$\eta_\mathrm{app} = \frac{\sigma}{\mathrm{d}v/\mathrm{d}x}\,, \tag{3.201}$$

which for the simplified case of constant force becomes

$$\eta_\mathrm{app} = \frac{\eta}{1 + \lambda \frac{\mathrm{d}v}{\mathrm{d}x}} = \frac{\eta}{1 + De}\,, \tag{3.202}$$

where De denotes the *Deborah*-number, which describes the ratio between relaxation time $\lambda = \eta/E$ and deformation time $(\mathrm{d}v/\mathrm{d}x)^{-1}$. Visco-elasticity appears as an artificial reduction of viscosity if the deformation is described in the manner above.

Ziabicki [184] discussed the solution of Eq. 3.200 for the special case of $E = \text{const.}$ and pointed out that after reaching the critical region some visco-elastic spinning instability occurs. Its intensity increases the more the ratio between the elastic and viscous properties of the material increases. Similar results were found by means of dynamic rheological measurements. For well spinnable polymers the ratio between storage modulus G' and loss modulus G'' must not exceed any critical value [269]. To illustrate the complexity of visco-elastic behaviour Fig. 3.69 (*left side*) shows the spinline velocity v vs. distance x from spinneret for three different ranges of elastic modulus: 1) the pure viscous case ($E \to \infty$), 2) the visco-elastic case describing the spinning behaviour of PA 6 more realistically by means of a deformation dependent modulus $E = E_0 f(\varepsilon_e)$ (Eq. 3.83), and 3) a visco-elastic case like 2) but with drastically reduced modulus (factor 0.01). The condition used for the calculation in all three cases was that both viscosity and modulus reach infinity at solidification distance and deformation finishes afterwards. The corresponding apparent viscosities Eq. 3.201 are shown at the right-hand side of Fig. 3.69. The DEBORAH number for the visco-elastic case 2)

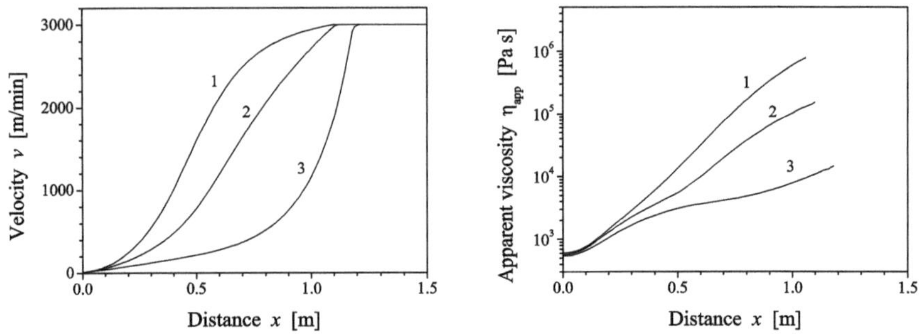

Fig. 3.69. The influence of visco-elasticity on spinning behaviour (calc.). *Left:* velocity v vs. distance x from spinneret exit, *right:* apparent elongational viscosity vs. distance, 1 – pure viscous behaviour ($E_1 \to \infty$), 2 – visco-elastic behaviour ($E_2 = E_0 f(\varepsilon_e)$) realistically describing melt spinning of PA 6, 3 – highly elastic behaviour ($E_3 = 0.01 E_2$); material PA 6, mass throughput $Q = 1$ g/min, take-up velocity $v_L = 3000$ m/min

in Fig. 3.69 varies from $De \approx 0.05$ near the spinneret to $De \approx 5$ near the solidification point. This means that the spinning behaviour changes from viscous at the beginning over visco-elastic to mostly elastic deformation at the end.

A high deformation rate may lead to another effect well known in rheology: the decrease of viscosity with increasing the deformation rate. This kind of nonlinear rheological behaviour is often described in shear melt rheology by a power-law equation or by the CROSS-CARREAU model which can be generalised for extensional deformation to any equation of the form

$$\eta(T, \dot{\varepsilon}) = \frac{\eta_0(T)}{1 + a\,\dot{\varepsilon}^{\,b}}\,, \tag{3.203}$$

or, respectively

$$\eta(T, \dot{\varepsilon}) = \frac{\eta_0(T)}{1 + a\left(\eta_0(T) \cdot \dot{\varepsilon}\right)^b}\,, \tag{3.204}$$

with parameters a, b.

The nonlinear effect of decreasing viscosity is demonstrated in Fig. 3.70 where the fibre formation was calculated again for the visco-elastic MAXWELL model (Eq. 3.47) but now with deformation rate depending upon viscosity Eq. 3.203. Because it is only a qualitative comparison the simplified relation with power exponent $b = 1$ was used. Other used parameters are the same as in the example for the purely viscous (case 1) and the visco-elastic (case 2) behaviours.

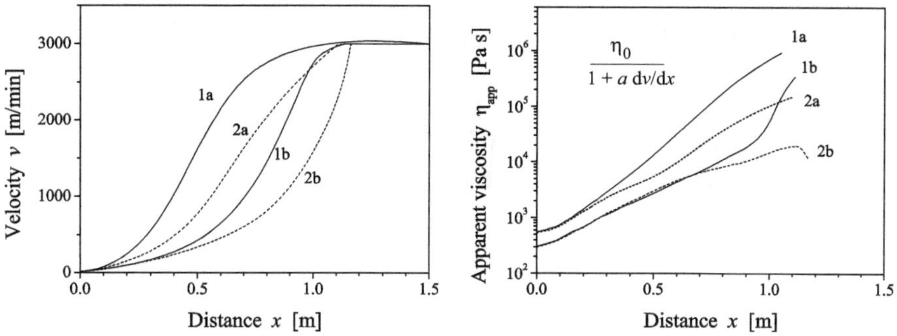

Fig. 3.70. The influence of nonlinear viscous and visco-elastic spinning behaviour (calc.), *left:* velocity v vs. distance x from spinneret exit, *right:* apparent elongational viscosity vs. distance, 1 – purely viscous behaviour ($E \to \infty$), 2 – visco-elastic behaviour ($E = E_0\, f(\varepsilon_e)$) realistically describing spinning of PA 6, (a) linear viscous behaviour, parameter $a = 0$, (b) nonlinear viscous behaviour, parameter $a = 0.1$; material PA 6, mass throughput $Q = 1$ g/min, take-up velocity $v_L = 3000$ m/min

Nonlinearity and visco-elasticity amplify each other because both show similar behaviour of decreasing the apparent viscosity while increasing the deformation rate. In Fig. 3.70 the beginning of failure at the end of graph (2b) can be seen, only prevented here by the abrupt end of the deformation because the viscosity and modulus are set to infinity at the solidification point. There is no solution for the calculation (within the frame of the used model equations) for any larger value of the nonlinearity parameter $a > 0.1$. The nonlinear and highly elastic behaviour can lead to fibre breakage if no rapid stabilisation effect follows. In melt spinning the stress induced crystallisation may overtake this role and acts as an additional hardener. Often the *neck*-like deformation observed is followed by stress induced crystallisation.

The different kinds of failure behaviour are shown in Fig. 3.71 (according to [270, 273]). Brittle fracture occurs if the stress exceeds the critical amount of the breaking stress. No special deformation appears at the point of rupture. Otherwise the ductile failure shows a typical thinning behaviour which may also lead to break, perhaps in a similar manner to the brittle fracture because the stress significantly increases. The thinning behaviour may be locally limited; this deformation is called *neck*-deformation. If the elongation or the elongation rate exceeds some critical value, the transition to 'catastrophic' failure may occur and can be explained with the special nonlinear visco-elastic rheological material behaviour. On the other hand, if there is any stabilizing effect like stress/elongation induced crystallisation the required breaking stress may rapidly increase and thus fibre breakage may be avoided.

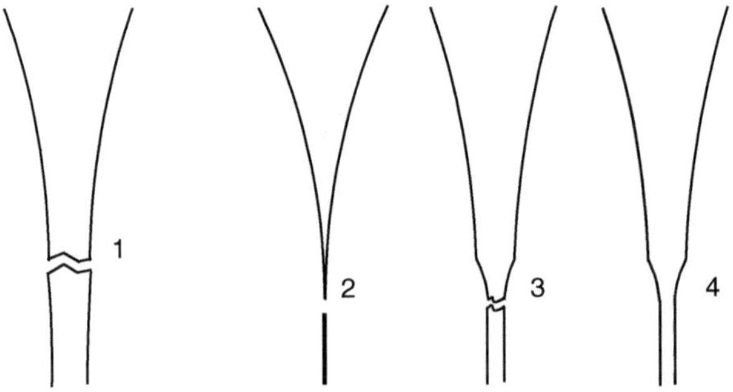

Fig. 3.71. Failure behaviour, 1 – brittle fracture, 2 – ductile failure, 3 – neck-like deformation, 4 – neck-like deformation followed by any stabilizing effect

Figure 3.72 shows some experimental results together with results from calculation, exemplary for high speed spinning of PA 6. The range from 3000 m/min to 7200 m/min and from 1.0 g/min to 3.0 g/min was investigated experimentally (the bold marked region in the diagram). Each point within the 'spinnability map' (according to [271–273]) represents a combination of take-up velocity v_L and draw down ratio $ddr = v_L/v_0 \propto v_L/Q$. Horizontal lines in this map represent constant draw down ratios and therefore constant finenesses. The straight lines passing the origin mark the states of constant mass throughput. For the experiments the take-up velocity was increased stepwise at constant mass throughput until fibre breakage occurred and no stable spinning process was further possible. The points (or more exactly: the small region) where the stable spinning behaviour turned into unstable behaviour mark the right border of the investigated region.

Additionally, the calculated lines of constant maximum deformation rate $\dot{\varepsilon} = \mathrm{d}v/\mathrm{d}x$ and the lines of constant stress σ_s at the solidification point are shown in the diagram.

The maximum possible draw down ratio decreases with increasing throughput and increasing take-up velocity. This means that in order to get fine filaments lower throughput and lower take-up speed is to be recommended with respect to spinning stability. It seems that for small throughputs the maximum deformation rate and the ductile failure behaviour are the limiting factors of spinnability. For sufficiently high mass throughputs and higher spinning speeds the melt spinning behaviour of PA 6 shows the typical neck-effect. Much higher deformation rates and also higher stresses are now possi-

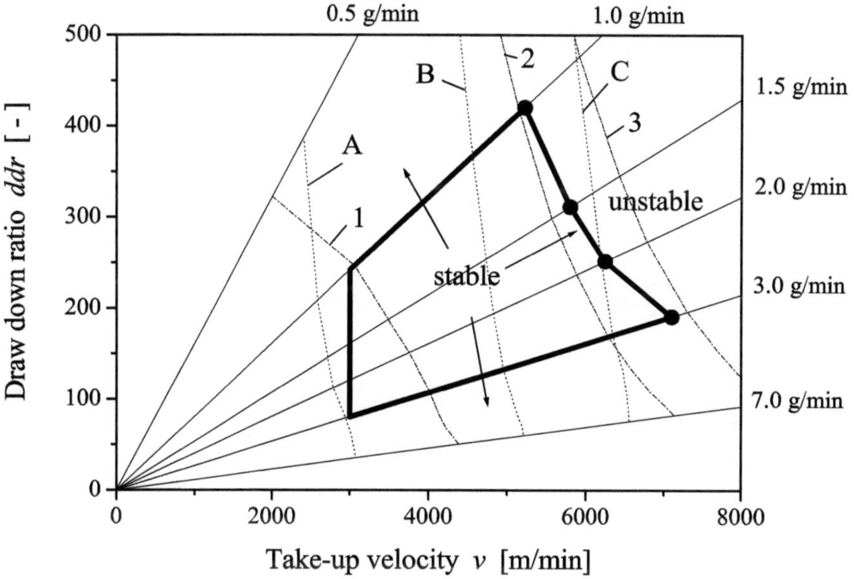

Fig. 3.72. Spinnability map for PA 6 melt spinning, lines of constant mass throughput (indicated); bold marked region: experimentally investigated, stable spinning conditions; points (•): limitation of spinnability, right from the border line no further spinning is possible; lines of constant maximum deformation rate (calculated): 1 – $\dot{\varepsilon} = 100\,s^{-1}$, 2 – $800\,s^{-1}$, 3 – $1200\,s^{-1}$, lines of constant stress at solidification point (calculated): A – $\sigma_s = 5\,\text{MPa}$, B – 10 MPa, C – 15 MPa, data source: Institute of Polymer Research Dresden

ble. It seems that the stress induced crystallisation that occurs after necking acts as a stabiliser and the failure then results from brittle fracture. Similar conclusions were drawn also by other authors [273]. The failure behaviour of melt spun polymers is an interesting field of fibre research with many unsolved problems. The question of spinnability includes problems of material behaviour (flow behaviour, rheology, crystallisation, phase transition, structure development, and so on) as well as measuring and analytical techniques and also engineering practice.

4. Dynamics of Fibre Formation Processes

4.1 Task

It is well-known that textile fibres are produced with essentially three basic technologies:

a) The separating, refining, strengthening and winding up of spinnable liquid mass streams. All organic and inorganic chemical fibres are to be subsumed to this group independently, if they are produced in a melt, dry or wet spinning process.
b) The separating, parallel join and twist of fibres. All fibre yarns are to be classified into this group independently, if the single fibres come from natural (animal, vegetable) or chemical sources. The latter case is mostly extending in front of a chemical fibre spinning process as seen in case a) with following cut process.
c) Cut of plain sheets of organic polymers into thin, tape like stripes (slit film yarn).

Each fibre formation process aims at the manufacturing of yarns with equal properties along to the yarn length axis. This means in conformity with the given definitions, that all product variables, which estimate the textile processing and wear properties of the yarn, should be as constant as possible. The case of effect yarn manufacture with consciously determined periodic or stochastic disturbed yarn structures along its length axis is an exception that should be mentioned. However, it will not be subject of the following considerations.

The processing of textile yarns and their wear behaviour is characterised by the product variable mass (fineness) along the yarn length axis or deformation resistance (elastic modulus) along the fibre length axis. These product variables oscillate around their averages caused by oscillations of raw materials and process variables. Therefore these product variables characterise the yarn unevenness, in which the fineness characterises the so-called outer yarn unevenness and the elastic modulus characterises the so-called inner yarn unevenness.

4.2 Melt Spinning of Polymers

4.2.1 Variable Fibre Fineness

The yarn finenesses and the yarn orientation are the most important among the different variables which describe the yarn quality. Therefore, the development of a mathematical model for these two yarn variables should be demonstrated here. At first, we start the investigations with the variable fibre fineness.

Cause-Effect-Scheme

The cause-effect relations of the process and the product variables for the target quantity yarn fineness should be demonstrated in the following. The recommended first step of the modelling process (registration and order of relevant process and product variables; see Sect. 2.5.1) can be carried out best through the elaboration of a cause-effect-scheme. The technological scheme of a melt spinning process is shown in Fig. 4.1.

It is to be remarked that this scheme is strongly simplified. It only contains the absolutely necessary tools and variables for our considerations. For instance thread guides, the oiling system and in some cases existing godets before the winder are not drawn. It is assumed that the heating system for the spinning die is an electrical resistance heating equipment. This is usual for laboratory equipment. Typical for the polymer melt spinning process is, that the thermoplastic melt (produced normally by means of an extruder) is fed to the single spinning positions along a melt distribution system by means of an exactly feeding volume conveyor tool for each (gear pump, spinning pump). After passing the spinning die (the tool, which distributes the melt stream into the number of filaments in the yarn) the single thin melt filaments are rapidly deformed, cooled and strengthened. At this complicated rheological and structural formation, shift processes take place in the filaments, which are caused directly or indirectly by the take-up velocity, created by the winder. For the target quantity or effect variable "fineness of the spun yarn" the process and product cause variables that are probably interesting at such a spinning position are shown in Fig. 4.1 as well.
Figure 4.2 shows the cause-effect-scheme for the target quantity fineness Tt_s (designed on this basis).

The cause-effect-arrows go from the cause to the effect. The box of the target quantity fineness is thickly framed, boxes of quantities at the process periphery are shaded (from these arrows only lead off). The fineness Tt_s is only caused under static conditions from the take-down (spinning) velocity at the output of the fibre formation distance v_s and from the throughput through the spinneret Q_s, which feeds the fibre formation distance at its input. The basic equation uses the suitable dimensions

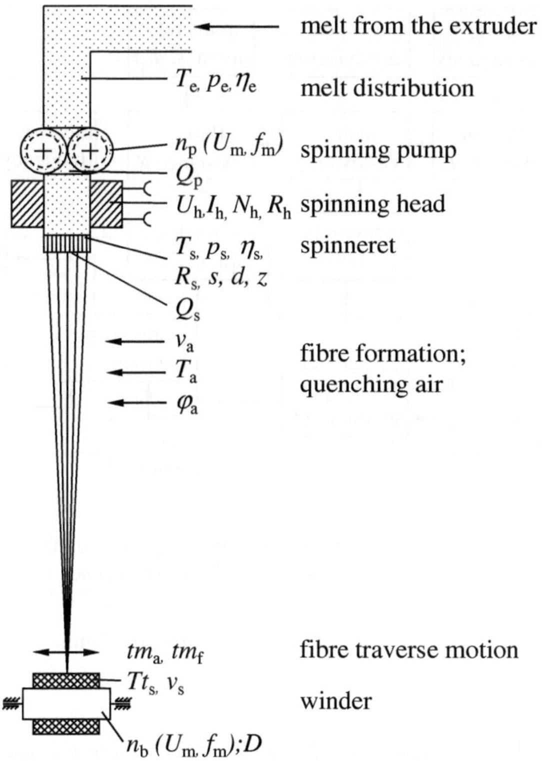

Fig. 4.1. Technological scheme (simplified) of a polymer melt spinning process

$$Tt_s[\text{tex}] = \frac{Q_s[\text{g/min}]}{v_s[\text{km/min}]} \tag{4.1}$$

It should be added, that changes of the quenching air velocity Δv_a below the die, and the necessary traverse motion at the bobbin of the winder, which is effected by the amplitude tm_a and the frequency tm_f of the thread guide for the traverse motion, also effect changes of the yarn fineness. However, they do not effect changes to the mean fineness. They only change the value of the fineness differentially. These 3 boxes are dotted frames. More detailed explanations to the latter are in Sects. 4.1.1.3 and 5.1.5.4.

The main cause variables for the fineness Tt_s, namely the throughput Q_s and the take-down (spinning) velocity v_s, can be traced back now regarding their cause process and product variables. The take-down velocity v_s is caused by the speed of the bobbin n_b, which is caused by the mains supply voltage U_m and the mains supply frequency f_m if an asynchronous drive motor is

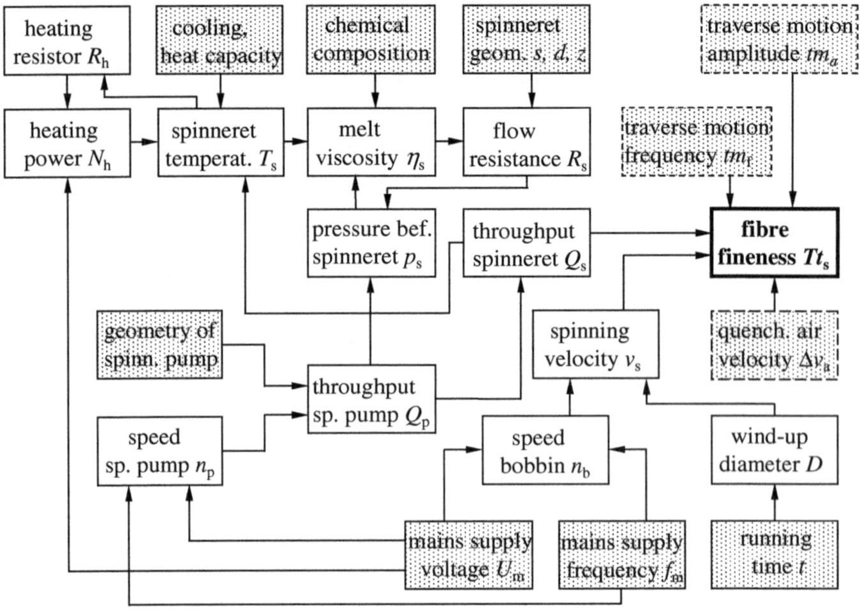

Fig. 4.2. Cause-effect-scheme for the target quantity fineness Tt_s of a polymer melt spinning process

used, and the wind-up diameter D. The latter of course increases with the running time t.

The throughput Q_s is only caused by the throughput of the spinning pump Q_p, which depends upon itself because of its geometrical design and its speed n_p. Back stream leakages of the spinning pump, which would mean $Q_s < Q_p$, are not regarded here. If the spinning pump would be driven by an asynchronous motor from the same mains supply U_m and f_m to Tt_s would be effectively doubled along the cause-effect-chains:

$$U_m, f_m \rightarrow n_b \rightarrow v_s \rightarrow Tt_s \text{ and}$$
$$U_m, f_m \rightarrow n_p \rightarrow Q_p \rightarrow Q_s \rightarrow Tt_s$$

The pressure before the spinning die p_s does not appear as a cause variable in regard to Q_s. It only depends, corresponding to the HAGEN-POISEUILLE-law for laminar flows in the tube, see [279], on the flow resistance inside the capillary holes of the die R_s and from Q_p. The flow resistance R_s depends on its part from the geometry of the capillary holes (length s, diameter d, number z) and from the melt viscosity η_s. The dependence of the melt viscosity η_s on the spinneret temperature T_s and on the chemical polymer composition is comprehensible by reason of simple basic physical laws. The dependence of T_s on the cooling conditions at the spinning die and on the heating power N_h

(itself depending on the mains supply voltage U_m and on the OHM'S heating resistance R_h) can be concluded with the same reasons.
In the next step it is necessary to set up the DEq. for each cause-effect relation or, if impossible, to investigate the dynamic signal transfer properties of the partial transfer systems. The signal transfer and signal interlacing character can be better represented by means of the so-called *functional block diagram*. In automatic control this is an often used scheme, which develops formally from the cause-effect-scheme by means of technological and *prior* physical knowledge. This only contains the change or oscillating parts of the process and product variables as their signals are connected together by cause-effect relations. The dynamic transfer properties are represented by the blocks, *"black boxes"*, which are unknown at the beginning of the analysis. This procedure will be demonstrated more fully in Sect. 4.3 with the example of "glass fibre spinning". In the following, the set up of the dynamic model will be demonstrated, which on the one hand describes the cause-effect relations between the cause variables throughput spinneret Q_s and the spinning velocity v_s and the fibre fineness Tt_s which is effected by these variables, on the other. The final goal of this procedure is to prepare technological statements about the disturbance transfer properties of the fibre formation distance.

Specified Differential Equation of a Fibre Formation Distance (simplified)

Figure 4.3 shows the fibre formation distance of the melt spinning process, only one monofilament fibre, which is reduced to the most necessary of elements and variables.

A melt stream is pressed through the capillary hole, cross section q_i, of the spinneret with the velocity v_i (input velocity or injection velocity into the fibre formation distance). At the length l between the spinneret and the take-up rolls with the output velocity v_s it is drawn, solidified and transported. The ready fibre with the cross section q_s esp. the fineness Tt_s appears at the take-up rolls. The relationship between the fibre fineness and the cross section is given by the density ϱ in the following manner:

$$Tt_s[\text{tex}] = \varrho[\text{g/cm}^3] \cdot q_s[\mu\text{m}^2] \cdot 10^{-3} \qquad (4.2)$$

The following relationship exists between the mass discharge per time (or throughput) Q_s and the input variables of the fibre formation distance:

$$Q_s = \varrho \cdot q_i \cdot v_i \qquad (4.3)$$

In Eq. 4.3 it is not distinguished between ϱ_{fibre} and ϱ_{melt}, because a constant factor exists between these two densities and an influence on the time and frequency oscillation relations will not be given. We split according to the

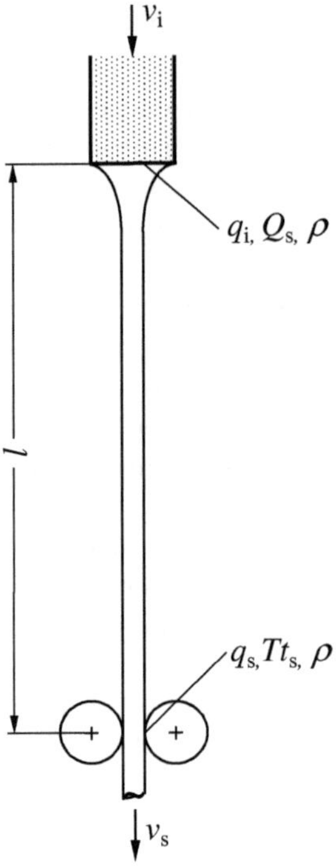

Fig. 4.3. Fibre formation distance (simplified) of the polymer melt spinning

agreement the variables, if necessary, into the mean value and the fluctuating part, consequently for instance:

$$v_i = v_{im} \pm \Delta v_i,$$
$$q_s = q_{sm} \pm \Delta q_s \text{ and so on}$$

The question is: Which Tt_s-fluctuations appear and if fluctuations of the variables v_i, q_i, ϱ, v_s or l appear? The question can only be answered on the basis of a dynamic fibre formation model. Ingeniously the starting point is the dynamic continuum equation (2.18).
Applied to the present fibre formation distance are:

$$\text{mass inflow/time} = q_i \cdot v_i \cdot \varrho$$
$$\text{mass discharge/time} = q_s \cdot v_s \cdot \varrho = Tt_s \cdot v_s$$
$$\text{change of stored mass} = \dot{q}_s \cdot l \cdot \varrho = \dot{T}t_s \cdot l$$

The assumed simplification at the formulation of the stored mass is, that the deformation range of the melt stream until it reaches its solidification point is not considered. At this point the diameter or the fineness of the spun fibre is reached. However, this range of the whole fibre formation distance is relatively small (~ 0.5...0.8 m), and the related mistakes do not prevent qualitatively correct results. An exact and quantitatively correct consideration of this range makes a correct mathematical solution impossible. In Sect. 3. the very complex processes which take place in the fibre formation distance are explained in more detail.

Using the expressions above the complete DEq. of the fibre formation distance can now be written as:

$$q_\mathrm{i} \cdot v_\mathrm{i} \cdot \varrho - Tt_\mathrm{s} \cdot v_\mathrm{s} - \dot{T}t_\mathrm{s} \cdot l = 0 \tag{4.4}$$

By marking the whole left side of the DEq. 4.4 with the letter $\varPhi$ and introducing the LAPLACE-operator $p = \frac{\mathrm{d}}{\mathrm{d}t}$, the DEq. is converted into the transformed quantic:

$$\varPhi = q_\mathrm{i} \cdot v_\mathrm{i} \cdot \varrho - Tt_\mathrm{s} \cdot v_\mathrm{s} - p \cdot Tt_\mathrm{s} \cdot l = 0 \tag{4.5}$$

Equation 4.5 represents a nonlinear DEq. first order, because all variables in the single terms, which can fluctuate, are multiplicatively connected together. Equation 4.5 can be linearised by means of the partial differentiation as follows:

$$\frac{\partial \varPhi}{\partial q_\mathrm{s}} \cdot \Delta q_\mathrm{s} + \frac{\partial \varPhi}{\partial v_\mathrm{s}} \cdot \Delta v_\mathrm{s} + \frac{\partial \varPhi}{\partial q_\mathrm{i}} \cdot \Delta q_\mathrm{i} + \frac{\partial \varPhi}{\partial v_\mathrm{i}} \cdot \Delta v_\mathrm{i} + \frac{\partial \varPhi}{\partial l} \cdot \Delta l + \frac{\partial \varPhi}{\partial \varrho} \cdot \Delta \varrho = 0 \tag{4.6}$$

The instruction of Eq. 4.6 means, that the whole DEq. 4.5 is to be derived partially with respect to each single variable of change. The mean value is to be set by the single derivation step for these variables which are not to be derived. The following linearised complete DEq. is achieved as the mathematical dynamic model of the fibre formation distance after the partial derivation and order of the single terms:

$$\begin{aligned}(v_\mathrm{sm} + p \cdot l_\mathrm{m}) \cdot \Delta Tt_\mathrm{s} + Tt_\mathrm{sm} \cdot \Delta v_\mathrm{s} - v_\mathrm{im} \cdot \varrho_\mathrm{m} \cdot \Delta q_\mathrm{s} \\ - q_\mathrm{im} \cdot \varrho_\mathrm{m} \cdot \Delta v_\mathrm{i} + p \cdot Tt_\mathrm{sm} \Delta l - q_\mathrm{im} \cdot v_\mathrm{im} \cdot \Delta \varrho = 0\end{aligned} \tag{4.7}$$

Equation 4.7 is a multilateral applicable dynamic model equation for fibre formation and fibre transport processes (see also Sect. 5.1). The performed linearisation (each term of the Eq. 4.7 contains only variables of change in time) is connected to the following consequences regarding the analysis: Equation 4.5 represents primarily a nonlinear relationship. This is imaginable as a spatial multidimensional *curved* sheet and it is approached by a *plane* tangential sheet in the technological operating point. It is determined by the mean values of the single variables. The linearised relation is better validated the nearer the analytical investigation remains at this technological operation point. That means that the linearised Eq. 4.7 is valid more exact the smaller the investigated change quantities are in relation to their mean values. In practice it should be kept for any change variable x:

$$\Delta x \leqq 0.1 \cdot x_\mathrm{m} \tag{4.8}$$

Solutions of the Differential Equation of the Fibre Formation Process

DEq. 4.7 is the calculation basis for the characterisation of the dynamic behaviour of the fibre formation distance. It is now possible to calculate the effects of Δv_{s}-, Δv_{i}-, Δq_{i}-, $\Delta \varrho$- and Δl- disturbances on the fineness changes ΔTt_{s}. Each change variable is ingeniously to be regarded separately.

Disturbance Δv_{s} (changes of the take-down velocity). The boundary condition for this case is:

$$\Delta v_{\mathrm{i}} = \Delta q_{\mathrm{i}} = \Delta \varrho = \Delta l = 0$$

Introducing this into the DEq. 4.7 and applicatiing the calculation and conversion algorithms (which were explained in detail in Sect. 2.5.2) results in:

dynamic transfer function:

$$G(p) = \frac{\Delta Tt_{\mathrm{s}}}{\Delta v_{\mathrm{s}}} = -\frac{Tt_{\mathrm{sm}}}{v_{\mathrm{sm}}} \cdot \frac{1}{1 + p \cdot \dfrac{l_{\mathrm{m}}}{v_{\mathrm{sm}}}} \tag{4.9}$$

The *complex frequency response* follows:

$$G(\mathrm{j}\omega) = \frac{\widetilde{\Delta Tt_{\mathrm{s}}}}{\widetilde{\Delta v_{\mathrm{s}}}} \cdot \mathrm{e}^{\mathrm{j}\varphi} = -\frac{Tt_{\mathrm{sm}}}{v_{\mathrm{sm}}} \cdot \frac{1}{1 + \mathrm{j}\omega \cdot \dfrac{l_{\mathrm{m}}}{v_{\mathrm{sm}}}} \tag{4.10}$$

Equation 4.10 can be split into the *amplitude frequency response* (simply also *amplitude response*)

$$|G(\mathrm{j}\omega)| = \left| \frac{\widetilde{\Delta Tt_{\mathrm{s}}}}{\widetilde{\Delta v_{\mathrm{s}}}} \right| = (-)\frac{Tt_{\mathrm{sm}}}{v_{\mathrm{sm}}} \left[1 + \left(\omega \cdot \frac{l_{\mathrm{m}}}{v_{\mathrm{sm}}} \right)^2 \right]^{-1/2} \tag{4.11}$$

and the *phase frequency response* (simply *phase response*)

$$\varphi(\omega) = \arctan \left[-\frac{\omega \cdot l_{\mathrm{m}}}{v_{\mathrm{sm}}} \right] - \pi \tag{4.12}$$

By means of Eqs. 4.9 and 2.41 the *time transient function* can be calculated as: (also *step response*)

$$\underline{\Delta Tt_{\mathrm{s}}}\overline{|\Delta v_{\mathrm{s}}} = -\Delta v_{\mathrm{s}} \cdot \frac{Tt_{\mathrm{sm}}}{v_{\mathrm{sm}}} \left[1 - \exp \left(-\frac{v_{\mathrm{sm}}}{l_{\mathrm{m}}} \cdot t \right) \right] \tag{4.13}$$

A collected qualitative and quantitative evaluation of the results of Eqs. 4.9 to 4.13 is given in the next Sect. *"Summarised Evaluation and Conclusions to the Solutions of the Differential Equation"*. However, already here it should be referred to the minus sign in the Eqs. 4.9, 4.10, 4.11, 4.13, which hints at the physical right dependence: A positive Δv_{s}-change effects a negative ΔTt_{s}-change. That means an *increase* of the take-down velocity effects, under constant other conditions, a *decrease* of the spun fibre fineness.

Disturbance Δv_{i} (changes of the input velocity). Boundary condition:

$$\Delta v_{\mathrm{s}} = \Delta q_{\mathrm{i}} = \Delta \varrho = \Delta l = 0$$

It is to be written in the same manner as described before:

dynamic transfer function:

$$G(p) = \frac{\Delta Tt_{\mathrm{s}}}{\Delta v_{\mathrm{i}}} = \frac{q_{\mathrm{i}}m \cdot \varrho_{\mathrm{m}}}{v_{\mathrm{sm}}} \cdot \frac{1}{1 + p \cdot \dfrac{l_{\mathrm{m}}}{v_{\mathrm{sm}}}} \tag{4.14}$$

or, because $q_{\mathrm{im}} \cdot \varrho_{\mathrm{m}} = Tt_{\mathrm{sm}} \cdot \dfrac{v_{\mathrm{sm}}}{v_{\mathrm{im}}}$

$$G(p) = \frac{\Delta Tt_{\mathrm{s}}}{\Delta v_{\mathrm{i}}} = \frac{Tt_{\mathrm{sm}}}{v_{\mathrm{im}}} \cdot \frac{1}{1 + p \cdot \dfrac{l_{\mathrm{m}}}{v_{\mathrm{sm}}}} \tag{4.15}$$

complex frequency response:

$$G(\mathrm{j}\omega) = \frac{\widetilde{\Delta Tt_{\mathrm{s}}}}{\widetilde{\Delta v_{\mathrm{i}}}} \cdot \mathrm{e}^{\mathrm{j}\varphi} = \frac{Tt_{\mathrm{sm}}}{v_{\mathrm{im}}} \cdot \frac{1}{1 + \mathrm{j}\omega \cdot \dfrac{l_{\mathrm{m}}}{v_{\mathrm{sm}}}} \tag{4.16}$$

amplitude frequency response (simply also *amplitude response*)

$$|G(\mathrm{j}\omega)| = \left| \frac{\widetilde{\Delta Tt_{\mathrm{s}}}}{\widetilde{\Delta v_{\mathrm{i}}}} \right| = \frac{Tt_{\mathrm{sm}}}{v_{\mathrm{im}}} \left[1 + \left(\omega \cdot \frac{l_{\mathrm{m}}}{v_{\mathrm{sm}}} \right)^2 \right]^{-1/2} \tag{4.17}$$

phase frequency response (simply *phase response*)

$$\varphi(\omega) = \operatorname{arc\,tan} \left[-\frac{\omega \cdot l_{\mathrm{m}}}{v_{\mathrm{sm}}} \right] \tag{4.18}$$

time transient function (or *step response*)

$$\underline{\Delta Tt_{\mathrm{s}}} | \overline{\Delta v_{\mathrm{i}}} = \Delta v_{\mathrm{i}} \cdot \frac{Tt_{\mathrm{sm}}}{v_{\mathrm{im}}} \left[1 - \exp\left(-\frac{v_{\mathrm{sm}}}{l_{\mathrm{m}}} \cdot t \right) \right] \tag{4.19}$$

A collected qualitative and quantitative evaluation of the result Eqs. 4.15 to 4.19 is given in the next Sect. *"Summarised Evaluation and Conclusions to the Solutions of the Differential Equation"*. The effects of Δv_{i}-disturbances are co-directional to the ΔTt_{s}-changes opposite to the described Δv_{s}-disturbances. That means an *increase* of the input velocity effects under constant other conditions, an *increase* of the spun fibre fineness as well. Besides it should also be hinted here, that the result equations of Sect. 2.5.2 (example drawing process at input velocity disturbances) are equivalent to the presented result equations of the fibre formation distance for Δv_{i}-disturbances. That means it corresponds in each case Eqs. 4.15 to 2.40, 4.16 to 2.36, 4.17 to 2.38, 4.18 to 2.39 and 4.19 to 2.46.

Disturbance Δq_{i} (changes of the input cross sectional area). Boundary condition:

$$\Delta v_{\mathrm{s}} = \Delta v_{\mathrm{i}} = \Delta \varrho = \Delta l = 0$$

Such a disturbance is actually unlikely for a melt spinning process (it would be more plausible to assume step-like or oscillating changes of the capillary hole diameter in the spinneret). Nevertheless, dynamic solution equations will be explained in the following for this disturbance model as well. In a row of other fibre formation and fibre processing processes Δq_{i}-disturbances are identical namely with changes of the fineness at the process input. A modified application of the here given result equations is easily possible. Examples will follow in later sections.

dynamic transfer function:

$$G(p) = \frac{\Delta Tt_{\mathrm{s}}}{\Delta q_{\mathrm{i}}} = \varrho_{\mathrm{m}} \cdot \frac{v_{\mathrm{im}}}{v_{\mathrm{sm}}} \cdot \frac{1}{1 + p \cdot \dfrac{l_{\mathrm{m}}}{v_{\mathrm{sm}}}} \tag{4.20}$$

or, because $v_{\mathrm{im}} \cdot \varrho_{\mathrm{m}} = \dfrac{Tt_{\mathrm{sm}} \cdot v_{\mathrm{sm}}}{q_{\mathrm{im}}}$

$$G(p) = \frac{\Delta Tt_{\mathrm{s}}}{\Delta q_{\mathrm{i}}} = \frac{Tt_{\mathrm{sm}}}{q_{\mathrm{im}}} \cdot \frac{1}{1 + p \cdot \dfrac{l_{\mathrm{m}}}{v_{\mathrm{sm}}}} \tag{4.21}$$

complex frequency response:

$$G(\mathrm{j}\omega) = \frac{\widetilde{\Delta Tt_{\mathrm{s}}}}{\widetilde{\Delta q_{\mathrm{i}}}} \cdot \mathrm{e}^{\mathrm{j}\varphi} = \frac{Tt_{\mathrm{sm}}}{q_{\mathrm{im}}} \cdot \frac{1}{1 + \mathrm{j}\omega \cdot \dfrac{l_{\mathrm{m}}}{v_{\mathrm{sm}}}} \tag{4.22}$$

amplitude frequency response (simply also *amplitude response*)

$$|G(\mathrm{j}\omega)| = \left| \frac{\widetilde{\Delta Tt_{\mathrm{s}}}}{\widetilde{\Delta q_{\mathrm{i}}}} \right| = \frac{Tt_{\mathrm{sm}}}{q_{\mathrm{im}}} \left[1 + \left(\omega \cdot \frac{l_{\mathrm{m}}}{v_{\mathrm{sm}}} \right)^2 \right]^{-1/2} \tag{4.23}$$

phase frequency response (simply *phase response*)

$$\varphi(\omega) = \text{arc tan}\left[-\frac{\omega \cdot l_\mathrm{m}}{v_\mathrm{sm}}\right] \tag{4.24}$$

time transient function (or *step response*)

$$\underline{\Delta Tt_\mathrm{s}}|\overline{\Delta q_\mathrm{i}} = \Delta q_\mathrm{i} \cdot \frac{Tt_\mathrm{sm}}{q_\mathrm{im}}\left[1 - \exp\left(-\frac{v_\mathrm{sm}}{l_\mathrm{m}} \cdot t\right)\right] \tag{4.25}$$

A collected qualitative and quantitative evaluation of the result of Eqs. 4.21 to 4.25 is given in the next Sect. *"Summarised Evaluation and Conclusions to the Solutions of the Differential Equation"*. But, it should already be said here, that the effected ΔTt_s-disturbances, caused by Δq_i-disturbances, are co-directional.

Disturbance $\Delta\varrho$ (changes of the density of fibre material). Boundary condition:

$$\Delta v_\mathrm{s} = \Delta v_\mathrm{i} = \Delta q_\mathrm{i} = \Delta l = 0$$

It can be concluded in the same manner as described before:

dynamic transfer function:

$$G(p) = \frac{\Delta Tt_\mathrm{s}}{\Delta\varrho} = \frac{q_\mathrm{im} \cdot v_\mathrm{im}}{v_\mathrm{sm}} \cdot \frac{1}{1 + p \cdot \dfrac{l_\mathrm{m}}{v_\mathrm{sm}}} \tag{4.26}$$

or, because $q_\mathrm{im} \cdot v_\mathrm{im} = \dfrac{Tt_\mathrm{sm} \cdot v_\mathrm{sm}}{\varrho_\mathrm{m}}$

$$G(p) = \frac{\Delta Tt_\mathrm{s}}{\Delta\varrho} = \frac{Tt_\mathrm{sm}}{\varrho_\mathrm{m}} \cdot \frac{1}{1 + p \cdot \dfrac{l_\mathrm{m}}{v_\mathrm{sm}}} \tag{4.27}$$

complex frequency response:

$$G(\mathrm{j}\omega) = \frac{\widetilde{\Delta Tt_\mathrm{s}}}{\widetilde{\Delta\varrho}} \cdot \mathrm{e}^{\mathrm{j}\varphi} = \frac{Tt_\mathrm{sm}}{\varrho_\mathrm{m}} \cdot \frac{1}{1 + \mathrm{j}\omega \cdot \dfrac{l_\mathrm{m}}{v_\mathrm{sm}}} \tag{4.28}$$

amplitude frequency response (simply also *amplitude response*)

$$|G(\mathrm{j}\omega)| = \left|\frac{\widetilde{\Delta Tt_\mathrm{s}}}{\widetilde{\Delta\varrho}}\right| = \frac{Tt_\mathrm{sm}}{\varrho}\left[1 + \left(\omega \cdot \frac{l_\mathrm{m}}{v_\mathrm{sm}}\right)^2\right]^{-1/2} \tag{4.29}$$

phase frequency response (simply *phase response*)

$$\varphi(\omega) = \text{arc tan}\left[-\frac{\omega \cdot l_{\text{m}}}{v_{\text{sm}}}\right] \tag{4.30}$$

time transient function (or *step response*)

$$\underline{\Delta T t_{\text{s}}|\overline{\Delta \varrho}} = \Delta\varrho \cdot \Delta\varrho \cdot \frac{T t_{\text{sm}}}{\varrho_{\text{m}}}\left[1 - \exp\left(-\frac{v_{\text{sm}}}{l_{\text{m}}} \cdot t\right)\right] \tag{4.31}$$

A co-directional dependence of $\Delta\varrho$-changes on effected $\Delta T t_{\text{s}}$-changes is also to be seen here. A collected qualitative and quantitative evaluation of the result of Eqs. 4.27 to 4.31 is given in the next Sect. *"Summarised Evaluation and Conclusions to the Solutions of the Differential Equation"*.

Disturbance Δl (changes of the length of the fibre formation distance). Boundary condition:

$$\Delta v_{\text{s}} = \Delta v_{\text{i}} = \Delta q_{\text{i}} = \Delta\varrho = 0$$

At first glance, this disturbance does not seem to be of any practical interest. But it is to be hinted, that the fibre influence by means of the dynamics of thread traverse motion at winders (especially at the godetless high speed melt spinning) is exactly equivalent to the change model *"length of the fibre formation distance"*. More details can be found in Sect. 5.1.5.

The following result equations are to be concludet:

dynamic transfer function:

$$G(p) = \frac{\Delta T t_{\text{s}}}{\Delta l} = -\frac{T t_{\text{sm}}}{l_{\text{m}}} \cdot \frac{p}{p + \dfrac{v_{\text{sm}}}{l_{\text{m}}}} \tag{4.32}$$

complex frequency response:

$$G(\text{j}\omega) = \frac{\widetilde{\Delta T t_{\text{s}}}}{\widetilde{\Delta l}} \cdot e^{\text{j}\varphi} = -\frac{T t_{\text{sm}}}{l_{\text{m}}} \cdot \frac{\text{j}\omega}{\text{j}\omega + \dfrac{v_{\text{sm}}}{l_{\text{m}}}} \tag{4.33}$$

amplitude frequency response (simply also *amplitude response*)

$$|G(\text{j}\omega)| = \left|\frac{\widetilde{\Delta T t_{\text{s}}}}{\widetilde{\Delta l}}\right| = (-)\frac{T t_{\text{sm}}}{l_{\text{m}}} \cdot \omega \cdot \left[\omega^2 + \left(\frac{v_{\text{sm}}}{l_{\text{m}}}\right)^2\right]^{-1/2} \tag{4.34}$$

phase frequency response (simply *phase response*)

$$\varphi(\omega) = \text{arc tan}\left[\frac{v_{\text{sm}}}{\omega \cdot l_{\text{m}}}\right] - \pi \tag{4.35}$$

time transient function (or *step response*)

$$\underline{\Delta Tt_{\text{s}}|\overline{\Delta l}} = -\Delta l \cdot \frac{Tt_{\text{sm}}}{l_{\text{m}}} \cdot \exp\left(-\frac{v_{\text{sm}}}{l_{\text{m}}} \cdot t\right) \tag{4.36}$$

Also these result Eqs. 4.32 to 4.36 will be collected and discussed in the following Sect. *"Summarised Evaluation and Conclusions to the Solutions of the Differential Equation"*

Summarised Evaluation and Conclusions to the Solutions of the Differential Equation

Three summarised unifying statements can be made by the comparing all result equations regarding the cause-effect relations between the different process and product variables Δv_{s}, Δv_{i}, Δq_{i}, $\Delta \varrho$ and Δl on the one hand and the product variable ΔTt_{s} on the other:

1. Statement

Four of the five dynamic transfer functions are equally constructed. (cp. 4.9, 4.15, 4.21 and 4.27). This fact leads of course to equally constructed, from it derived functions, as complex frequency response, amplitude frequency response, phase frequency response and step response. This applies to the disturbances Δv_{s}, Δv_{i}, Δq_{i} and $\Delta \varrho$. This dynamic behaviour is characterised by it in the transfer function, that the LAPLACE-operator p exists only once linear in one of two terms of the denominator. Such a behaviour is called proportional action with delay of first order. It is typical for such a system behaviour, that residual changes (esp. step-like) of the cause variables (if they continue long enough) effect residual changes of the effect variables (ΔTt_{s}), in which the quantities of cause and effect changes are proportional to each other. Delay of first order means, that only *one* time constant and only *one* exponential function (coming from only *one* differential quotient in the DEq.) determine the dynamic transient process.

2. Statement

The dynamic transfer function for the disturbance Δl (Eq. 4.32) is deflectively constructed. It actually possesses the same denominator compared with the named transfer functions in the first statement, but the LAPLACE-operator p appears once more linear in the numerator. This behaviour is called differential action with delay of first order. It is typical for such a system behaviour, that changes of the effect variables (ΔTt_s) reduce to zero again, if the changes(esp. step-like) of the cause variables only continue long enough. Delay of first order means in this case, that likewise only *one* time constant and only *one* exponential function determine the subsidence of the effect to zero (a transient process as well).

3. Statement

The dynamic behaviour of a fibre formation distance is dominated by the quantity l_m/v_{sm}. This was also the case in Sect. 2.5.2 (there for the drawing process) for the system time constant T_c (see Eq. 2.47). The time constant T_c is readable from the exponent of the e-functions in the time transient functions of Eqs. 4.13, 4.19, 4.25, 4.31 and 4.36.
The critical frequency f_c can be determined likewise by l_m and v_{sm} corresponding to Eq. 2.49 (v_{sm} corresponds to v_{om}).

Considering the first statement it is possible to get common solutions for the four disturbance causes Δv_s, Δv_i, Δq_i and $\Delta \varrho$ regarding their effects to the fineness ΔTt_s. Uniform, normalised solutions can be determined for the complex frequency response, the amplitude frequency response, the phase frequency response and the time transient function (step response), which are valid for all of the four disturbance causes in similar manners.

Considering Eqs. 2.48 and 2.49 one gets from the Eqs. 4.10, 4.16, 4.22 and 4.28 the *common normalised complex frequency response*, in which ω is substituted by f:

$$\begin{aligned} G[j(f/f_c)] &= -\frac{\widetilde{\Delta Tt_s/Tt_{sm}}}{\widetilde{\Delta v_s}/v_{sm}} \cdot e^{j\varphi} = \frac{\widetilde{\Delta Tt_s/Tt_{sm}}}{\widetilde{\Delta v_i}/v_{im}} \cdot e^{j\varphi} \\ &= \frac{\widetilde{\Delta Tt_s/Tt_{sm}}}{\widetilde{\Delta q_i}/q_{im}} \cdot e^{j\varphi} = \frac{\widetilde{\Delta Tt_s/Tt_{sm}}}{\widetilde{\Delta \varrho}/\varrho_m} \cdot e^{j\varphi} \\ &= \frac{1}{1 + j(f/f_c)} \end{aligned} \tag{4.37}$$

Considering Eqs. 2.48 and 2.49 one gets from the Eqs. 4.11, 4.17, 4.23 and 4.29 in the same manner the *common normalised amplitude frequency response*:

$$\begin{aligned} |G[j(f/f_c)]| &= \left|\frac{\widetilde{\Delta Tt_s/Tt_sm}}{\widetilde{\Delta v_s}/v_{sm}}\right| = \left|\frac{\widetilde{\Delta Tt_s/Tt_sm}}{\widetilde{\Delta v_i}/v_{im}}\right| \\ = \left|\frac{\widetilde{\Delta Tt_s/Tt_sm}}{\widetilde{\Delta q_i}/q_{im}}\right| &= \left|\frac{\widetilde{\Delta Tt_s/Tt_sm}}{\widetilde{\Delta \varrho}/\varrho_m}\right| = [1 + (f/f_c)^2]^{-1/2} \end{aligned} \tag{4.38}$$

The *transfer locus* of this normalised complex frequency response is shown in Fig. 4.4.

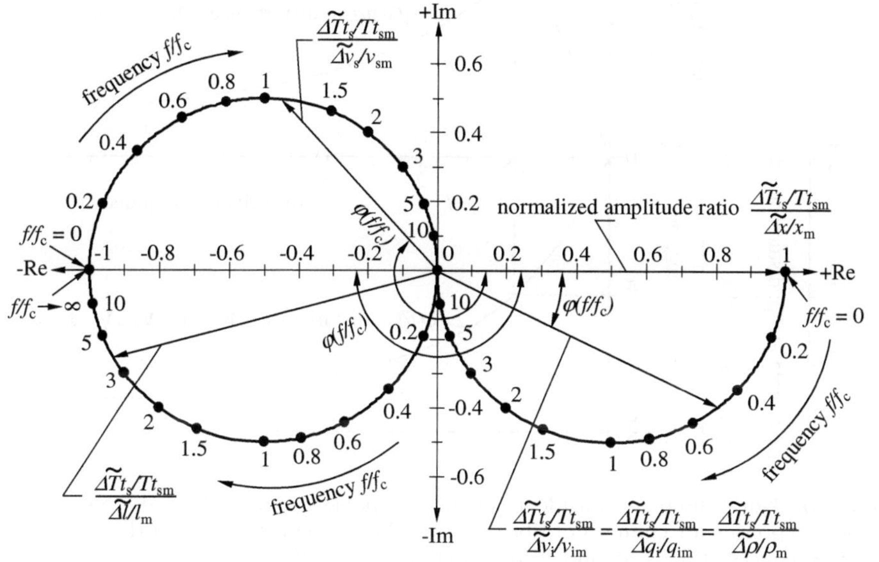

Fig. 4.4. Transfer locus of the normalised complex frequency responses for changes of fibre fineness ΔTt_s in consequence of different disturbance causes (Eqs. 4.37 and 4.42)

It should be considered that, at the modulus generation the negative sign for the disturbance Δv_s disappears. The normalised amplitude frequency response is presented in Fig. 4.5.

The *common normalised phase frequency response* of the same single phase frequency responses of Eqs. 4.18, 4.24 and 4.30 is:

$$\varphi(f) = \text{arc}\tan[-f/f_c] \tag{4.39}$$

and for 4.12:

$$\varphi(f) = \text{arc}\tan[-f/f_c] - \pi \tag{4.40}$$

Both are also included in Fig. 4.5. At last the *common normalised step responses* can be concluded from Eqs. 4.13, 4.19, 4.25 and 4.29 considering Eq. 2.48:

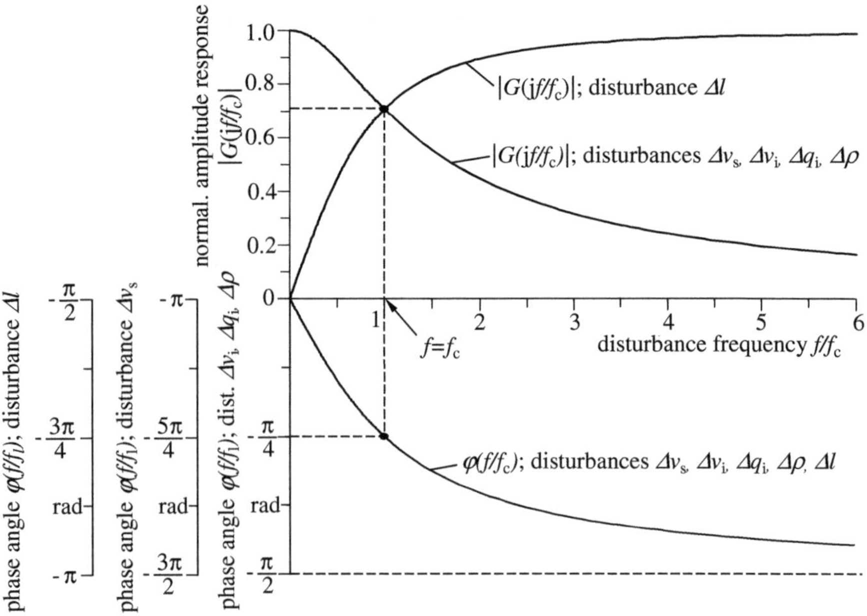

Fig. 4.5. Normalised amplitude frequency responses and normalised phase frequency responses for changes of fibre fineness ΔTt_s in consequence of different disturbance causes (Eqs. 4.38 and 4.43 as well as 4.39, 4.40 and 4.44)

$$\begin{aligned}\frac{-\Delta Tt_s/Tt_{sm}\overline{|\Delta v_s}}{\Delta v_s/v_{sm}} &= \frac{\Delta Tt_s/Tt_{sm}\overline{|\Delta v_i}}{\Delta v_i/v_{im}} \\ = \frac{\Delta Tt_s/Tt_{sm}\overline{|\Delta q_i}}{\Delta q_i/q_{im}} &= \frac{\Delta Tt_s/Tt_{sm}\overline{|\Delta \varrho}}{\Delta \varrho/\varrho_m} \\ &= \mathbf{1 - \exp(-\boldsymbol{t}/\boldsymbol{T}_c)}\end{aligned} \tag{4.41}$$

The common normalised step response is shown in Fig. 4.6. It is characteristic for this type of disturbances, that the effect to the fibre fineness is smaller the smaller the disturbance frequency f is. In a way steady state Δl-changes (frequency $f \approx 0$) do not result in fineness changes (see Figs. 4.4 and 4.5), whereas the frequency and amplitude frequency responses rapidly approach the maximum amplification factor 1 for $f > f_c$.
This differential action appears in the step response. The fineness shift is first of all a maximum after the imprint of the cause step Δl and after that decreases to zero according to e^{-t/T_c}. The critical frequency f_c or the time constant of the system T_c are the significant sizes for quantitative estimations here as well.

Quantitative conditions can be deduced easily, for instance from the developed relationships for the design of yarn traverse motion systems at wind-up devices. The goal is in this case a minimising of the fineness changes ΔTt_s,

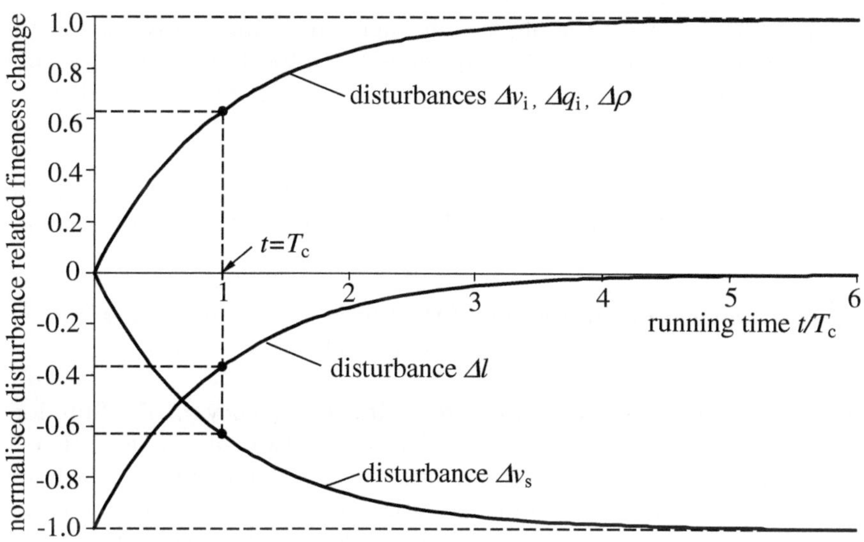

Fig. 4.6. Normalised step response for changes of fibre fineness ΔTt_s in consequence of different disturbance causes (Eqs. 4.41 and 4.45)

caused by the permanent effecting Δl-changes (see example presentation in Sect. 5.1.5). A high critical frequency f_c (that means short fibre formation distances l_m and high take-up velocities v_{om}) would be preferred for the disturbance type Δl alone. However, these are conditions which are connected with high amplifications of the other dealt with disturbance causes. At this point it is necessary that compromises be settled.

More remarks to the phase frequency responses of the dealt with disturbance causes (Fig. 4.5):

The disturbances Δv_i, Δq_i, and $\Delta \varrho$ result in the running behind of the phase shift angles of the effect oscillations against the periodic cause oscillations (they start for $f/f_c = 0$ with $\varphi = 0$ and end for $f/f_c \to \infty$ with $\varphi = -\pi/2$). Whereas, the disturbance Δv_s causes (settled by the negative coupling to the effect oscillation ΔTt_s - see the minus signs before the functions in Eqs. 4.9 until 4.11 and 4.13) a running behind of the phase shift angle which is additionally shifted about $-\pi$. That means it starts for $f/f_c = 0$ with $\varphi = -\pi$ and ends for $f/f_c \to \infty$ with $\varphi = -3\pi/2$ (see also Eq. 4.12).

A differential action with delay of first order (in the present case Δl-disturbances) normally results in a running before of the phase shift angles which start at $f/f_c = 0$ with $\varphi = +\pi/2$ and end for $f/f_c \to \infty$ with $\varphi = 0$. Because the coupling between the Δl-disturbances and the effected ΔTt_s-disturbances is also negative (minus signs before the functions in

Eqs. 4.32...4.34 and 4.36) the phase shift angles are shifted in the same way about $-\pi$ (see also Eq. 4.35). The result is the altogether running behind of the phase shift angles, which appear as $\varphi = -\pi/2$ (for $f/f_c = 0$) until $\varphi = -\pi$ (for $f/f_c \to \infty$).

These specialities for Δv_s- and Δl-disturbances are considered in Fig. 4.5 by different ordinates for the common dotted drawn phase frequency response curve.

Further explanations to the use of the phase frequency response for the analysis of specific problems will be given in Sect. 5.1.5.

The system time constant T_c and the critical frequency f_c of a fibre formation distance respectively a fibre formation line (we will see later that it is also valid for many yarn processing lines in the same manner), which are characterised by a length l_m, can be investigated by means of the nomogram in Fig. 4.7. These lines are characterised by a length of l_m which passes the fibre or yarn with the output or take-up velocity v_{om}. The nomogram is the graph of the definition of Eqs. 2.47 and 2.49 and allows for a quick estimation of T_c and f_c. l_m and v_{om} (outer ladders) are to be lined rectilinearly and T_c and f_c can be read from the point of intersection with the middle ladder. The drawn straight lines (1) and (2) represent the concrete application examples that have been described in Sects. 2.5.2 and 5.1.5.

As one can see, the ordinates of Figs. 4.5 and 4.6 are divided appropriately into the dimensionless ratio of relative effect-cause-changes, and the abscissa is divided into the also dimensionless ratio f/f_c or t/T_c. The ordinate values assert, by which factor a percentage cause change is to be multiplied in order to get the effected fineness change. This factor is at the most 1 for small disturbance frequencies (periodic disturbances) or for infinite long times (step-like disturbances). If the frequency of the disturbances is exactly equal to the critical frequency of the fibre formation distance ($f = f_c$) then the factor is $1/\sqrt{2} \approx 0.71$ and decreases with greater disturbance frequency quickly to zero.
Normally disturbances should only influence the aim variable as little as possible. Taking this into consideration the technological operating point of the fibre formation distance demands that the time constant T_c should be as much as possible and concerning the critical frequency f_c as little as possible. It is then guaranteed, that disturbances of small frequency no longer have any considerable penetrance to the aim variable fibre fineness. It is to be reached, if either the length of the fibre formation distance l_m can be enlarged and/or the take-up (spinning) velocity v_{sm} can be reduced. The latter is not usually desired. But this dynamic model reflection shows that the instable process behaviour at increased velocity is objectively foundable on the one

hand. It also shows on the other the limits of constructive and technological compromises, which can be made if necessary.

The cause-effect-transmission of the disturbance model Δl is the exact opposite of the four dealt with disturbance models above. The normalised complex frequency response can be concluded from Eq. 4.33 considering Eqs. 2.48 and 2.49:

$$G(\mathrm{j}f) = \frac{\widetilde{\Delta T t_\mathrm{s}/T t_\mathrm{sm}}}{\widetilde{\Delta l_\mathrm{s}}} \cdot e^{\mathrm{j}\varphi} = -\frac{j(f/f_\mathrm{c})}{1 + j(f/f_\mathrm{c})} \tag{4.42}$$

The transfer locus of this normalised complex frequency response is also included in Fig. 4.4.
One can get the amplitude frequency response from Eq. 4.34 in the same manner considering Eqs. 2.48 and 2.49:

$$|G(\mathrm{j}f)| = \left| \frac{\widetilde{\Delta T t_\mathrm{s}/T t_\mathrm{sm}}}{\widetilde{\Delta l_\mathrm{s}}} \right| = (-)(f/f_\mathrm{c}) \left[1 + (f/f_\mathrm{c})^2\right]^{-1/2} \tag{4.43}$$

The normalised phase frequency response is according to 4.35:

$$\varphi(f) = \mathrm{arc}\ \tan(f_\mathrm{c}/f) - \pi \tag{4.44}$$

Normalised amplitude and phase frequency responses are shown in Fig. 4.5. The minus sign of the complex frequency response is suppressed in the modulus representation of the amplitude frequency response. At quantitative evaluations it has to be considered of course, that a positive Δl-change correlates with a negative $\Delta T t_\mathrm{s}$-change and vice versa.
At last from Eq. 4.36 we get the step response as:

$$\frac{\underline{\underline{\Delta T t_\mathrm{s}/T t_\mathrm{sm}|\overline{\Delta l}}}}{\Delta l/l_\mathrm{m}} = -\exp(-t/T_\mathrm{c}) \tag{4.45}$$

This normalised step response is shown in Fig. 4.6.

It is characteristic of this kind of disturbance, that the effect to the fibre fineness is smaller if the frequency f of disturbances is smaller. In a way steady state Δl-changes (frequency $f \approx 0$) result in no fineness changes whatsoever (see Figs. 4.4 and 4.5), whereas the frequency and amplitude frequency responses quickly reach the maximum amplification factor 1 for $f > f_\mathrm{c}$.

This differential action is expressed in the presentation of the step response (Fig. 4.6) so that an imprinted step Δl causes first of all a maximum fineness shift, which then decreases to zero according to $\exp(-t/T_\mathrm{c})$. The critical frequency f_c or the system time constant T_c are significant quantities also for quantitative estimations in such cases.

The developed relationships can be used for instance as the calculation basis for the design of traverse motion systems at winders. It is possible to give quantitative conditions for a reduction of the fineness changes ΔTt_s caused by the permanent effecting Δl-changes (see example in Sect. 5.1.5). A high critical frequency f_c would be preferred for the disturbance type Δl, this means a short fibre formation distance l_m and a high take-up velocity v_{om} would be favourable. However, these are just conditions for a high amplification of the other dealt with disturbance causes as named before. The necessity of compromises is obvious at this point.

A few remarks to the phase frequency responses of the dealt with disturbance causes (Fig. 4.5):

The disturbances $\Delta v_i, \Delta q_i$ and $\Delta \varrho$ produce running behind phase shift angles versus the causing oscillations ($f/f_c = 0$ starts with $\varphi = 0$ and ends for $f/f_c \to \infty$ with $\varphi = -\pi/2$). Whereas, the disturbance Δv_s (involved by the negative coupling to the effected oscillation ΔTt_s - see minus signs before the dynamic functions of Eqs. 4.9 to 4.11 and 4.13) produces a running behind phase shift angle which is shifted additionally by $-\pi$. This means, it starts with $f/f_c = 0$ with $\varphi = -\pi$ and ends with $f/f_c \to \infty$ with $\varphi = -3\pi/2$ (see also Eq. 4.12).

A differential action with delay of first order (as being submitted by the disturbance Δl) normally produces running before phase shift angles which start with $f/f_c = 0$ with $\varphi = +\pi/2$ and ends with $f/f_c \to \infty$ with $\varphi = 0$. But, also because a negative coupling is given for Δl-disturbances to the effected ΔTt_s-disturbances (see minus signs before the dynamic functions of Eqs. 4.32 to 4.34 and 4.36) the phase shift angles are also shifted by $-\pi$ (see also Eq. 4.35). Therefore, running behind phase shift angles appear altogether which start with $\varphi = -\pi/2$ (for $f/f_c = 0$) and end with $\varphi = -\pi$ (for $f/f_c \to \infty$). These specialities by Δv_s- and Δl-disturbances are considered in Fig. 4.5 by different ordinates for the dotted lined common phase frequency response. More detailed explanations to the use of the phase frequency response for special problems will be given in Sect. 5.1.5.

The system time constant T_c and the critical frequency f_c of a fibre formation distance (we will see later, that it is also valid for many other fibre processing distances) can be found out by means of the nomogram Fig. 4.7. This converts the Eqs. 2.47 and 2.49 into a quickly utilisable manner. To this the length of the fibre formation distance l_m and the take-up velocity v_{om} (outer nomogram ladders) are connected rectilinearly and in the cross point with the middle ladder the T_c and/or f_c can be read. This is done for instance for two applications, dealt with in Sects. 2.5.2 and 5.1.5 (straight lines (1) and (2)).

At last some remarks to the question of the model-mathematical treatment of different process and product variable disturbances which simultane-

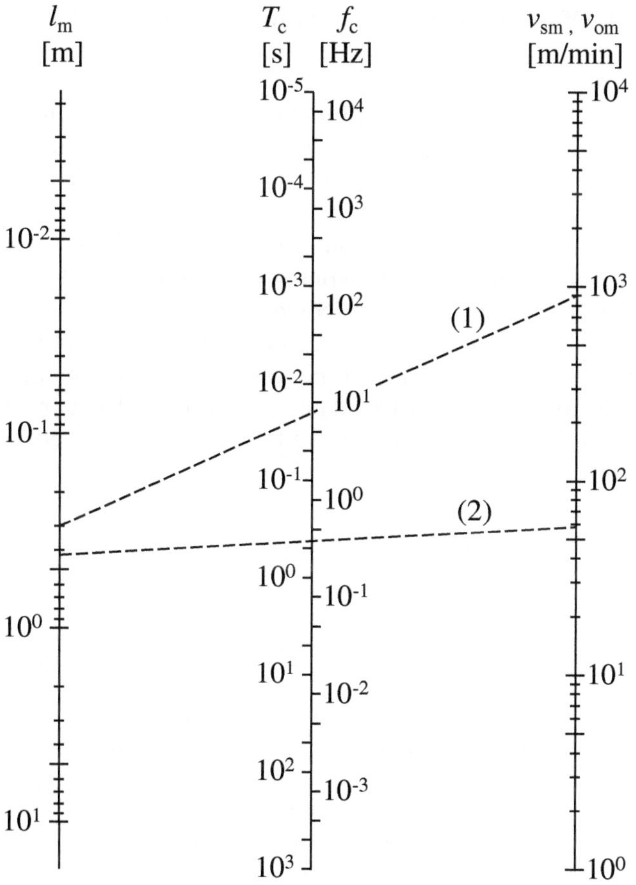

Fig. 4.7. Nomogram to the estimation of the time constant T_c and of the critical frequency f_c for fibre formation and fibre processing processes;
example (1): drawing process; v_{om}=900 m/min, l_m=0.3 m, T_c=0.02 s, f_c=8 Hz
example (2): length of the yarn input distance on a twister; v_{om}=58 m/min, l_m=0.42 m, T_c=0.43 s, f_c=0.37 Hz

ously appear and effect fibre fineness changes. In principle, the superposition principle can be applied to the single developed solution equations (transfer function, complex frequency response and time transient function) for the single disturbance of the DEq. 4.7. However, the following has be taken into consideration:

a) The related step response (for instance Eqs. 4.13 and 4.19, if step-like disturbances Δv_s and Δv_i are at hand) are added in consideration of the time range. If the different disturbances do not start at the same time $t = 0$ then the corresponding time shift of the one transient function to the other is to be considered. Of course a graphic addition of the transient functions is also possible.

b) The related dynamic transfer functions or complex frequency responses (for instance Eqs. 4.9 and 4.15 or 4.10 and 4.16, if periodic disturbances $\widetilde{\Delta v_s}$ and $\widetilde{\Delta v_i}$ are at hand) are to be added corresponding to the rules of the addition of complex numbers (vector addition) in consideration of the frequency range. If the exciting frequency ω is not the same for the viewed disturbances then ω_1 and ω_2 must be put in separately to both basic equations of the complex frequency responses (corresponding to $p_1 = \mathrm{j}\omega_1$ and $p_2 = \mathrm{j}\omega_2$ to both basic equations of the transient functions). The amplitude frequency responses and the phase frequency responses can only be calculated by means of the complex added complex frequency responses and not by means of the simple addition of the amplitude and phase frequency responses from the both single disturbances.

4.2.2 Variable Fibre Orientation

Cause-Effect-Scheme

A very important product variable of a melt spun fibre is the achieved orientation of the macromolecules along the fibre length axis, consequently along the main tension direction during the fibre formation and elongation processes. The orientation determines the textile-physical properties of the fibre decisively. This fibre orientation determines the breaking tensile force F_b and the breaking elongation ε_b. It should be characterised by the orientation elongation ε_o which the fibre has suffered, starting from the complete unsettled isotropic state of the thermoplastic melt. A measurement for ε_o is the birefringence of the fibre Δ_o at the exit of a special fibre formation distance.

The following a-priori knowledges from the literature [280, 281] are important and are concerned with the qualitative and quantitative relationship between the imprinted orientation elongation ε_o and the causing process and product variables in the fibre formation process:

a) Thermoplastic fibre polymers possess a maximum imprintable orientation elongation ε_{omax}. This is independent of the kind and the quantity of as to time consecutive elongation steps (spinning, draw and tensile testing or high speed spinning, post draw and tensile testing and so on) and its value is nearly constant for an appointed fibre polymer.

If the achieved elongation ε_{oi} of a single elongation step i is expressed as the natural logarithm, consequently:

$$\varepsilon_{oi} = \ln\left[\frac{Tt_{ii}}{Tt_{oi}}\right]$$

then it is valid with good exactness for the fibre material PA

$$\varepsilon_{oi} = \sum_{i=1}^{n} \varepsilon_{oi} \approx 2$$

and for the fibre material PET

$$\varepsilon_{oi} = \sum_{i=1}^{n} \varepsilon_{oi} \approx 2.3$$

b) In the first step of such a consecutive chain of elongation steps (this is the spinning step) the fibre has reached the orientation elongation ε_{o1}. This is dependent upon the tensile stress (fineness related tensile force) R_{T_g} which appears in the fibre at the glass transition temperature T_g (for PA 6 nearly 50 °C, for PET nearly 80 °C). For temperatures $< T_g$ the fibre is "ready", this means Tt_s is reached and on the following take-up way the fibre will not be further elongated plastically.

c) R_{T_g} is the tensile stress (fineness related tensile force) at the point $< T_g$. Therefore, it is necessary to analyze the tensile force F_{T_g} at this point because it is valid:

$$R_{T_g} = F_{T_g}/Tt_s \tag{4.46}$$

As informed in [280] and [281] F_{T_g} is improved essentially by an air friction component F_{drag} and an acceleration component F_{inert}, in which both of these components have totally different quantitative importance in the classical spinning ($v_s \leq 1200$ m/min) and in the high speed spinning ($v_s \geq 3000$ m/min). With completing this it should be remarked, that in the classical spinning process the rheological initial force F_{rheo} should be taken into consideration.

d) The air friction force F_{drag} is dependent upon the throughput through the spinneret Q_s, the spinning velocity v_s and the distance between the spinneret and the point at which the fibre reaches the glass transition temperature T_g. This spinneret distance l_{T_g} is essentially dependent upon the throughput through the spinneret Q_s, the spinneret temperature T_s and the cooling conditions (surrounding air velocity, surrounding air temperature and surrounding air humidity).

e) The acceleration force F_{inert} is dependent upon the spinning velocity v_s as well as upon the throughput through the spinneret Q_s.

The verbally described dependences are summarised as shown in the technological scheme Fig. 4.8 and the cause-effect-scheme Fig. 4.9.

Looking at Fig. 4.9 it is evident that the fibre fineness Tt_s is one of the two relevant product variables which effects the tensile stress (fineness related tensile force) R_{T_g}. The latter effects the target quantity fibre orientation

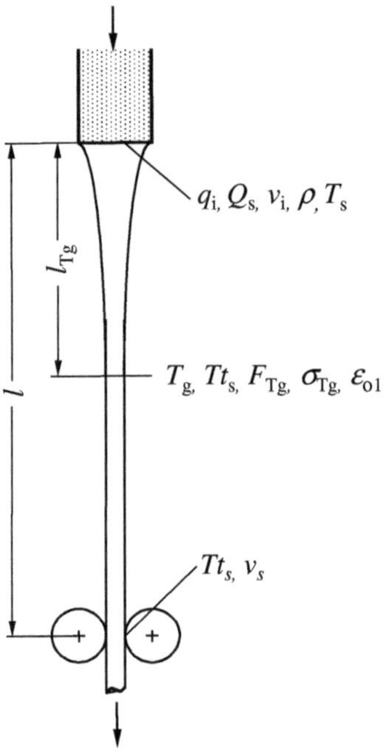

Fig. 4.8. Scheme of fibre formation at the melt spinning of polymers. Solidification point at the glass transition temperature T_g in the distance l_{T_g}

through the elongation orientation ε_{o1}. It would be possible to integrate the whole cause-effect-scheme Fig. 4.2 into Fig. 4.9 for the variable fibre fineness Tt_s. Nevertheless, only the important process variables Q_s and v_s (which directly effect the fineness Tt_s) have been inserted in Fig. 4.9 because these additionally influence the tensile stress (fineness related tensile force) R_{T_g} through the tensile force F_{Tg}.

Estimation to the Disturbance Transfer

In the following it should be attempted to estimate (without detailed derivation) the dynamic disturbance transmission to the fibre orientation ε_{o1} considerating the derived detailed relations from Sect. 4.2.1.

All disturbance quantities which influence Q_s (these are Δv_i, Δq_i and $\Delta \varrho$), effect by means of the ΔTt_s-changes (itself effected by the well-known dynamic transfer equations) also changes of force (ΔR_{T_g})- and with this of orientation elongation ($\Delta\varepsilon_{o1}$)-changes with the same time and frequency behaviour. ΔQ_s-changes additionally cause by means of the F_{T_g}-branch changes ΔR_{T_g} (see Fig. 4.9) in the same manner. The necessarily following change of the stored melt mass in the fibre formation distance with the distance l_{T_g} (which changes too) is decisively for the transfer behaviour of the acceleration force changes as well as for the air friction force changes. The spinning velocity v_s and the distance l_{T_g} are there, on the other hand, the time constant determining quantities for the acceleration and air friction forces. The same

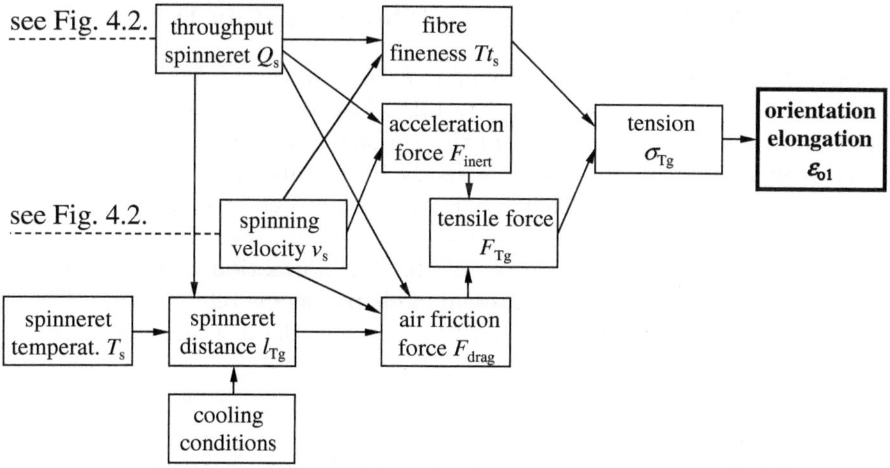

Fig. 4.9. Cause-effect-scheme for the target quantity orientation elongation ε_{o1} at the melt spinning of polymers, especially at the high speed spinning process ($v_s \geq 3000$ m/min)

statement is also valid for the effect of spinning velocity disturbances Δv_s to changes of the orientation elongation $\Delta\varepsilon_{o1}$. The steady state amplification factors are indeed different to estimate for the different disturbance types: Changes of the spinneret throughput ΔQ_s (and its effecting disturbances) effect in each case changes of the fineness Tt_s and also of the tensile force F_{Tg} with equal sign. The effects referred to R_{T_g} and ε_{o1} are comparatively small, because they will be compensated by means of the quotient according to Eq. 4.46. The amplification factor will be small.

Spinning velocity changes Δv_s on the other hand effect changes of the fineness ΔTt_s and also of the tensile force ΔF_{Tg} with unequal sign. The quotient according to Eq. 4.46 for R_{T_g} and its effect to ε_{o1} will be large, and the amplification factor will also be large.

Consequently, spinning velocity changes Δv_s in principle to value are much more influential than spinneret throughput changes ΔQ_s (which are caused for instance by unevenly operating gear pumps, effecting changes of input velocity Δv_i) with view to the generation of structural unevennesses of a melt spun fibre.

The same quantitative data can be authoritative therefore for the velocity of disturbance transmissions to the fibre orientation (and with it to fibre or yarn length with undefined structural properties) than they have been obtained with the dynamic fineness changes.

4.2.3 Complex Proceedings in the Fibre Formation Distance

A *"Specified Differential Equation of a Fibre Formation Distance (simplified)"* has been developed under Sect. 4.2.1. This equation does not consider the first fibre formation range from the spinneret to the solidification point of the melt which characterises the ready fibre state. This actual range of the fibre formation is in principle a range of the highest complex dynamic proceedings even by an undisturbed steady state process. Dynamic change proceedings of all product variables of the fibre formation take place inside this range in a very short time interval. This passage is referred to in Chap. 3, especially to the whole Sect. 3.1 which contains detailed investigations to these complex processes.

4.3 Glass Fibre Spinning; Variable Fibre Fineness

4.3.1 Cause-Effect-Scheme

Another melt spinning material of a totally different source and with different physical and chemical structure is glass. This inorganic fibre material obtained a separate importance for technical-textile applications (especially as a reinforcing fibre material) and (because of its inflammability) for decorative textiles in the last 50 years. Some process dynamic questions for this technological fibre formation process will be discussed in the following sections. Specific further specialities for the methodical practice in the physical analysis of a given technological situation will be explained in more detail later.

The technological scheme of the glass fibre spinning process is shown in Fig. 4.10. [1]

The missing of a spinning pump is evident in comparison with the melt spinning process of polymers (Fig. 4.1). The spinneret is the bottom of a metallic Pt-Rh-oven which is connected serially in the secondary circuit of an electrical heating current transformer. The oven with the spinneret is, in this manner, practically an OHM's heating resistance. The throughput Q_s will be achieved by means of the hydrostatic pressure of the glass melt which stands over the spinneret with the glass level h. A volume feeding spinning pump would not be thinkable here for higher than 1200°C of the oven temperature. The pressure in front of the spinneret holes p_s is a process

[1] Additionally is to remark, that the so-called two steps process is demonstrated here. Glass marbles (single mass from 10 until 15 g) are fed into the spinneret oven. This is typical for small production or single lab equipments. In large production is to find generally the one step process. The spinneret oven is fed there directly by molten glass mass coming from the glass melt tub and the marble phase is avoided.

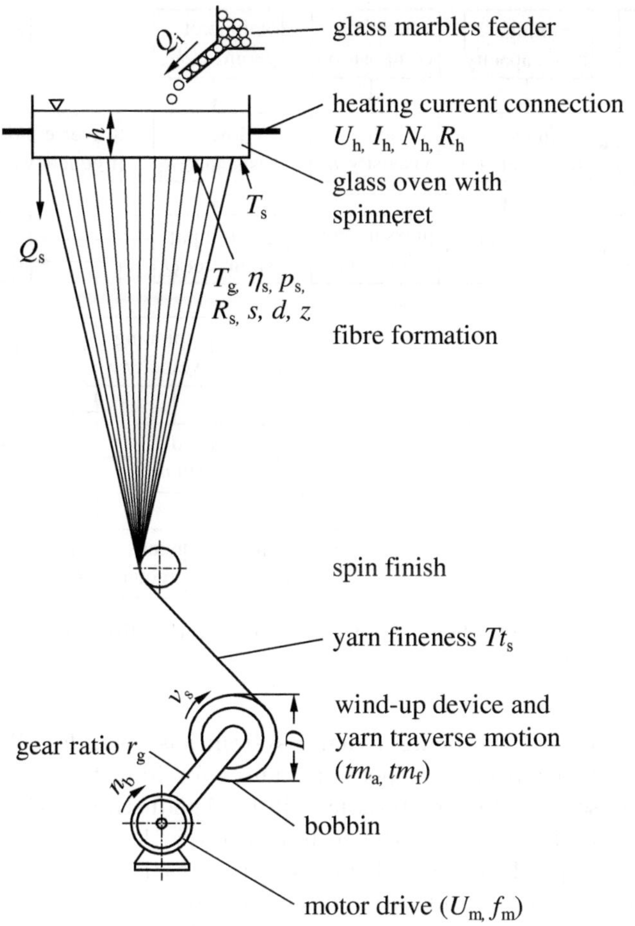

Fig. 4.10. Simplified technological scheme of the glass fibre spinning

variable which directly effects the spinneret throughput Q_s concerning the HAGEN-POISEUILLE-law for viscous melts (in the polymer spinning process, see Fig. 4.2, the pressure p_s was a secondary variable and did not control Q_s and following Tt_s directly!). The valid cause-effect-scheme of the glass spinning process is shown in Fig. 4.11 for the target quantity fineness Tt_s.

These specifics can be learned from Fig. 4.11 in comparison to Fig. 4.2. Another essential specific is that the spinneret temperature T_s (it is the main cause variable for the melt viscosity η_s which effects the flow resistance R_s in the spinneret holes) now directly causes the spinneret throughput Q_s and following the fineness Tt_s.

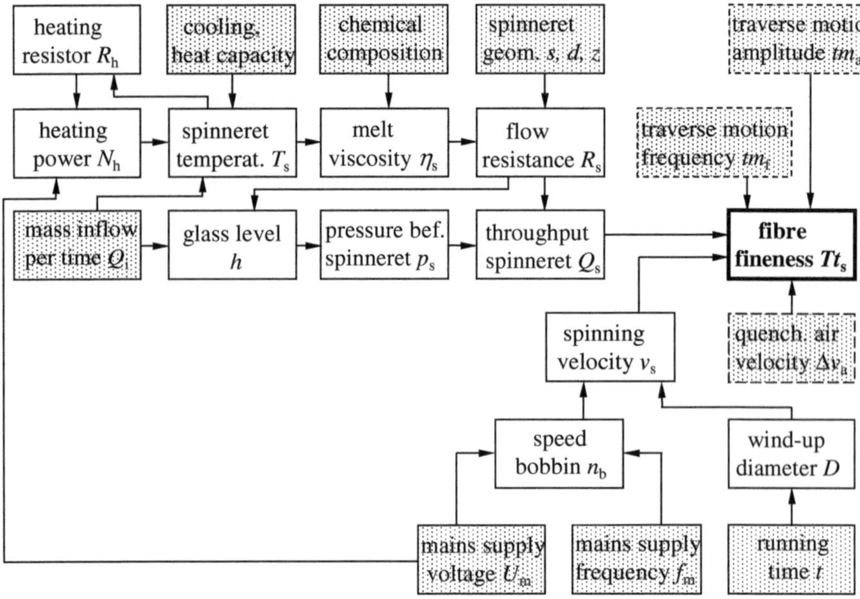

Fig. 4.11. Cause-effect-scheme for the target quantity fibre fineness Tt_s by the glass fibre spinning

The dynamics of the being at hand system has been described in detail in [282, 283] in perspective of possible causes for fibre fineness changes. In the following some few characteristics will be demonstrated which are based upon the general strategy of the technical-physical analysis recommended in Sect. 2.5.1. Specifically it will be clear that each concrete technological situation enforces the consideration or elaboration of new product or process specific a-priori knowledges. An additional product variable at the input of the spinneret oven is the glass mass inflow per time unit Q_i. This variable carries out effects to the glass level h and the spinneret temperature T_s. All other not else named variables of the cause-effect-scheme are explained in Sect. 4.2.1.

4.3.2 Functional Block Diagram

The fibre formation distance possesses, in the glass fibre spinning process, a few other properties than in the polymer spinning in relation to the disturbance transmission. The fibre fineness is actually realised by means of the glass mass inflow per time unit Q_i as well as the spinning velocity v_s. However, the glass fibre is already ready in a distance of 3 to 5 cm below the spinneret holes, because the glass melt does not have a viscous and elastic elongation power. At this point the spinning velocity v_s and the fineness Tt_s are reached. The whole distance between this extremely die near deformation zone until the take-up bobbin is a pure transport distance in which fineness changes do not take place further. The transmission range of disturbances is reduced to the named short fibre formation distance below the die. This possesses a very small time constant T_c in consequence of the small stored glass mass and the short length from this fibre formation distance. This means further that all disturbances Δv_s and ΔQ_s will be transmitted practically undamped to ΔTt_s. The dynamic properties of the fibre formation distance can be described in the first approximation by means of the steady state model.[2]

Therefore, the single cause-effect relations which produce either ΔQ_s or Δv_s changes must be investigated to the description of the dynamics relating to the product variable fibre fineness. A similar practice would also be necessary for a further complete dynamic "backward" view of an adequate polymer spinning equipment. However, this has been neglected in Sect. 4.2.1, because a spinning pump enforces the spinneret throughput Q_s and the dynamic investigation in the Q_s, as well as in the v_s branch, results in principle in the investigation of the dynamic transmission properties of the electromotoric drives. This will be given now anyway for the lied before glass fibre spinning process in the v_s branch.

The functional block diagram Fig. 4.12 of the cause-effect relations can be designed now on the basis of Fig. 4.11. This supposes partly the mathematical calculation of physical relationships, partly the experimental fixation of the transmission behaviour of single transfer elements. These details are not given here. They are described completely in [282], [283]. As one can see, from the outside imprintable cause quantities to fineness changes ΔTt_s are possible concerning the dynamic transfer functions G_1 to G_{18} which the transmission effects from changes of the mains supply voltage ΔU_m, changes of the input mass per time unit ΔQ_{i1}, changes of the mains supply frequency Δf_m and changes of the wind-up diameter ΔD, symbolically marked in step response in the single transfer elements of the functional block diagram characterise the transmission behaviour of the concerned element.

A specific of the signal chain is the transfer elements G_2 to G_6 and G_{10} to G_{12} in the present functional block diagram. G_2 to G_6 characterise the (linearised) relationships between a change of a heating voltage ΔU_h and the effected change of the spinneret temperature ΔT_s as a standard example of the electrical heating of an OHM's resistance generally.

G_{10} considers that a change of the glass melt temperature ΔT_g effects a change of the output mass per time unit ΔQ_{s1} as well as (about the transfer

[2] Strictly speaking the transport distance from the end of the deformation zone until the take-up bobbin is the dynamic transmission distance which had to be included in the considerations. This generates however a phase shift only between cause and effect and not a change of the amplitude ratio of cause and effect.

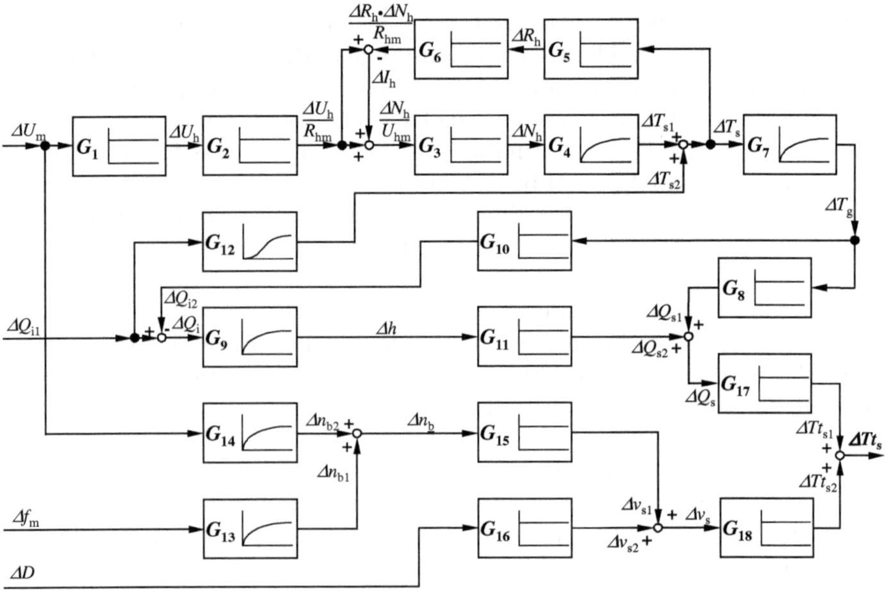

Fig. 4.12. Functional block diagram for the target quantity change fibre fineness ΔTt_s by the glass fibre spinning

element G_9) a change of the glass level Δh. G_{12} considers that a change of the input mass per time unit ΔQ_{i1} effects not only a change of the glass level Δh but also a change of the glass melt temperature ΔT_g. G_{10} and G_{12} consider the existing couplings between the temperature and the glass level regime of the spinneret oven.

The most important sizes of the theoretical and experimental investigated glass fibre spinning equipment which characterise the technological regime are collected in Table 4.1, the dynamic transfer functions G_1 to G_{18} are collected in Table 4.2.[3]

Both tables give an impression concerning the physical and technological constants and relationships which are necessary to know for the solution of the given task on the other hand. A better understanding of the following quantitative result interpretations should be obtained herewith.

[3] If it is unambiguously the question of dynamic transfer functions it is often written instead of $G_i(p)$ shortened G_i. This is used sometimes in the following.

Table 4.1. Compilation of technological and physical constants of the glass fibre melt spinning equipment at the investigated technological operating point

Data to complex	Designation	Symbol	Quantity and dimension
Spinneret oven	number of holes	z	100
	surface of glass melt	A_s	20000 mm^2
	transfer factor of heating transformer	k	$1.023 \cdot 10^{-2}$
	mains supply voltage	U_m	220 V
	heating voltage	U_hm	2.25 V
	heating current	I_hm	2.06 kA
	heating power	N_hm	4.64 kW
	heating (OHMic) resistance of spinneret oven	R_hm	$1.092 \cdot 10^{-3}\,\Omega$
	spinneret temperature	T_sm	1223 °C
	glass melt temperature	T_gm	1223 °C
	glass level	h_m	90 mm
	throughput per spinneret	Q_sm	20 g/min
Winder	speed of bobbin	n_b	2550 min^{-1}
	wind-up diameter	D_m	0.15 m
	gear ratio	r_g	1
	mains supply frequency	f_m	50 Hz
	spinning velocity	v_sm	1200 m/min
	fibre fineness	Tt_sm	16.7 tex
	bobbin formation time	T_bf	10 min
Static amplification factors	transfer element G_4	K_S	100 °C/kW
	transfer element G_8	$K_{\overline{\mathrm{S}}}$	$0.25\ \frac{\mathrm{g}}{\mathrm{min}}/$°C
	transfer element G_9	$K(T_0)$	$0.22\ \frac{\mathrm{g}}{\mathrm{min}}/\mathrm{mm}$
	transfer element G_{12}	K_K	-1.5 °C$/\frac{\mathrm{g}}{\mathrm{min}}$
	transfer element G_{13}	K_U	51 min^{-1}/Hz
	transfer element G_{14}	$K_{\overline{\mathrm{U}}}$	2.32 min^{-1}/V
Delay time constants	transfer element G_4	T_H	140 s
	transfer element G_7	$T_{\overline{\mathrm{H}}}$	67 s
	transfer element G_9	T_h	13500 s
	transfer element G_{12}	T_K1, T_K2	52 s, 386 s
	transfer element G_{13}, G_{14}	T_U	0.9 s
Physical constants	resistance-temperature coefficient of the Pt-Rh-spinneret oven	α_rt	$1.55 \cdot 10^{-4}$/°C
	glass density	ϱ_m	2.5 g/cm^3

Table 4.2. Compilation of the dynamic transfer functions of all functional block diagram elements corresponding to Fig. 4.12

Transfer element G_i	Dynamic transfer function $G_i(p)$
G_1	$\frac{\Delta U_h}{\Delta U_m} = k$
G_2	$\frac{\Delta U_h / R_{hm}}{\Delta U_h} = \frac{1}{R_{hm}}$
G_3	$\frac{\Delta N_h}{\Delta N_h / U_{hm}} = U_{hm}$
G_4	$\frac{\Delta T_{s1}}{\Delta N_h} = \frac{K_S}{1 + p \cdot T_H}$
G_5	$\frac{\Delta R_h}{\Delta T_s} = R_{hm} \cdot \alpha_{rt}$
G_6	$\frac{\Delta R_h \cdot I_{hm}/R_{hm}}{\Delta R_h} = \frac{I_{hm}}{R_{hm}}$
G_7	$\frac{\Delta T_g}{\Delta T_s} = \frac{1}{1 + p \cdot T_{\overline{H}}}$
G_8	$\frac{\Delta Q_{s1}}{\Delta T_g} = K_{\overline{S}}$
G_9	$\frac{\Delta h}{\Delta Q_i} = \frac{1}{K\,(T_0)} \cdot \frac{1}{1 + p \cdot T_h}$
G_{10}	$\frac{\Delta Q_{i2}}{\Delta T_g} = K_{\overline{S}} = G_8$
G_{11}	$\frac{\Delta Q_{s2}}{\Delta h} = K(T_0)$
G_{12}	$\frac{\Delta T_{s2}}{\Delta Q_{i1}} = \frac{K_K}{(1 + p \cdot T_{K1})\,(1 + p \cdot T_{K2})}$
G_{13}	$\frac{\Delta n_{b1}}{\Delta f_m} = \frac{K_U}{1 + p \cdot T_U}$
G_{14}	$\frac{\Delta n_{b2}}{\Delta U_m} = \frac{K_{\overline{U}}}{1 + p \cdot T_U}$
G_{15}	$\frac{\Delta v_{s1}}{\Delta n_b} = \pi \cdot D_m \cdot r_g$
G_{16}	$\frac{\Delta v_{s2}}{\Delta D} = \pi \cdot n_{bm} \cdot r_g$
G_{17}	$\frac{\Delta Tt_{s1}}{\Delta Q_s} = \frac{1}{v_{sm}}$
G_{18}	$\frac{\Delta Tt_{s2}}{\Delta v_s} = -\frac{Q_{sm}}{v_{sm}^2}$

It means in Table 4.2:

ΔQ_{i1}	change of the input mass per time unit
ΔQ_s	change of the output mass per time unit
ΔU_h	change of the heating voltage
ΔI_h	change of the heating current
ΔN_h	change of the heating power
ΔR_h	change of the heating (ohmic) resistance
ΔU_m	change of the mains supply voltage
Δh	change of the glass level
Δn	change of the speed of bobbin motor
Δf_m	change of the mains supply frequency
Δv_s	change of the spinning velocity
ΔD	change of the wind-up diameter
ΔT_s	change of the spinneret temperature
ΔT_g	change of the glass melt temperature
ΔTt_s	change of the fibre fineness

4.3.3 Evaluation and Results

The awaited dynamic changes of the glass fibre fineness ΔTt_s can be calculated now on the basis of Fig. 4.12 and the relationships which are given in Tables 4.1 and 4.2. The time transient function as well as the complex frequency response can be used. However, the dynamic transfer functions should be preferred to the calculation because a handling, as of items and factors, on the basis of the functional block diagram is possible. The fineness change ΔTt_{s2} caused by a change of the mains supply frequency Δf_m through the drive of the wind-up bobbin is to be calculated simply for instance by means of the multiplication of the dynamic transfer functions in the cause-effect-chain from Δf_m to ΔTt_{s2}, also

$$\frac{\Delta Tt_{s2}}{\Delta f_m} = G_{13} \cdot G_{15} \cdot G_{18} = -\frac{\pi \cdot K_U \cdot D_m \cdot r_g \cdot Q_{sm}}{v_{sm}^2 (1 + p \cdot T_U)} \tag{4.47}$$

All interesting total transfer equations can be built up step-like if the following basic rule is considered: Each output size of a transfer element in the functional block diagram is equal to the product of its input size and transfer function of the concerned transfer element, also for instance $\Delta Q_{s1} = \Delta T_g \cdot G_8$ or $\Delta h = \Delta Q_i \cdot G_9$. This rule is also valid, if single process variables are held constant by means of an additional automatic control. Their changes in size would be then zero (ideal imagination). The following single elements in the functional block diagram would also be assumed to be zero when starting from this points, because their input sizes are missing. If T_s were to be controlled, for instance, then ΔT_s would be zero in the ideal case and the transfer elements G_7 and G_8 would be inoperative in practice. This would mean that changes ΔQ_{s1} could not exist anymore and fineness changes could only be induced by means of glass level changes or changes of the wind-up diameter or changes of the speed of the bobbin motor.

Nine different disturbance transfer functions G_{z1} to G_{z9} of the fibre fineness are presented in Table 4.3 which have been calculated on the basis of the dynamic transfer functions G_1 to G_{18} for 9 different disturbance variants. These can be analysed quantitatively after inputting the equation expressions for G_1 until G_{18} given in Table 4.2. The last and further disturbance transfer functions of other process variables (for instance glass level or spinneret temperature) and all resulting amplitude frequency responses, phase frequency responses and step response can be found by the interested reader in [282], [283].

Table 4.3. Compilation of the dynamic disturbance transfer functions of the fibre fineness at the glass fibre spinning

Cause variable disturb.	Additionally const. held variable	Disturbance transfer function $G_{zi}(p)$
ΔU_m	not one	$G_{z1}(p) = \dfrac{\Delta Tt_s}{\Delta U_m} = \dfrac{2G_1G_2G_3G_4G_7G_{17}(G_8 - G_9G_{10}G_{11})}{1 + G_3G_4G_5G_6} + G_{14}G_{15}G_{18}$
ΔU_m	n_b	$G_{z2}(p) = \dfrac{\Delta Tt_s}{\Delta U_m} = \dfrac{2G_1G_2G_3G_4G_7G_{17}(G_8 - G_9G_{10}G_{11})}{1 + G_3G_4G_5G_6}$
ΔU_m	n_b and h	$G_{z3}(p) = \dfrac{\Delta Tt_s}{\Delta U_m} = \dfrac{2G_1G_2G_3G_4G_7G_{17}}{1 + G_3G_4G_5G_6 - G_7G_{10}G_{12}}$
ΔU_m	h	$G_{z4}(p) = \dfrac{\Delta Tt_s}{\Delta U_m} = \dfrac{2G_1G_2G_3G_4G_7G_{17}}{1 + G_3G_4G_5G_6 - G_7G_{10}G_{12}} + G_{14}G_{15}G_{18}$
ΔQ_{i1}	not one	$G_{z5}(p) = \dfrac{Tt_s}{\Delta Q_{i1}} = \dfrac{G_7G_{12}G_{17}(G_8 - G_9G_{10}G_{11})}{1 + G_3G_4G_5G_6} + G_9G_{11}G_{17}$
ΔQ_{i1}	T_s	$G_{z6}(p) = \dfrac{Tt_s}{\Delta Q_{i1}} = G_9G_{11}G_{17}$
ΔU_m	T_s	$G_{z7}(p) = \dfrac{Tt_s}{\Delta U_m} = G_{14}G_{15}G_{18}$
Δf_m	not one	$G_{z8}(p) = \dfrac{Tt_s}{\Delta f_m} = G_{13}G_{15}G_{18}$
ΔD	not one	$G_{z9}(p) = \dfrac{Tt_s}{\Delta D} = G_{16}G_{18}$

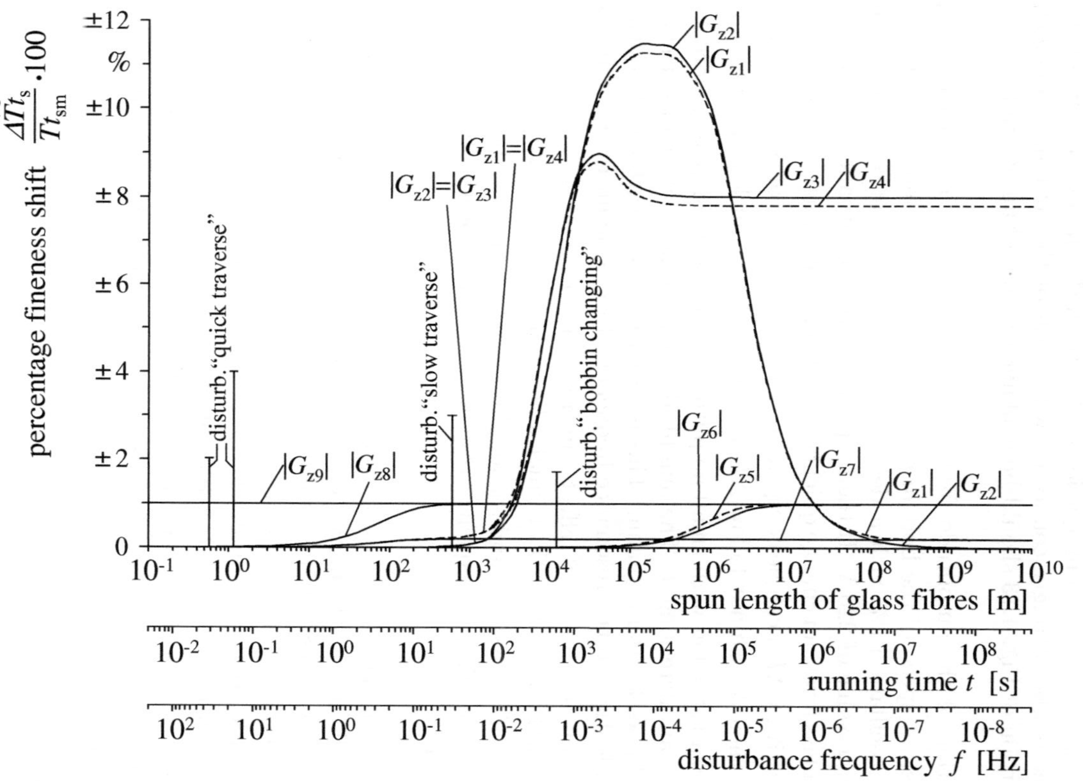

Fig. 4.13. Amplitude frequency responses of percentage fineness changes $(\widetilde{\Delta Tt_s / Tt_{sm}})100$ by means of changes of different disturbance quantities of 1% by the glass fibre spinning, v_{sm}=1200 m/min

The overview representation Fig. 4.13 is only a demonstration of amplitude frequency responses of the fineness $|G_{z1}(jf)|$ to $|G_{z9}(jf)|$, shortened $|G_{z1}|$ to $|G_{z9}|$.

The oscillations percent of the fineness are drawn on the ordinate which would appear at a one percent sinusoidal disturbance of $\widetilde{\Delta U_m}$, $\widetilde{\Delta Q_{i1}}$, $\widetilde{\Delta f_m}$ or $\widetilde{\Delta D}$ appropriately of the selected disturbance variant. Three interacting measurements are drawn on the abscissa (valid for the special numerical example according to Table 4.1). The first abscissa measurement presents the meter yarn spun from the start time point, the second presents the running time according to the applied spinning velocity, the third the disturbance frequency. The latter is also strongly connected with both of the first measurements. The assertion of Fig. 4.13 should be explained by use of an example.

Assertion of the amplitude frequency response $|G_{z2}|$: Disturbance $\widetilde{\Delta U_m}$ with constant speed of the bobbin motor ($\widetilde{\Delta n_b} = 0$). The mains supply voltage shall oscillate ±1% with a determined frequency sinusoidal around its steady state value of 220 V. If the frequency of the disturbance is very small then the glass level can change quickly enough and the change $\widetilde{\Delta Q_s}$ (caused by $\widetilde{\Delta T_s}$ respectively $\widetilde{\Delta T_g}$) can be compensated. If the frequency of the distrubance increases then $\widetilde{\Delta h}$ cannot follow $\widetilde{\Delta T_s}$ respectively $\widetilde{\Delta T_g}$ quickly enough and a $\widetilde{\Delta Tt_s}$ appears starting with small amplitude values. This increases more and more with an increasing f. The amplitude curve stands out against the abscissa and goes to its maximum. This is reached by a disturbance frequency which is identical to the resonance frequency of the system. In this range an oscillation of $\widetilde{\Delta U_m} = \pm 1\%$ effects an oscillation of $\widetilde{\Delta Tt_s} = \pm 10.8\%$. If f increases furthermore then T_g cannot follow quickly enough (involved by the heat inertia of the system) and the amplitudes $\widetilde{\Delta Tt_s}$ will again be smaller and smaller. The amplitude frequency response again asymptotically approaches the abscissa.

In the first abscissa measurement it is readable, that the mains supply voltage oscillations in the named critical frequency range operates to the fineness in a length range of 10^3 until $5 \cdot 10^7$ m yarn. This increase is therefore precarious, because the $\widetilde{\Delta U_h}$ (caused by $\widetilde{\Delta U_m}$) can be throughout ± 5% (in a non-automatic controlled process). Fineness oscillations of ± 54% would be the result! The conclusion of the dynamic analysis is that the process must be stabilised by means of an automatic control. However, $\widetilde{\Delta Tt_s}$ inside of a yarn length from 10^3 m would not be recordable independently of the quantities of $\widetilde{\Delta U_m}$ respectively $\widetilde{\Delta U_h}$.

The further disturbance inflows can be explained and estimated in an adequate manner using the other curves of the diagram. Four additional single

$\widetilde{\Delta Tt_s}$-oscillation parts are drawn in Fig. 4.13 (perpendicular straight lines to the abscissa), which are caused by the yarn winding advance motion and the wind-up diameter changes along a bobbin changing. These four disturbances only exist at four defined frequencies. The length of these straight lines represents the maximum possible fineness oscillations caused by these disturbances.

Figure 4.14 shows the yarn traverse motion element, which is generally used in the glass fibre wind-up process. It consists of a rotating axle with spacious buckled wires at which the yarn slips off. This special traverse motion element generates three distinguished oscillation frequencies. Analytical details shall not be explained here because they are connected to the special traverse motion element with its geometrical and kinematic specifities.

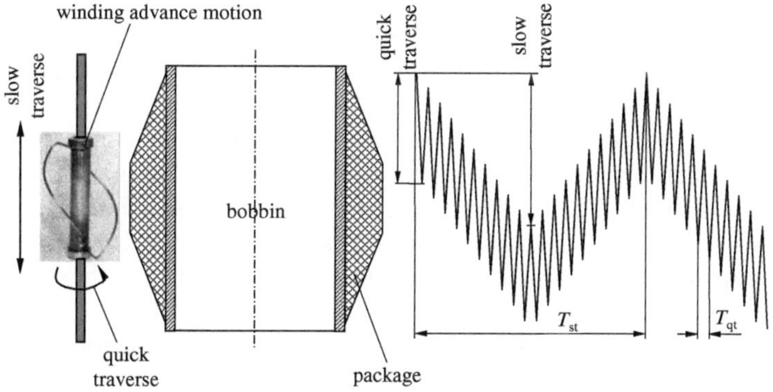

Fig. 4.14. Scheme of the package construction by means of winding advance motion on the glass fibre spinning

The until now given explanations show that practically each disturbance kind in the spinning process will generate changes or oscillations of the fibre fineness. These disturbances of the process variables are the real causes of the fibre or yarn fineness unevenness. This is valid not only for the glass fibre spinning process but for fibre formation processes in general. The dynamical analysis of the whole process allows assertions which cause variables as disturbance sizes are suitable for an appointed target quantity and in which quantity and/or frequency range this will happen.

The reversed question to the conditions for a test proof of appointed disturbance causes in the ready spun yarn will be cleared in Sect. 4.5. We will come back to the summarised presentation of Fig. 4.13 there once more.

4.4 Dynamics of Fibre Formation in the Spun Yarn Spinning Process

4.4.1 Task

The classical technological basic principle to the manufacture of spun staple yarn exists as you may know in different dissolution (beater), parallelisation (roller top card or stationary flat card), and refining (drafting, doubling) process steps of unarranged fibres, mostly pressed in fibre bales, of different length (staple length) to a roving (flyer) which is the input sliver for the last spinning process step, mostly realised by means of a ring frame. The dynamic of the process steps of yarn formation has been investigated and described (only partly or even empirically) in the past frequently under the preparation or realisation view of an automatic control system to influence the fineness unevenness of the produced rovings for the yarn manufacture. Especially the papers are to be mentioned here which present a sufficient and founded dynamic analysis of the automatic controlled process step by means of the time transient function or the dynamic transfer function description [284] to [289].

Two different types of the dynamic system behaviour can be met at the appropriate partial processes: the dead time behaviour [4] and the proportional action with delay [5]. The first is typical for some partial systems of passages in roller top cards and stationary flat cards, the second is for all passages in drafting systems which drafts imprint into the product. The basic equations and some conclusions will be given in the following for both types. Estimations in principle and a further approach to the whole problem should be possible for special tasks.

4.4.2 Dynamic Transfer Behaviour of Carding Engines

Each fibre spinning process pursues the goal to spin yarns with a fineness unevenness as small as possible. Two general strategies can help to reach this goal (besides a further development of the machine technique):
a) by means of a greater number of roving operation passages. This was the only possibility before the invention of automatic controlled drafting systems. The general law underlies this method that a yarn will be more even the more it will be doubled and drafted in the spinning process. However, this is connected as you know with a high expense of machines and (not unimportant)

[4] Changes of a cause variable at the input of a system appears true after a determined time (the dead time) at the output as changes of the effected variable (in the examples of this book it is always a transport time of the treated product). See also Sect. 5.1.3.

[5] To the definition of a proportional action with delay see the first statement of the *"Summarised Evaluation and Conclusions to the Solutions of the Differential Equation"* in Sect. 4.2.1

a lot of working power is necessary.
b) by means of an automatic controlled drafting system in at least one drafting passage in which the draft ratio changes automatically according to the continuously measured fineness changes of the input (open loop control) with or output (closed loop control) slivers. This method allows for drafting and doubling passages to economise in the spinning process plane by an unchanged quality of the ready yarn.

The fineness unevenness of the layed before fibre material (coming from the roller top card or from the stationary flat card as pile, fleece or sliver) is important for both strategies now. At this point, at the latest, the question appears as to the dynamic disturbance transmission on these machines and vice versa to the evenness power for uneven feeded fibre mats, fibre fleeces, fibre flocks and the like.[6]

The dynamic transfer properties of roller top and stationary flat cards have been described previously in the Russian special literature [285] to [287]. The essential ideas are explained in the following which lead to the dynamic transfer function.

Roller Top Card

The technological scheme is shown in Fig. 4.15. Unordered and undissolved fibres will be fed into the machine, picked up from the main drum and conveyed successively to a row of q worker-angle-stripper-pairs i (which $i = 1, 2, 3 \ldots q$). In the end the fibres will be processed (dissolved and parallelised) and in each case delivered up to the main drum back. After the run through all work elements the web will be taken up from the main drum and taken off as a sliver. Nevertheless, not all fibres (coming from the main drum) at each work station (consisting of a worker-angle-stripper-pair) go to the concerned worker-angle-stripper-pair but only the Kth part ($0 < K < 1$). The $(1 - K)^{\text{th}}$ part remains on the main drum and will be transported immediately to the next worker-angle-stripper-pair. The, from the worker-angle-stripper-pair picked up, fibres go back to the main drum after the processing or transport time T_{d}. Here also only a part of the fibres (coming from the main drum) will be picked up in fact too at the removal point of the web. This part shall be named K_{a}, whereas the $(1 - K_{\text{a}}^{\text{th}})$ part on the main drum will be mixed to the input of the machine once more after the run of the transport time T_{d2}. The transport time from the flake feeding until the first worker-angle-stripper-pair shall be named T_{d2}. It is assumed, to

[6] At this point the following hint: The dynamic transfer behaviour of the cotton beater (as well-known, directly extending in front of the cotton stationary flat card) has been described fully in [288]. Details must be exempted here. These would exceed the present investigations because these are technologically to far from the fibre formation.

simplify matters, that the coefficient of the fibre distribution K is the same for all worker-angle-stripper-pairs. If this is not the case (in the practice experimental to fix only) different K should be put into the following equations.

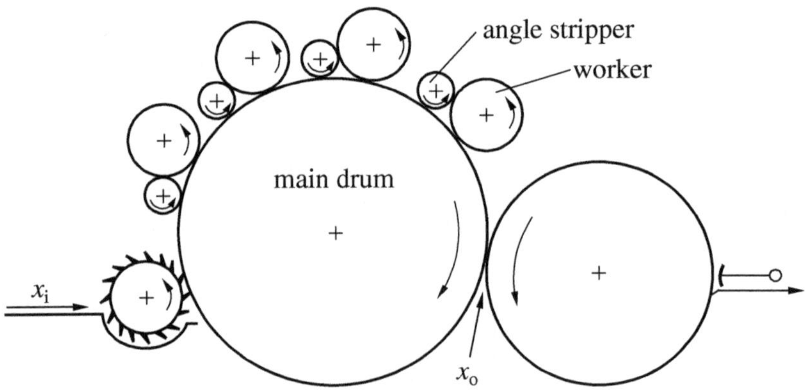

Fig. 4.15. Technological scheme of a roller top card; x_i, x_o input and output fibre mass per time unit

The dynamic transfer function of a roller top card can be developed now by means of the given a-priori knowledge. All quantities are to be taken in equally as changing quantities around the mean averages of an adjusted steady state technological operation point.

The disturbance of the fibre mass Δx_i (caused by uneven flock feed) comes on the main drum at the input of the machine according to Fig. 4.16. This Δx_i will be mixed with the not fully taken up fibre mass Δx_{i1} from the exit of the machine.

The sum passes the dead time element $\exp(-p \cdot T_{d1})$.

$$\Delta x_{i2} = \Delta x_i + \Delta x_{i1} \tag{4.48}$$

This exponential function is the dynamic transfer function of a pure transport or dead time line. The fed fibre mass from the machine entrance appears at the first worker-angle-stripper-pair:

$$\Delta x_1 = \Delta x_{i2} \cdot \exp(-p \cdot T_{d1}) \tag{4.49}$$

The functional block diagram Fig. 4.17 is qualified for the deviation of the dynamic transfer function of a worker-angle-stripper-pair.

The fibre mass Δx_1 (coming from the main drum) and the fibre mass Δx_2 (which is fed back from the processing) will be mixed and the sum Δx will be processed either once more according to the fibre distribution coefficient

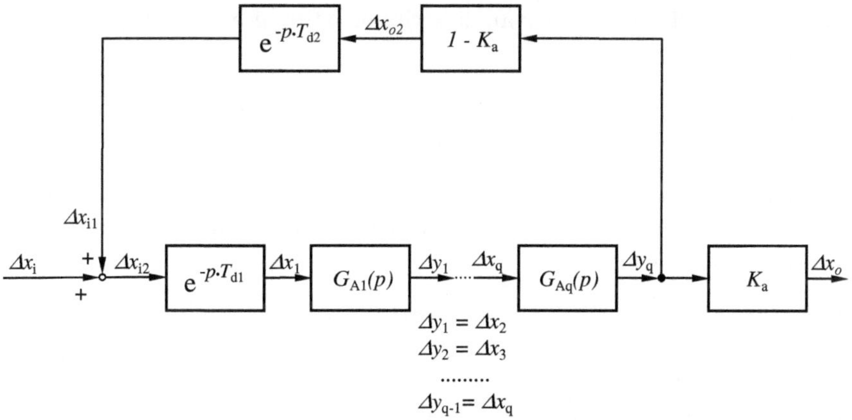

Fig. 4.16. Functional block diagram of the roller top card; Δx_i, Δx_o changes of the input and output fibre mass per time unit

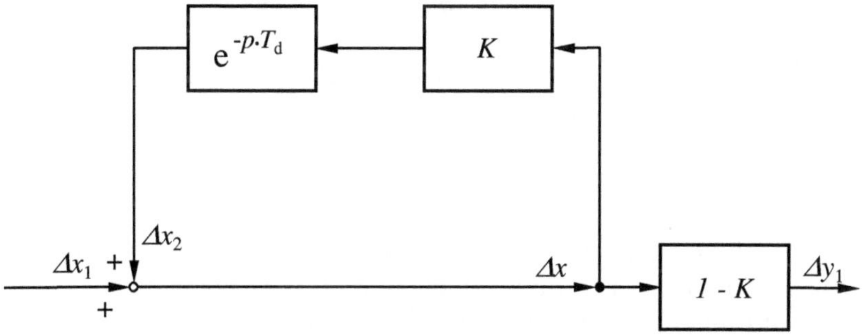

Fig. 4.17. Functional block diagram of the first worker-angle-stripper-pair on the roller top card; Δx_1, Δy_1 changes of the input and output fibre mass per time unit

K (branch with K) or to the main drum transmitted for the next worker-angle-stripper-pair as output fibre mass per time unit Δy_1. The transport time transfer element $\exp(-p \cdot T_d)$ is situated in the branch K. The following equations are readable as:

$$\Delta x = \Delta x_1 + \Delta x_2$$

$$\Delta x_2 = \Delta x \cdot K \cdot \exp(-p \cdot T_{d1})$$

$$\Delta y_1 = \Delta x \cdot (1 - K)$$

From this it is possible to calculate the dynamic transfer function:

$$G_{W1}(p) = \frac{\Delta y_1}{\Delta x_1} = \frac{1 - K}{1 - K \cdot \exp(-p \cdot T_d)} \tag{4.50}$$

Because each output Δy_i of a worker-angle-stripper-pair is simultaneously equal the input Δx_{i+1} of the next worker-angle-stripper-pair the dynamic transfer function of all q worker-angle-stripper-pairs of a roller top card can be written as:

$$G_{\mathrm{W1}}(p) \cdot G_{\mathrm{W2}}(p) \cdots G_{\mathrm{Wq}}(p) = \frac{\Delta y_{\mathrm{q}}}{\Delta x_1} = \left[\frac{1-K}{1-K \cdot \exp(-p \cdot T_{\mathrm{d}})}\right]^q \tag{4.51}$$

The dynamic transfer function of the whole roller top card can be calculated now considering Eq. 4.51 and the following equations, which are readable from Fig. 4.16 as:

$$\Delta x_{\mathrm{o}} = \Delta y_{\mathrm{q}} \cdot K_{\mathrm{a}}$$

$$\Delta x_{\mathrm{i2}} = \Delta x_{\mathrm{i}} + \Delta x_{\mathrm{i1}}$$

$$\Delta x_{\mathrm{i1}} = \Delta y_{\mathrm{q}} \cdot (1 - K_{\mathrm{a}}) \cdot \exp(-p \cdot T_{\mathrm{d2}})$$

$$\Delta x_1 = \Delta x_{\mathrm{i2}} \cdot \exp(-p \cdot T_{\mathrm{d1}})$$

The final result for the whole roller top card is at last:

$$G_{\mathrm{rc}}(p) = \frac{\Delta x_{\mathrm{o}}}{\Delta x_{\mathrm{i}}} = \frac{K_{\mathrm{a}} \cdot \exp(-p \cdot T_{\mathrm{d1}})}{\left[\frac{1-K \cdot \exp(-p \cdot T_{\mathrm{d}})}{1-K}\right]^q - (1-K_{\mathrm{a}}) \cdot \exp[-p(T_{\mathrm{d1}} + T_{\mathrm{d2}})]} \tag{4.52}$$

The complex frequency response as we know it is obtained when the operator p is substituted by the complex frequency $\mathrm{j}\omega$. Data about the dynamic disturbance transmission properties and about the evenness power of the roller top card are only general and extrapolated from results of possible concrete realised machines. Prerequisites for a quantitative analysis are the knowledge of the fibre distribution coefficients K and K_{a} and the dead times T_{d}, T_{d1} and T_{d2}, which can be appointed only experimentally. Quantitative analyses by means of graphic methods of the vector addition and inversion say [285]:

a) The evenness power of the roller top card is only given for input disturbances of the fed fibre mass per time unit, if the disturbance frequency is large. The cycle duration for a full oscillation must be smaller than the whole transport time of the fibres from the card entrance to its exit. This time can be nearly estimated with $T_{\mathrm{d1}} + q \cdot T_{\mathrm{d}}$.

b) The evenness power of changes, which the conditions under a) actually fulfill, will not take place if the cycle duration of the disturbances is equal or if an integer part multiple of the dead time of a worker-angle-stripper-pair T_{d}. Then namely, the maxima from fibres which come back after the time T_{d} coincide with maxima of the next, after the next and so on disturbance changes.

c) The evenness power is the better the more worker-angle-stripper-pairs are situated on the main drum periphery from the entrance of the card to its exit (trivial assertion) and/or the more uneven these are designed regarding its transport dead time. Because T_d arises from the diameter of the roller and from their speed (that means from their periphery and their periphery velocity) two possibilities of variation are given in principle. This means physically-obviously that resonance conditions should be at another disturbance frequency for each worker-angle-stripper-pair to avoid whole resonance conditions for all disturbance frequencies.

Stationary Flat Card

In comparison to the roller top card with discreet situated working elements on the main drum (worker-angle-stripper-pairs) these melt together to a closed technological unit in form of a flat clothing (Fig. 4.18).

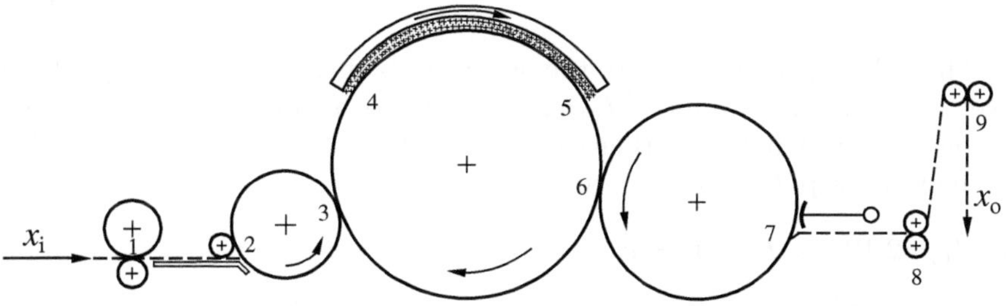

Fig. 4.18. Technological scheme of a stationary flat card; ranges 1 - 2, 7 - 8 and 8 - 9 draft of the web; ranges 2 - 3, 3 - 4, 5 - 6 and 6 - 7 transport of the web; range 4 - 5 treatment of the web

Referring to [287] the following dynamic transfer function has been derived and experimentally confirmed for the stationary flat card:

$$G_\mathrm{fc}(p) = \frac{\Delta x_\mathrm{o}}{\Delta x_\mathrm{i}} = \frac{1}{V} \cdot \frac{\exp(-p \cdot T_\mathrm{d})}{1 + (T_\mathrm{c1} + T_\mathrm{c2}) \cdot p + T_\mathrm{c1} T_\mathrm{c2} \cdot p^2} \tag{4.53}$$

The single symbols mean:

Δx_i, Δx_o changes of the input and output fibre mass

V whole draft of the fibre mass$=\dfrac{v_\mathrm{output}}{v_\mathrm{input}}$ (ranges 1 - 2, 7 - 8, 8 - 9)

T_d sum of all transport times of the fibres (ranges 2 - 3, 3 - 4, 5 - 6, 6 - 7)

T_c1, T_c2 time constants, describing the dynamic of the fibre redeposition power of the whole flat clothing, which is a fibre mass storage with permanent fibre exchange

With that (Eq. 4.53) the dynamic model of a stationary flat card shows a dead time behaviour (exponential function in the numerator) with delay of second order.

Complex frequency response (4.54), amplitude frequency response (4.55) and phase frequency response (4.56) can be calculated on the basis of Eq. 4.53 as follows:

$$G_{\mathrm{fc}}(\mathrm{j}\omega) = \frac{\widetilde{\Delta x_{\mathrm{o}}}}{\widetilde{\Delta x_{\mathrm{i}}}} \cdot \mathrm{e}^{\mathrm{j}\varphi} = \frac{1}{V} \cdot \frac{\exp(-\mathrm{j}\omega \cdot T_{\mathrm{d}})}{1 + (T_{\mathrm{c1}} + T_{\mathrm{c2}}) \cdot \mathrm{j}\omega + T_{\mathrm{c1}}T_{\mathrm{c2}} \cdot (\mathrm{j}\omega)^2} \tag{4.54}$$

$$|G_{\mathrm{fc}}(\mathrm{j}\omega)| = \left| \frac{\widetilde{\Delta x_{\mathrm{o}}}}{\widetilde{\Delta x_{\mathrm{i}}}} \right| = \frac{1}{V} \left[(1 - T_{\mathrm{c1}}T_{\mathrm{c2}}\omega^2)^2 + \omega^2 (T_{\mathrm{c1}} + T_{\mathrm{c2}})^2 \right]^{-1/2} \tag{4.55}$$

$$\varphi(\omega) = \mathrm{arc\ tan} \left[-\frac{(1 - T_{\mathrm{c1}}T_{\mathrm{c2}}\omega^2) \cdot \sin(\omega T_{\mathrm{d}}) + \omega (T_{\mathrm{c1}} + T_{\mathrm{c2}}) \cdot \cos(\omega T_{\mathrm{d}})}{(1 - T_{\mathrm{c1}}T_{\mathrm{c2}}\omega^2) \cdot \cos(\omega T_{\mathrm{d}}) - \omega (T_{\mathrm{c1}} + T_{\mathrm{c2}}) \cdot \sin(\omega T_{\mathrm{d}})} \right] \tag{4.56}$$

The following equivalent, to the output web of the card referred distances gives both time constants as $T_{\mathrm{c1}} \widehat{=} 2.12$ m, $T_{\mathrm{c2}} \widehat{=} 0.78$ m in [287] for a specific cotton stationary flat card. Hence they follow the time constants after division by the web output velocity v_{o} to:

$$T_{\mathrm{c1}} = 2.12/v_{\mathrm{o}},\ T_{\mathrm{c2}} = 0.78/v_{\mathrm{o}}$$

The times will be derived in the dimension s, if v_{o} is put in the dimension m/s.

The amplitude frequency response according to Eq. 4.55 is shown in Fig. 4.19 and actually not versus the disturbance frequency ω but versus the wavelength λ_{o} in the output web. The used conversion relation can be read as:

$$\omega = 2 \cdot \pi \cdot v_{\mathrm{o}}/\lambda_{\mathrm{o}} \tag{4.57}$$

The awaited evenness effect of the stationary flat card can be estimated from this amplitude frequency response. The normalised to the mean values x_{om} and x_{im} related fluctuation parts $\widetilde{\Delta x_{\mathrm{o}}}$ and $\widetilde{\Delta x_{\mathrm{i}}}$ are represented in the ordinate.

The quotient

$$\frac{\widetilde{\Delta x_{\mathrm{o}}}/x_{\mathrm{om}}}{\widetilde{\Delta x_{\mathrm{i}}}/x_{\mathrm{im}}}$$

is obtained considering the relation

$$V = \frac{v_{\mathrm{om}}}{v_{\mathrm{im}}} = \frac{x_{\mathrm{im}}}{x_{\mathrm{om}}}$$

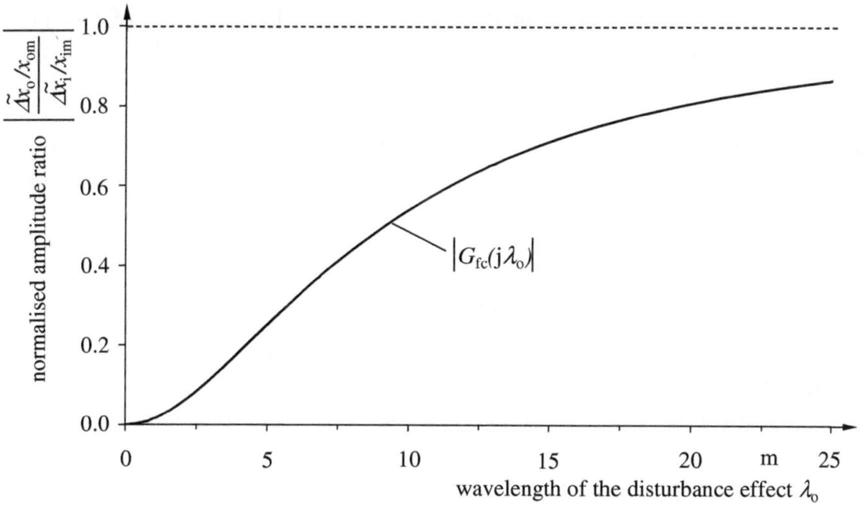

Fig. 4.19. Normalised amplitude frequency response by changes of the output fibre mass per time unit $\widetilde{\Delta x_o}$ caused by changes of the input fibre mass per time unit $\widetilde{\Delta x_i}$ of a stationary flat card

Figure 4.19 shows that a worth-while evenness of input fluctuations $\widetilde{\Delta x_i}$ are only to be awaited to a web length of 5 m, an even good evenness to a web length of 10 m and no worth-while evenness for a web length of more than 15 m. The pertinent wavelength of the disturbance cause in the coming in sliver λ_i of the stationary flat card ($\widetilde{\Delta x_i}$-course) is simple to calculate by using the quotient of λ_o and the draft V:

$$\lambda_i = \lambda_o / V$$

Finally it is to be remarked that the preceding considerations relate only to the fibre mass distribution along the web. Uneven distributions across to the transport direction have been excluded from the considerations because they must be described preferably with other methods. Some hints are represented moreover in [285].

4.4.3 Dynamic Transfer Behaviour of Drafting Zones in Drafting Systems

Specified Differential Equation of a Drafting Zone and its Disturbance Transfer

The drafting or drawing process is the most important processing step for the refinement of the slivers at the roving operation and at the final spinning machines. The dynamic of these processes is not to describe easy by means of a mathematical model if the statistical discontinuity of the geometrical single fibre properties (above all their length) are to be taken into consideration by the investigations. An impression of the complication (and also unfortunately the unwieldy handling) of a complete model description is shown in [289] and [290]. Nevertheless, it is enough to use the following simplified DEq. and its solutions for the outlet of design lines for automatic control tools of drafting systems or also only for the estimation how the fineness fibre of a sliver will in summary be influenced in such a drafting zone.

On this occasion it is assumed, that

- the fibre number in the sliver cross section at the drafting zone exit is greater than (> 100) and
- the roll setting is $1.5 \cdot l_{\mathrm{f}} \leq l \leq 2 \cdot l_{\mathrm{f}}$, in which l_{f} is the middle staple length of the fibres.

The latter prerequisite means, that neither the drafting forces should increase to high (under limit) nor should "swimming" fibres appear in a greater number which can no longer guarantee the cohesion of the free, not conducted, sliver section.

According to Fig. 4.20 the process and product variables which are necessary for the derivation of the DEq. can be defined in a similar manner as with the drawing zone of the drawing process (Fig. 2.3). They are somewhat the same:

v_{i}	input velocity
v_{o}	output velocity
z_{i}	number of fibres in the cross section of the coming in sliver
z_{o}	number of fibres in the cross section of the coming out sliver
l	roll setting
l_{f}	fibre length
Tt_{f}	fibre fineness
Tt_{i}	fineness of the coming in sliver $= z_{\mathrm{i}} \cdot Tt_{\mathrm{f}}$
Tt_{o}	fineness of the coming out sliver $= z_{\mathrm{o}} \cdot Tt_{\mathrm{f}}$

The dynamic basic Eq. 2.23 for the exchange processes in a mass storage system is similarly valid for the drafting zone. With this it is now possible to immediately describe the dynamic transfer behaviour with the general linearised DEq. 4.7. Single product variables, which apply to the yarn with endless fibres, must be adapted and defined to the present sliver structure. The linearised DEq. which fits the drafting zone of the sliver draft can be defined, if the following variables of Eq. 4.7 are substituted: the middle input cross sectional area q_{im} and its change Δq_{i} by the mean number of fibres in the cross sectional area of the input sliver z_{im} and its change Δz_{i}. Furthermore the mean density of the fibre material ϱ_{m} and its change $\Delta \varrho$, here no more consisting of a continuum mass, by the mean fineness of the fibres Tt_{fm} and

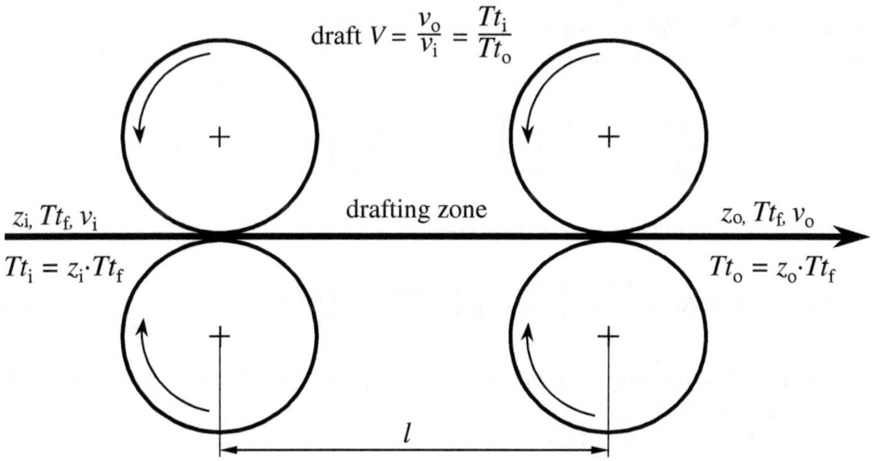

Fig. 4.20. Technological scheme of a one step sliver drafting system

its change ΔTt_f:

$$\begin{aligned}(v_{om} + p \cdot l_m) \cdot \Delta Tt_o + Tt_{om} \cdot \Delta v_o - v_{im} \cdot Tt_{fm} \cdot \Delta z_i \\ -z_{im} \cdot Tt_{fm} \cdot \Delta v_i + p \cdot Tt_{om} \cdot \Delta l - z_{im} \cdot v_{im} \cdot \Delta Tt_f = 0\end{aligned} \tag{4.58}$$

The dynamic transfer function of the fineness Tt_o of the yarn at the drafting system output can be given on the basis of Eq. 4.58 which was demonstrated similarly in Sect. 4.2.1. The dynamic functions complex frequency response, amplitude frequency response, phase frequency response and step response for the changes of Δv_o, Δv_i, Δz_i, ΔTt_f or Δl to the effected ΔTt_o-changes can now be derived. Probably, the most interesting dynamic functions (amplitude and step responses) for quantitative estimations are given as follows accordingly to Eqs. 4.13, 4.17, 4.19, 4.23, 4.25, 4.29 and 4.31. The functions for Δl-changes are lost on this occasion because these are not of any interest in drafting systems. The roll setting is assumed as constant and designed as mean value l_m:

disturbance Δv_o (changes of the output velocity)
amplitude frequency response

$$|G(j\omega)| = \left| \frac{\widetilde{\Delta Tt_o}}{\widetilde{\Delta v_o}} \right| = (-)\frac{Tt_{om}}{v_{om}} \left[1 + \left(\omega \cdot \frac{l_m}{v_{om}} \right)^2 \right]^{-1/2} \tag{4.59}$$

time transient function (step response)

$$\underline{\Delta Tt_o}\overline{|\Delta v_o} = -\Delta v_o \cdot \frac{Tt_{om}}{v_{om}} \left[1 - \exp\left(-\frac{v_{om}}{l_m} \cdot t \right) \right] \tag{4.60}$$

disturbance Δv_i *(changes of the input velocity)*
amplitude frequency response

$$|G(\mathrm{j}\omega)| = \left|\frac{\widetilde{\Delta Tt_\mathrm{o}}}{\widetilde{\Delta v_\mathrm{i}}}\right| = \frac{Tt_\mathrm{om}}{v_\mathrm{im}}\left[1 + \left(\omega \cdot \frac{l_\mathrm{m}}{v_\mathrm{om}}\right)^2\right]^{-1/2} \tag{4.61}$$

time transient function (step response)

$$\underline{\Delta Tt_\mathrm{o}}|\overline{\Delta v_\mathrm{i}} = \Delta v_\mathrm{i} \cdot \frac{Tt_\mathrm{om}}{v_\mathrm{im}}\left[1 - \exp\left(-\frac{v_\mathrm{om}}{l_\mathrm{m}} \cdot t\right)\right] \tag{4.62}$$

disturbance Δz_i *(changes of the number of fibres in the cross section of the coming in sliver)*
amplitude frequency response

$$|G(\mathrm{j}\omega)| = \left|\frac{\widetilde{\Delta Tt_\mathrm{o}}}{\widetilde{\Delta z_\mathrm{i}}}\right| = \frac{Tt_\mathrm{om}}{z_\mathrm{im}}\left[1 + \left(\omega \cdot \frac{l_\mathrm{m}}{v_\mathrm{om}}\right)^2\right]^{-1/2} \tag{4.63}$$

time transient function (step response)

$$\underline{\Delta Tt_\mathrm{o}}|\overline{\Delta z_\mathrm{i}} = \Delta z_\mathrm{i} \cdot \frac{Tt_\mathrm{om}}{z_\mathrm{im}}\left[1 - \exp\left(-\frac{v_\mathrm{om}}{l_\mathrm{m}} \cdot t\right)\right] \tag{4.64}$$

disturbance ΔTt_f *(changes of the fibre fineness)*
amplitude frequency response

$$|G(\mathrm{j}\omega)| = \left|\frac{\widetilde{\Delta Tt_\mathrm{o}}}{\widetilde{\Delta Tt_\mathrm{f}}}\right| = \frac{Tt_\mathrm{om}}{Tt_\mathrm{fm}}\left[1 + \left(\omega \cdot \frac{l_\mathrm{m}}{v_\mathrm{om}}\right)^2\right]^{-1/2} \tag{4.65}$$

time transient function (step response)

$$\underline{\Delta Tt_\mathrm{o}}|\overline{\Delta Tt_\mathrm{f}} = -\Delta Tt_\mathrm{f} \cdot \frac{Tt_\mathrm{om}}{Tt_\mathrm{fm}}\left[1 - \exp\left(-\frac{v_\mathrm{om}}{l_\mathrm{m}} \cdot t\right)\right] \tag{4.66}$$

In all cases it is the question of proportional action with a delay of first order (in Sect. 4.2.1 already described for fibre formation distances) as one can see and Eqs. 4.59 to 4.62 are identical with Eqs. 4.11, 4.13, 4.17 and 4.19. Also the phase frequency responses (here not given once more) are identically with the previously given Eqs. 4.12, 4.18, 4.24 and 4.30 for the appropriate disturbances. This result means, that the normalised complex frequency responses of Fig. 4.4, the normalised amplitude and phase frequency responses of Fig. 4.5 and the normalised step response of the Fig. 4.6 are valid for the estimating of the dynamic behaviour and its graphic presentation of a sliver drafting zone in the same manner. Only the following equivalent relations are to be considered: Δq_i, q_im of the polymer fibre formation correspond to

Δz_i, z_{im} of the sliver drafting zone and $\Delta\varrho$, ϱ_m of the polymer fibre formation correspond to ΔTt_f, Tt_{fm} of the sliver drafting zone. On the other hand, Δv_o, v_{om}, Δv_i, v_{im} and Δl, z_m are fully identical for both different process steps of the fibre, respectively, yarn formation. The discussion in Sect. 4.2.1 is also consequently valid for the disturbance transmission of drafting zones in drafting systems according to the fineness of the outrunning drafted yarn. An additional hint: The disturbance sizes ΔTt_f and, of even greater importance, Δz_i characterise of course fineness disturbances of the inrunning sliver ΔTt_i whose evenness power chances can be estimated specifically through means of the appropriate (homogeneously constructed) amplitude frequency responses 4.63 and/or 4.65. A good evenness power effect through means of an increased system time constant $T_c = l_m/v_{om}$ is indeed limited at a sliver drafting system, because the distance of the drafting zone l_m cannot be selected (dependent on the staple length of the single fibres) as large as one would wish. The general observable smaller fineness fibre of long staple spun yarns compared to short staple spun yarns is to be found objectively through the viewpoint of the dynamic system. Because, longer staples enforce imperatively longer roll settings with greater time constants T_c (and smaller critical frequencies f_c) which are more effectively dampening for disturbances.

Disturbance Transfer of a Drafting System with Two Successive Drafting Zones

Disturbance Δz_i (changes of the fibre number in the input sliver). Sliver drafting systems are often created in multi stages. It is the question of how disturbances will be transmitted from the input to the different stages and how these appear finally on the output. A drafting system with two successive drafting zones according to Fig. 4.21 will be investigated in the following.

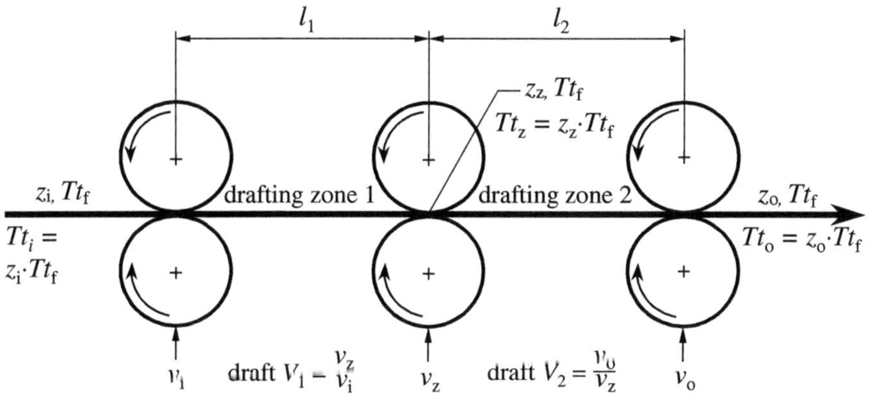

Fig. 4.21. Technological scheme of a two steps sliver drafting system

A sliver with the fineness Tt_i will be drafted to the fineness Tt_o along the drafting zones 1 and 2. The whole draft results as one knows to $V_1 \cdot V_2$ and the steady state equation is valid:

$$Tt_0 = \frac{Tt_\text{i}}{V_1 \cdot V_2} = Tt_\text{i} \cdot \frac{v_\text{i}}{v_\text{o}} \tag{4.67}$$

disturbance variant 1

There should exist a fineness disturbance at the input as a change of the fibre number in the input sliver cross sectional area Δz_i. How will ΔTt_o appear at the output of the second drafting zone? The following process and product variables are to be defined additionally compared with the previous section: The dynamic transfer functions for the drafting zones 1 and 2 must

- v_z velocity of the middle drafting godet pair (output velocity of the drafting zone 1 and simultaneously input velocity of the drafting zone 2)
- z_z number of fibres in the cross section of the coming out sliver from the drafting zone 1 and simultaneously of the coming in sliver into the drafting zone 2
- l_1 length of the drafting zone 1
- l_2 length of the drafting zone 2
- Tt_z sliver fineness at the output of the drafting zone 1 and the input of the drafting zone 2 $= z_\text{z} \cdot Tt_\text{f}$

be derived at first from the dynamic basic Eq. 4.58 which correspond to the question.
This is for the drafting zone 1:

$$G_1(p) = \frac{\Delta Tt_\text{z}}{\Delta z_\text{i}} = \frac{Tt_\text{zm}}{z_\text{im}} \cdot \frac{1}{1 + p \cdot \dfrac{l_\text{1m}}{v_\text{zm}}} \tag{4.68}$$

One considers that the output fineness of the drafting zone 1 is Tt_z and not Tt_o which at first appears at the output of the drafting zone 2. Therefore, the quantities ΔTt_o and Tt_om have been substituted by the quantities ΔTt_z and Tt_zm in Eq. 4.58 before the Eq. 4.68 has been derived.
The drafting zone 2 is valid as:

$$G_2(p) = \frac{\Delta Tt_\text{o}}{\Delta z_\text{z}} = \frac{Tt_\text{om}}{z_\text{zm}} \cdot \frac{1}{1 + p \cdot \dfrac{l_\text{2m}}{v_\text{om}}} \tag{4.69}$$

If the fineness of the fibres Tt_fm is also constantly valid as
$Tt_\text{zm} = z_\text{zm} \cdot Tt_\text{fm}$ and $\Delta Tt_\text{z} = \Delta z_\text{z} \cdot Tt_\text{fm}$
Therefore Eq. 4.68 is also to be written as:

$$G_1(p) = \frac{\Delta z_\mathrm{z} \cdot Tt_\mathrm{fm}}{\Delta z_\mathrm{i}} = \frac{z_\mathrm{zm} \cdot Tt_\mathrm{fm}}{z_\mathrm{im}} \cdot \frac{1}{1 + p \cdot \dfrac{l_\mathrm{1m}}{v_\mathrm{zm}}} \tag{4.70}$$

The dynamic transfer function of the successive systems results from the product of the transfer functions of the single systems. If one applies this law to Eqs. 4.69 and 4.70 the following will be obtained for the two steps drafting system:

$$G_\mathrm{D}(p) = G_1(p) \cdot G_2(p) = \frac{\Delta Tt_\mathrm{o}}{\Delta z_\mathrm{i}} = \frac{Tt_\mathrm{om}}{z_\mathrm{im}} \cdot \frac{1}{\left(1 + p \cdot \dfrac{l_\mathrm{1m}}{v_\mathrm{zm}}\right)\left(1 + p \cdot \dfrac{l_\mathrm{2m}}{v_\mathrm{om}}\right)} \tag{4.71}$$

The *complex frequency response* follows from this:

$$G_\mathrm{D}(\mathrm{j}\omega) = \frac{\widetilde{\Delta Tt_\mathrm{o}}}{\widetilde{\Delta z_\mathrm{i}}} \cdot e^{\mathrm{j}\varphi} = \frac{Tt_\mathrm{om}}{z_\mathrm{im}} \cdot \frac{1}{\left(1 + \mathrm{j}\omega \cdot \dfrac{l_\mathrm{1m}}{v_\mathrm{zm}}\right)\left(1 + \mathrm{j}\omega \cdot \dfrac{l_\mathrm{2m}}{v_\mathrm{om}}\right)} \tag{4.72}$$

Equation 4.72 can be split into the *amplitude frequency response*

$$|G_\mathrm{D}(\mathrm{j}\omega)| = \left|\frac{\widetilde{\Delta Tt_\mathrm{o}}}{\widetilde{\Delta z_\mathrm{i}}}\right| = \frac{Tt_\mathrm{om}}{z_\mathrm{im}} \cdot \frac{\sqrt{\left[1 - \omega^2 \cdot \dfrac{l_\mathrm{1m} \cdot l_\mathrm{2m}}{v_\mathrm{zm} \cdot v_\mathrm{om}}\right]^2 + \omega^2 \left[\dfrac{l_\mathrm{1m}}{v_\mathrm{zm}} + \dfrac{l_\mathrm{2m}}{v_\mathrm{om}}\right]^2}}{\left[1 + \left(\omega \cdot \dfrac{l_\mathrm{1m}}{v_\mathrm{zm}}\right)^2\right]\left[1 + \left(\omega \cdot \dfrac{l_\mathrm{2m}}{v_\mathrm{om}}\right)^2\right]} \tag{4.73}$$

and into the *phase frequency response*

$$\varphi(\omega) = \arctan\left[-\omega \cdot \frac{\dfrac{l_\mathrm{1m}}{v_\mathrm{zm}} + \dfrac{l_\mathrm{2m}}{v_\mathrm{om}}}{1 - \omega^2 \cdot \dfrac{l_\mathrm{1m} \cdot l_\mathrm{2m}}{v_\mathrm{zm} \cdot v_\mathrm{om}}}\right] \tag{4.74}$$

Now the *time transient function* to be calculated from Eq. 4.71 is analogous to Eq. 2.41:

$$\begin{aligned}\underline{\Delta Tt_\mathrm{o}}|\overline{\Delta z_\mathrm{i}} = \Delta z_\mathrm{i} \cdot \frac{Tt_\mathrm{om}}{z_\mathrm{im}} &\left[1 - \frac{v_\mathrm{om} \cdot l_\mathrm{1m}}{v_\mathrm{om} \cdot l_\mathrm{1m} - v_\mathrm{zm} \cdot l_\mathrm{2m}} \cdot \exp\left(-\frac{v_\mathrm{zm}}{l_\mathrm{1m}} \cdot t\right)\right] \\ - \Delta z_\mathrm{i} \cdot \frac{Tt_\mathrm{om}}{z_\mathrm{im}} &\cdot \frac{v_\mathrm{zm} \cdot l_\mathrm{2m}}{v_\mathrm{zm} \cdot l_\mathrm{2m} - v_\mathrm{om} \cdot l_\mathrm{1m}} \cdot \exp\left(-\frac{v_\mathrm{om}}{l_\mathrm{2m}} \cdot t\right)\end{aligned} \tag{4.75}$$

Equations 4.72 to 4.75 do not mediate a clear assertion about the dynamic properties of the two steps drafting system without of course an appropriate quantitative analysis. The dynamic transfer properties should be demonstrated using a numerical example. It should be assumed:

v_{im} = 1 m/min
v_{om} = 36 m/min
v_{zm} = 1.5 m/min (V_1=1.5; V_2=24) or 6 m/min (V_1=6; V_2=6)
Tt_{im} = 1000 tex
z_{im} = 3600
Tt_{fm} = 0.28 tex
Tt_{om} = 27.8 tex
z_{om} = 100
l_{2m} = 45 mm
l_{1m} = 50 mm or 40 mm

The disturbance should be step-like or sinusoidal $\Delta z_i = 180$ respectively $\widetilde{\Delta z_i} = \pm 180$.

The time constants T_{cd1} and T_{cd2} and the critical frequencies f_{c1} and f_{c2} of the drafting zones 1 and 2 according to Eqs. 2.47 and 2.49 are:

		Drafting zone 1		Drafting zone 2	
l_{1m} [mm]	v_{zm} [m/min]	T_{cd1} [s]	f_{c1} [Hz]	T_{cd2} [s]	f_{c2} [Hz]
50	1.5	2.00	0.080	0.075	2.12
50	6	0.50	0.318	0.075	2.12
40	1.5	1.60	0.099	0.075	2.12
40	6	0.40	0.398	0.075	2.12

The normalised presentations of the complex frequency response, the amplitude frequency response and the step response are similarly valid if disturbances Δz_i, ΔTt_f or also Δv_i appear at the input. The result interpretation is qualitatively and quantitatively the same for these product and process variables regarding the effect to the output fineness ΔTt_o. The phase frequency response is of course similarly valid for all three disturbances.

The transfer locus of the complex frequency response, the amplitude frequency response and the phase frequency response of a two steps sliver drafting system are shown in Figs. 4.22, 4.23 and 4.24. The amplitude and phase diagrams include in each case through agreement two additional curves (drawn dotted) of a one step drafting zone. The appropriate step response is shown in Fig. 4.25.

The following statements can be taken from the diagrams:

a) The evenness power effect is more exact the more unsymmetrical the whole draft is divided between the drafting zones 1 and 2. And indeed, the

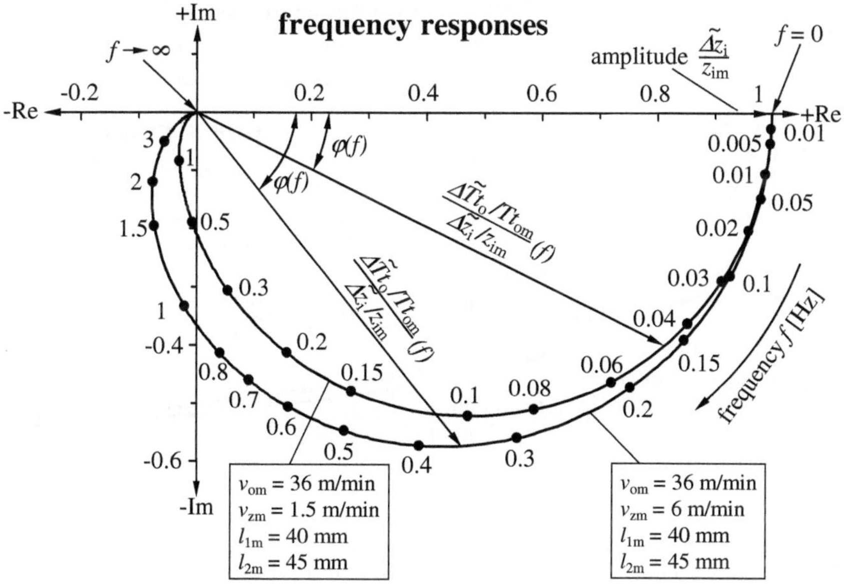

Fig. 4.22. Transfer locuses of the normalised complex frequency responses of fineness changes $\widetilde{\Delta Tt_o}$ in the coming out sliver caused by changes of the number of fibres in the coming in sliver $\widetilde{\Delta z_i}$ at a two steps sliver drafting system

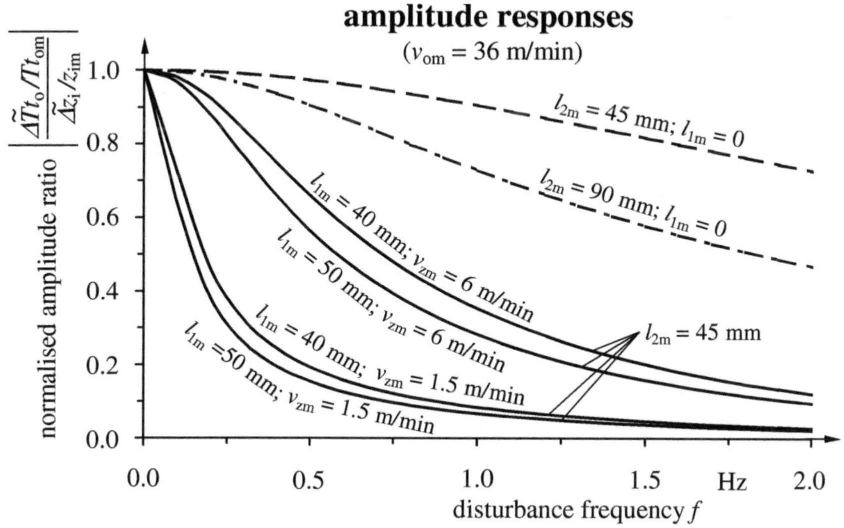

Fig. 4.23. Normalised amplitude frequency responses of fineness changes $\widetilde{\Delta Tt_o}$ in the coming out sliver caused by changes of the number of fibres in the coming in sliver $\widetilde{\Delta z_i}$ at a two steps sliver drafting system

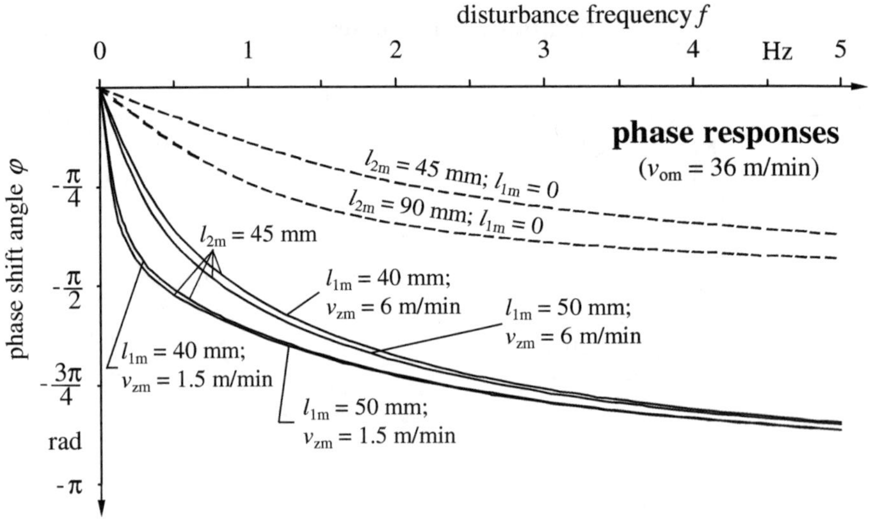

Fig. 4.24. Normalised phase frequency responses of fineness changes $\widetilde{\Delta Tt_o}$ in the coming out sliver caused by changes of the number of fibres in the coming in sliver $\widetilde{\Delta z_i}$ at a two steps sliver drafting system

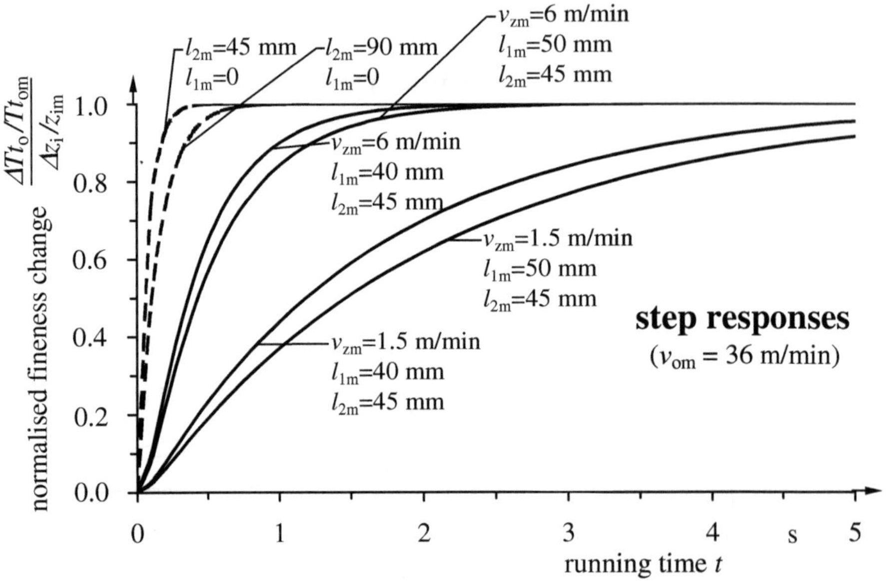

Fig. 4.25. Normalised step response of the fineness changes ΔTt_o of the coming out sliver caused by changes of the number of fibres in the coming in sliver Δz_i at an one and two steps sliver drafting system

greater part of the draft should be created in the drafting zone 2. This is to be seen by the amplitude frequency response courses which fall down to the abscissa at already clearly smaller disturbance frequencies for $v_z = 1.5$ m/min than for $v_z = 6$ m/min. This tendency can also be seen in the step transfer behaviour. The curves of $v_z = 1.5$ m/min clearly increase more slowly than those for $v_z = 6$ m/min.

b) A change of the first drafting zone length by 10 mm (from a starting point of 40 or 50 mm) does not effect aggravating changes of the dynamic behaviour. The longer drafting zone 1 improves the dampening behaviour for disturbances only insignificantly. A longer drafting zone 2 by same or similar quantities is not practically of influence to the dynamic transmission behaviour (not drawn here because appropriate diagram curves are nearly identical).

c) The comparison with the dropped drawn curves for the one step drafting system (following from the two steps system if l_{1m} is set to zero; compare also Eqs. 4.73 and 4.75 with 4.63 and 4.64) shows that a multi step drafting system (here demonstrated using the two steps system) a more effective evenness power is allowed than a one step drafting system for the fineness of the output sliver in a greater frequency range. In our example the two steps drafting system is able to suppress input disturbances with > 1 Hz to the fineness effectively if the draft is divided skillfully. This would mean for the output sliver in our example that a short periodic unevennesses of sliver length < 60 cm would not appear anymore in a troublesome manner. The same effect would occur for the comparable short one step drafting system only at > 15 Hz, appropriate < 9 m sliver length.

Disturbance Δv_z (velocity changes of the middle drafting roll pair). There should exist step-like or periodic disturbances of the velocity of the middle drafting godet pair v_z. How will ΔTt_o appear at the output of the drafting system?

disturbance variant 2

The mean value of the output fineness Tt_{om} will not change even after longtime disturbances of v_z, because in this case the whole draft does not change steady state. But short time ΔTt_o-effects will appear. These consist of two primary effective transmission processes: Δv_z effects at first a change of the fineness Tt_z at the output of the drafting zone 1, which is simultaneously an input fineness disturbance of the drafting zone 2. Second, Δv_z effects an input velocity disturbance of the drafting zone 2. ΔTt_o consists of the two parts ΔTt_{o1} and ΔTt_{o2} which are consequently to be added according to the superposition law. ΔTt_{o1} describes the disturbance transmission of ΔTt_z to

ΔTt_{o} and ΔTt_{o2} describes the disturbance transmission of Δv_{z} to ΔTt_{o}.

The dynamic transfer behaviour is to be derived from the dynamic transfer functions of the single part drafting zones.

effect of Δv_{z} to ΔTt_{z}:

$$G_1(p) = \frac{\Delta Tt_{\text{z}}}{\Delta v_{\text{z}}} = -\frac{Tt_{\text{zm}}}{v_{\text{zm}}} \cdot \frac{1}{1 + p \cdot \dfrac{l_{\text{1m}}}{v_{\text{zm}}}} \tag{4.76}$$

effect of ΔTt_{z} to ΔTt_{o1}:

$$G_2(p) = \frac{\Delta Tt_{\text{o1}}}{\Delta Tt_{\text{z}}} = \frac{Tt_{\text{om}}}{Tt_{\text{zm}}} \cdot \frac{1}{1 + p \cdot \dfrac{l_{\text{2m}}}{v_{\text{om}}}} \tag{4.77}$$

effect of Δv_{z} to ΔTt_{o1} = product of Eqs. 4.76 and 4.77:

$$G_3(p) = \frac{\Delta Tt_{\text{o1}}}{\Delta v_{\text{z}}} = -\frac{Tt_{\text{om}}}{v_{\text{zm}}} \cdot \frac{1}{\left(1 + p \cdot \dfrac{l_{\text{1m}}}{v_{\text{zm}}}\right)\left(1 + p \cdot \dfrac{l_{\text{2m}}}{v_{\text{om}}}\right)} \tag{4.78}$$

effect of Δv_{z} to ΔTt_{o2}:

$$G_4(p) = \frac{\Delta Tt_{\text{o2}}}{\Delta v_{\text{z}}} = \frac{Tt_{\text{om}}}{v_{\text{zm}}} \cdot \frac{1}{1 + p \cdot \dfrac{l_{\text{2m}}}{v_{\text{om}}}} \tag{4.79}$$

The total ΔTt_{o} is the sum of ΔTt_{o1} and ΔTt_{o2}. The total transfer function $G_{\text{D}}(p)$ is for this disturbance case the sum of $G_3(p)$ and $G_4(p)$ according to Eqs. 4.78 and 4.79:

$$G_{\text{D}}(p) = \frac{\Delta Tt_{\text{o1}}}{\Delta v_{\text{z}}} = \frac{Tt_{\text{om}}}{v_{\text{zm}}} \cdot \frac{p \cdot \dfrac{l_{\text{1m}}}{v_{\text{zm}}}}{\left(1 + p \cdot \dfrac{l_{\text{1m}}}{v_{\text{zm}}}\right)\left(1 + p \cdot \dfrac{l_{\text{2m}}}{v_{\text{om}}}\right)} \tag{4.80}$$

Equation 4.80 shows that a differential action with delay of second order lies before (no permanent fineness shift after a steady state disturbance). It is the starting point for the equations of the frequency, amplitude, phase and step response which can be derived according to the already cited manifold rules of Sect. 2.5.2.

complex frequency response

$$G_{\text{D}}(\text{j}\omega) = \frac{\widetilde{\Delta Tt_{\text{o1}}}}{\widetilde{\Delta v_{\text{z}}}} = \frac{Tt_{\text{om}}}{v_{\text{zm}}} \cdot \frac{\text{j}\omega \cdot \dfrac{l_{\text{1m}}}{v_{\text{zm}}}}{\left(1 + \text{j}\omega \cdot \dfrac{l_{\text{1m}}}{v_{\text{zm}}}\right)\left(1 + \text{j}\omega \cdot \dfrac{l_{\text{2m}}}{v_{\text{om}}}\right)} \tag{4.81}$$

amplitude frequency response

$$|G_{\mathrm{D}}(\mathrm{j}\omega)| = \left| \frac{\widetilde{\Delta Tt_{\mathrm{o}}}}{\widetilde{\Delta v_{\mathrm{z}}}} \right|$$

$$= \frac{Tt_{\mathrm{om}}}{v_{\mathrm{zm}}} \cdot \frac{\omega \cdot \dfrac{l_{1\mathrm{m}}}{v_{\mathrm{zm}}} \sqrt{\left[1 - \omega^2 \cdot \dfrac{l_{1\mathrm{m}} \cdot l_{2\mathrm{m}}}{v_{\mathrm{zm}} \cdot v_{\mathrm{om}}}\right]^2 + \omega^2 \left[\dfrac{l_{1\mathrm{m}}}{v_{\mathrm{zm}}} + \dfrac{l_{2\mathrm{m}}}{v_{\mathrm{om}}}\right]^2}}{\left[1 + \left(\omega \cdot \dfrac{l_{1\mathrm{m}}}{v_{\mathrm{zm}}}\right)^2\right] \left[1 + \left(\omega \cdot \dfrac{l_{2\mathrm{m}}}{v_{\mathrm{om}}}\right)^2\right]} \qquad (4.82)$$

phase frequency response

$$\varphi(\omega) = \mathrm{arc\ tan} \left[\frac{1 - \omega^2 \cdot \dfrac{l_{1\mathrm{m}} \cdot l_{2\mathrm{m}}}{v_{\mathrm{zm}} \cdot v_{\mathrm{om}}}}{\omega \left(\dfrac{l_{1\mathrm{m}}}{v_{\mathrm{zm}}} + \dfrac{l_{2\mathrm{m}}}{v_{\mathrm{om}}} \right)} \right] \qquad (4.83)$$

step response

$$\underline{\Delta Tt_{\mathrm{o}}|\overline{\Delta v_{\mathrm{z}}}} = \Delta v_{\mathrm{z}} \cdot \frac{Tt_{\mathrm{om}}}{v_{\mathrm{zm}}} \cdot \frac{\exp\left(-\dfrac{l_{1\mathrm{m}}}{v_{\mathrm{zm}}} \cdot t\right) - \exp\left(-\dfrac{l_{2\mathrm{m}}}{v_{\mathrm{om}}} \cdot t\right)}{1 - \dfrac{l_{2\mathrm{m}} \cdot v_{\mathrm{zm}}}{l_{1\mathrm{m}} \cdot v_{\mathrm{om}}}} \qquad (4.84)$$

The transfer locus of the frequency, amplitude and phase frequency responses of a two steps sliver drafting system for the disturbance Δv_{z} are shown in Figs. 4.26, 4.27 and 4.28. The step response is shown in Fig. 4.29.

Some statements can also be given here in principle:

a) Extremely short periodic as well as long periodic, fluctuations of the velocity of the middle drafting godet pair also effect the output fineness dampened strength (see amplitude frequency response curves). The range of the maximum disturbance transmission in the example is situated between about 0.5 and 1 Hz. Relative periodic Δv_{z}-changes have nearly the same great ΔTt_{o}-changes and are actually for strong unsymmetric draft partition ($v_{\mathrm{zm}} = 1.5$ m/min) somewhat greater than for less unsymmetric draft partition ($v_{\mathrm{zm}} = 6$ m/min).

b) This disturbance type Δv_{z} does not cause permanent fineness shifts after a step-like disturbance (see step response). The old fineness mean value Tt_{om} will be reached after a sufficient long time (here > 5 s) also if the disturbance continues. The drafting system further operates with increased

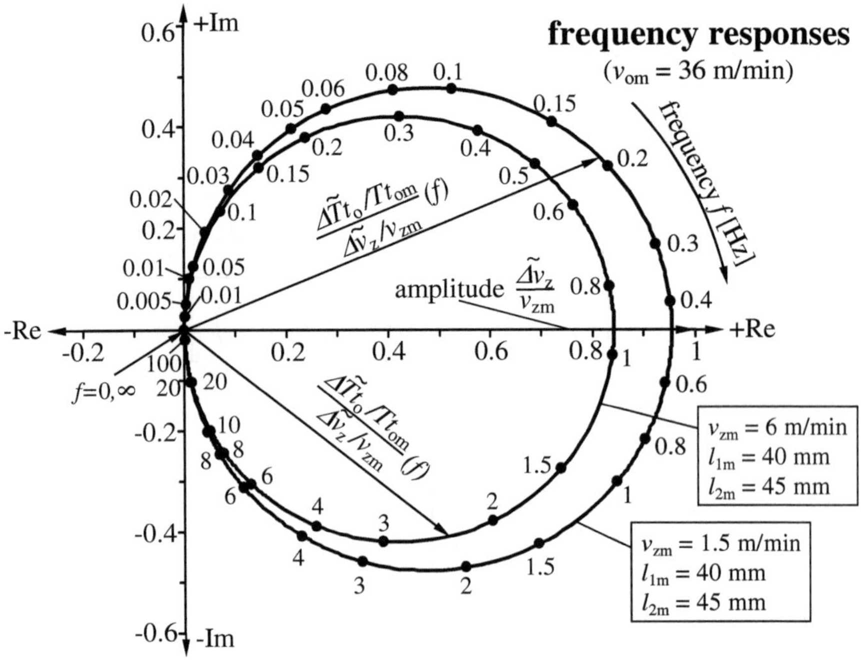

Fig. 4.26. Transfer locuses of the normalised complex frequency responses of fineness changes $\widetilde{\Delta Tt_o}$ in the coming out sliver caused by velocity changes $\widetilde{\Delta v_z}$ of the middle godets at a two steps sliver drafting system

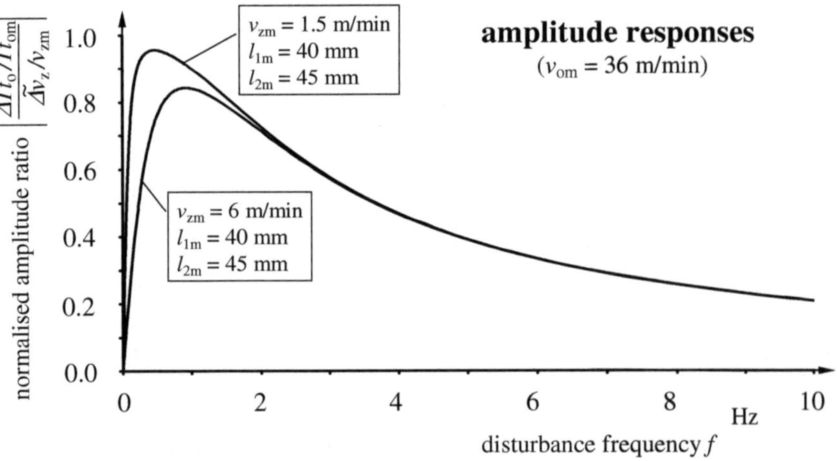

Fig. 4.27. Normalised amplitude frequency responses of fineness changes $\widetilde{\Delta Tt_o}$ in the coming out sliver caused by velocity changes $\widetilde{\Delta v_z}$ of the middle godets at a two steps sliver drafting system

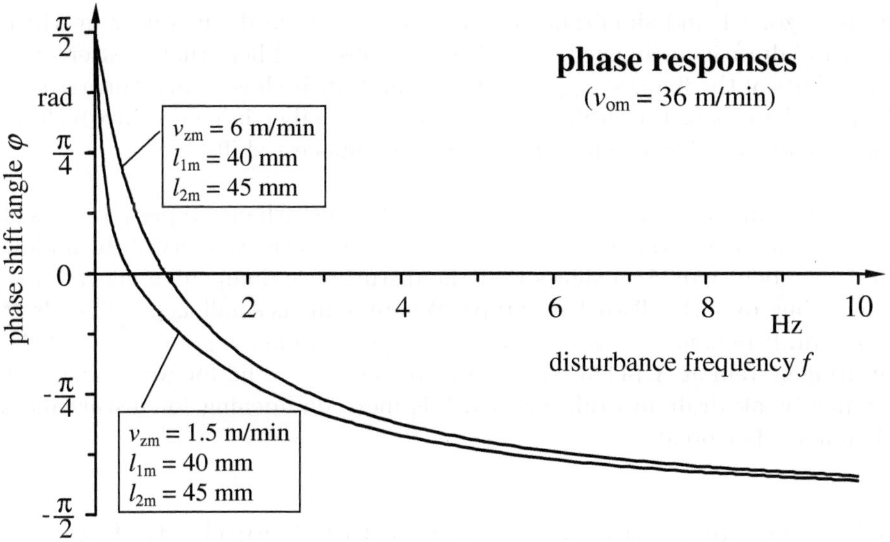

Fig. 4.28. Normalised phase frequency responses of fineness changes $\widetilde{\Delta Tt_o}$ in the coming out sliver caused by velocity changes $\widetilde{\Delta v_z}$ of the middle godets at a two steps sliver drafting system

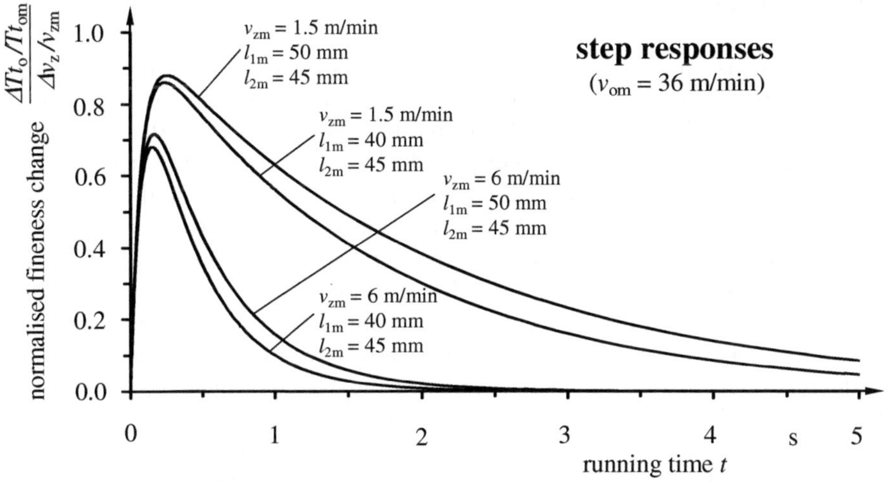

Fig. 4.29. Normalised step responses of the fineness changes ΔTt_o of the coming out sliver caused by changes of the number of fibres in the coming in sliver Δv_z at a one and two steps sliver drafting system

draft in zone 1 and simultaneously with decreased draft in zone 2 in which the total draft is unchanged. It is also to be observed here that greater short time shifts of the fineness appear if the total draft is chosen unsymmetrically. Length changes of the drafting zone 1 have a smaller influence in which an increase of l_{1m} effects a small increase of the fineness shift.

c) A result comparison of the demonstrated disturbance types (changes of the fibre number of the input sliver Δz_i or changes of the velocity of the middle drafting godet pair Δv_z) shows that the disturbance dampening effect varies depending upon the disturbance type. Whereas an, as small as possible, draft of the draft in zone 1 is an advantage for a good evenness power effect of the drafting system at Δz_i-changes the inverted effect is valid for Δv_z-changes. A greater break draft in drafting zone 1 is more dampening for disturbances than a smaller break.

4.5 Necessary Measuring and Gauge Lengths to the Proof of Dynamic Disturbances in Yarns

The presentation until now has shown that dynamic disturbances of different kinds (periodic, aperiodic, at different technological operating points) of different process and product variables effect yarn disturbances along different yarn lengths. It should be questioned as to which basic totality of (at one working position) manufactured yarn length is to be included into a continuous or discontinuous test in the textile test lab. What cut length is to be selected for the continuous single tests to decisively describe dynamic disturbance effects in the yarn?

One can use a well-known rule of thumb for the electric measurement technique to technically prove test disturbances by means of discontinuous measurements. This is possible if the frequency of the tracing disturbance is known and if this frequency is suitable for a measurable effect to the yarn variable (for instance the fineness) according to the transfer and dampening properties of the system. The latter mentioned are to estimate simply by means of the amplitude frequency response or also roughly for the dominating system time constant T_c or for the critical frequency f_c.

The named rule of the electric measurement technique means that the carrier frequency of the amplitude modulation (a well-known procedure of the analogous transmission of electrical measuring signals) must be at least five times greater than the highest awaiting frequency of the measuring size. If this condition is fulfilled then a true to the form reproduction and transmission is given of the dynamic measuring size. This means, in other words, that the cycle duration for an oscillation of the carrier frequency must be included at least five times in the cycle duration of the shortest awaiting oscillation of the measuring size [5].

This rule (applied to the presented problem of the gauge and measuring length estimation) means that the yarn length (which has been exactly manufactured or processed during a full oscillation of the disturbance) is to be divided into at least five equal parts for further (for instance statistical) estimations.

If the disturbance frequency f and the take-down velocity v_o are known, then the *wavelength* of the disturbance λ_f in the fibre or yarn amounts to:

$$\lambda_f = \frac{v_o}{f} \tag{4.85}$$

The *necessary cut length* L_{cl} for the named single tests can be derived from Eq. 4.85 as follows:

$$L_{cl} = \frac{1}{n} \cdot \frac{v_o}{f} = \frac{\lambda_f}{n} \tag{4.86}$$

condition: $n \geq 5$

It is not enough to only select as *necessary gauge length* L_{gl} the yarn length of one cycle duration of the disturbance. Rather it is recommended that the basic totality of the gauge length L_{gl} to select be at least so large that five full disturbance oscillations are included. The *necessary gauge length* L_{gl} for fibre or yarn testings can be recommended consequently to:

$$L_{gl} = n \cdot \frac{v_o}{f} = n \cdot \lambda_f \tag{4.87}$$

condition: $n \geq 5$

According to Eqs. 4.86 and 4.87, 25 continuous single tests of yarn pieces with the cut length L_{cl} would be necessary to prove an expected disturbance frequency. Typical disturbance frequency ranges and the necessary yarn lengths L_{gl} and L_{cl} for their proof (which can be derived from the wavelength of a full disturbance oscillation λ_f in the shown manner) are collected in Table 4.4 for a row of disturbance causes in the polymer and glass spinning processes. Questions about the named aspects could not be answered if fixed standard gauge lengths or cut lengths were used (fineness testing 100 m or stress strain testing with specimen length of 0.5 m).

However, if the named conditions are fulfilled the following remarks are to be added according to the assertion of statistical characteristics as quadratic dispersion or variance and coefficient of variation. These characteristics do not express the absolute shifts around the mean values of the concerned variables in consequence to their integral calculation laws. If conditions are being observed for the decision of the yarn lengths L_{gl} and L_{cl} according to

Eqs. 4.86 and 4.87 (to investigate an appointed disturbance cause of known frequency f) then the coefficient of variation will show sinusoidal shifts of approximately 0.7 times around the mean value only (exactly $\sqrt{2}/2$ times) for real existing maximum shift amplitudes.

It has been shown in Sect. 2.5.2 that the disturbance frequency f of process and product variables essentially effect the interesting product variables of the final product only if
- the dynamic transmission behaviour is shown with a considerable transfer or amplification factor and
- the disturbance frequency f is located in an appointed range, which depends on the critical frequency f_c of the process line.

The last condition results in a practical undamped dynamic transmission of a cause disturbance to the effect variable by a *proportional action with delay of first order*

$$f \leq f_c \tag{4.88}$$

It should be remembered that proportional action with delay of first order takes place in the fibre formation of melt spinning for the transmission of Δv_o-, Δv_i-, $\Delta \varrho$- or Δq_i-disturbances to ΔTt_o-shifts.

If *differential action with delay of first order* lies before, on the contrary then the condition reads for undamped disturbance transmission as:

$$f \geq f_c \tag{4.89}$$

Differential action with delay of first order exist for instance for Δl-disturbances, in melt spinning and its effect to ΔTt_o-shifts.

In Table 4.5 time constants T_c and critical frequencies f_c (calculated by means of Eqs. 2.47 and 2.49 are combined on the basis of the output velocity v_{om} and length of the process line l_m) for some typical yarn formation and processing lines. The collection can support estimations to possible effects of periodic disturbances with known frequency f for variables of the yarn at the output of a yarn formation or processing line.

Table 4.4. Disturbance causes, disturbance frequencies and disturbance wavelengths in the fibre or yarn of melt spinning processes; necessary gauge and cut lengths for the prove of disturbances (v_{om} = take-down velocity in m/min)

Cause of the disturbances at the spinning equipment	Disturbance frequency f [Hz]	Wavelength in the fibre or yarn λ_f [m]	Necess. gauge length L_{gl} [m]	Necess. cut length L_{cl} at discontinuous testing [m]
Melt spinning of polymers				
Bobbin changing	$(1.4...5.6) \cdot 10^{-4}$	$(12...3) \cdot 10 \cdot v_{om}$	$\geq (6..1.5) \cdot 10^2 \cdot v_{om}$	$\leq (24...6) \cdot v_{om}$
Melt temperature	$(1.7...17) \cdot 10^{-3}$	$(10...1) \cdot v_{om}$	$\geq (50..5) \cdot v_{om}$	$\leq (20...2) \cdot 10^{-1} \cdot v_{om}$
Oiling disk	$(1.7...3.3) \cdot 10^{-1}$	$(10...5) \cdot 10^{-2} \cdot v_{om}$	$\geq (5..2.5) \cdot 10^{-1} \cdot v_{om}$	$\leq (2...1) \cdot 10^{-2} \cdot v_{om}$
Spinning pump	$(1.7...6.7) \cdot 10^{-1}$	$(10...2.5) \cdot 10^{-2} \cdot v_{om}$	$\geq (5..1.25) \cdot 10^{-1} \cdot v_{om}$	$\leq (20...5) \cdot 10^{-3} \cdot v_{om}$
Quenching air	$(5...50) \cdot 10^{-1}$	$(33...3.3) \cdot 10^{-3} \cdot v_{om}$	$\geq (16.7...1.7) \cdot 10^{-2} \cdot v_{om}$	$\leq (67...6.7) \cdot 10^{-4} \cdot v_{om}$
Traverse motion	$(1...10) \cdot 10^{-1}$	$(17...1.7) \cdot 10^{-4} \cdot v_{om}$	$\geq (85...8.5) \cdot 10^{-4} \cdot v_{om}$	$\leq (33...3.3) \cdot 10^{-5} \cdot v_{om}$
Melt spinning of glass				
Bobbin changing	$(5.7...16.7) \cdot 10^{-4}$	$(3...1) \cdot 10 \cdot v_{om}$	$\geq (15..5) \cdot 10 \cdot v_{om}$	$\leq (6...2) \cdot v_{om}$
Main power (melt temp.)	$\leq 4 \cdot 10^{-1}$	$\geq 4.2 \cdot 10^{-2} \cdot v_{om}$	$\geq 2.1 \cdot 10^{-1} \cdot v_{om}$	$\leq 8.4 \cdot 10^{-3} \cdot v_{om}$
Throughput spinneret				
- without T_s-control	$\leq 1 \cdot 10^{-3}$	$\geq 1.7 \cdot 10^1 \cdot v_{om}$	$\geq 8.5 \cdot 10^1 \cdot v_{om}$	$\leq 3.4 \cdot v_{om}$
- with T_s-control	$\leq 1 \cdot 10^{-4}$	$\geq 1.7 \cdot 10^2 \cdot v_{om}$	$\geq 8.5 \cdot 10^2 \cdot v_{om}$	$\leq 3.4 \cdot 10 \cdot v_{om}$
- quick traverse	$\approx 1.751 \cdot 10^1$	$\approx 9.5 \cdot 10^{-4} \cdot v_{om}$	$\geq 4.75 \cdot 10^{-3} \cdot v_{om}$	$\leq 1.9 \cdot 10^{-4} \cdot v_{om}$
	$\approx 3.5 \cdot 10^1$	$\approx 4.8 \cdot 10^{-4} \cdot v_{om}$	$\geq 2.4 \cdot 10^{-3} \cdot v_{om}$	$\leq 9.6 \cdot 10^{-5} \cdot v_{om}$
- slow traverse	$\approx 3.3 \cdot 10^{-2}$	$\approx 5 \cdot 10^{-1} \cdot v_{om}$	$\geq 2.5 \cdot v_{om}$	$\leq 1 \cdot 10^{-1} \cdot v_{om}$

Table 4.5. Quantities of time constants and critical frequencies of different fibre formation and fibre processing process lines

Process step	Output velocity of the process step v_{om} [m/s]	Formation length of the process line l_m [m]	Time constant of the process line T_c [s]	Critical frequency of the process line f_c [Hz]
Man-made fibre manufacturing				
Spinning classically and spin-draw-winding	$(1.5...2) \cdot 10$	4...6	$(2...4) \cdot 10^{-1}$	$(8...4) \cdot 10^{-1}$
Spinning, high speed spinning	$(5...10) \cdot 10$	2...4	$(2...8) \cdot 10^{-2}$	8...2
Draw, classically	$(1.2...1.5) \cdot 10$	$3.5...4.5 \cdot 10^{-1}$	$(2.3...3.8) \cdot 10^{-2}$	6.9...4.2
Draw, spin-draw-winding	$(5...10) \cdot 10$	$4.5...5.5 \cdot 10^{-1}$	$(4.5...11) \cdot 10^{-3}$	$(3.5...1.5) \cdot 10$
False twisting, texturing	2.5...10	1...2	$(1...8) \cdot 10^{-1}$	$(16...2) \cdot 10^{-1}$
Yarn manufacturing				
Draw, slivers and rovings	1...8	$(1...5) \cdot 10^{-1}$	$(1.3...50) \cdot 10^{-2}$	12.2...0.32
Draw, ring frame	$(3...10) \cdot 10^{-1}$	$(5...20) \cdot 10^{-2}$	$(5...67) \cdot 10^{-2}$	$(32...2.4) \cdot 10^{-1}$

5. Dynamics of Fibre Processing Processes

5.1 Dynamics of the Fibre Transport

5.1.1 Task

Each fibre manufacture and processing process is connected to the dynamic basic principle of the transport. Neither fibre formation nor fibre processing are possible without continuously running down transport operations, which are realised as a rule, by means of rotations for specific machine tools. Fibres or yarns are not able to pick up axial pressure forces. Therefore, a continuous fibre transport is only possible if a tensile force is produced and permanently maintained in the transported fibre or yarn. A controlled refining of the yarn (for instance in the draw process of man-made fibres) can even be effected by means of this tensile force besides the pure transport and processing (for instance twisting, package winding, sectional warping, texturing). On the one hand, specially man-made fibres are very sensitive to tensile force fluctuations during their processing. On the other, some disturbed process and product variables (caused for instance by thread guide elements, preparation oils, unbalanced machine tools, fluctuating fibre material itself) permanently influence the quantity of the tensile force. It is obvious that the dynamic analysis and the dynamic modelling of the basic process fibre transport is from fundamental and primary importance for the process analysis of the technological fibre formation and processing processes.
Some elaborated model ideas to this set of problems are presented in the following. These are valid for small deformation ranges (elongation demand $< 10\%$) in which current linearity is between tensile force and elongation in the first approximation. Nonlinear deformation processes (with changing fibre deformation characteristics along the fibre length or with essential plastic deformation parts) are excluded.
The investigations can be used as a basis for the treatment of the following problems:

- *process synthesis oriented* (for instance statements to favourable or unfavourable arrangement and design of machine tools in the thread line),
- *process analysis oriented* (for instance influence of thread guide elements to the thread unevenness, reaction of measuring sensors to the thread) or even

- *thread test oriented* (for instance questions in relationship with the frictional test or with the unevenness measuring of threads).

Besides our own detailed paper to fibre transport modelling [291] the reader should be referred to the complementing three papers dealing with the problem of running elastic fabrics between driven rolls [292]- [294]; [295] is also recommended.
The analysis of any continuous running thread line in a fibre formation, fibre processing or fibre testing machine or apparatus shows that three different thread strain lines (arranged in different manners of series) can be found: These are the *delay thread line*, the *dead time thread line* and the *friction thread line.*
The dynamic modelling of any realised thread transport process therefore supposes knowledge and solutions of the dynamic description equations (that means the DEq.) for the three named typical thread transport strain lines.

5.1.2 Dynamic Model to the Description of a Delay Thread Line

A *delay thread line* is in the present relationship a part of a transported, under a tensile force situated, thread with the length l which runs into this part with the velocity v_{i} and the fineness Tt_{i} and comes out of this part with the velocity v_{o} and the fineness Tt_{o}. Here: $v_{\mathrm{o}} > v_{\mathrm{i}}$ (trivial condition for the improvement of a tensile force) is valid and each fineness change comes off evenly at each point along this part. The last prerequisite is a simplification for the transported thread in the settled limited strain range which the thread attributed rubber elastic behaviour.

Figure 5.1 (already known in a similar form, compare Figs. 2.5 and 4.20) shows a possible practical realisation for the given definition. At this, one can also imagine, the backing-off point of a thread from a cop which is transported along a determined free way distance into the rotating input element of a thread processing machine.

The linearised DEq. (the mathematical dynamic model) of such a delay thread line can be deduced naturally from the basic Eqs. 2.23 and 4.7 to:

$$(v_{\mathrm{om}}+p\cdot l_{\mathrm{m}})\cdot\Delta Tt_{\mathrm{o}}+Tt_{\mathrm{om}}\cdot\Delta v_{\mathrm{o}}-v_{\mathrm{im}}\cdot\Delta Tt_{\mathrm{i}}-Tt_{\mathrm{im}}\cdot\Delta v_{\mathrm{i}}+p\cdot Tt_{\mathrm{om}}\cdot\Delta l=0 \quad (5.1)$$

Equation 5.1 can be transmitted easily to the dynamic transfer function (repeatedly explained in Chap. 4 before) from which all further dynamic functions in the frequency and time range are calculable.
It is easy to state that after some practice in the handling of DEqs. in the operator style of writing and their solution functions that the dynamic behaviour of a delay thread line is practically identical to the already fully described behaviour of the fibre formation distance in melt spinning polymers for the product variable fibre fineness (see also Sect. 4.2.1). All resulting equations and their normalised presentations (Figs. 4.4 to 4.6) are valid, similarly, for the transport delay line. It is simple to substitute the variables q_{i}, q_{im} and Δq_{i}

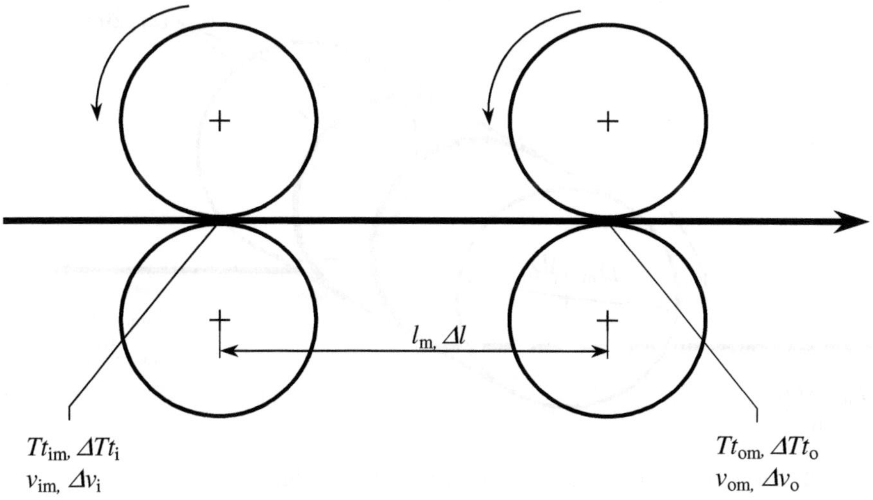

Fig. 5.1. Technological scheme of a delay thread line

by Tt_i, Tt_{im} and ΔTt_i in Sect. 4.2.1. $\Delta \varrho = 0$ is to be set additionally because this variable is not of interest here.

In the following practical example of Sect. 5.1.5 we will fall back upon Eq. 5.1 and its application will be demonstrated.

5.1.3 Dynamic Model to the Description of a Dead Time Thread Line

A *dead time thread line* is in the present relationship a part of a transported, under a tensile force situated, thread with the length l which runs into this part with the velocity v_i and the fineness Tt_i and comes out of this part with the velocity v_o and the fineness Tt_o. Here: $v_o = v_i$ is valid and the thread does not suffer fineness changes inside this part.
This definition says that the input thread fineness appears unchanged at the output after running through this part (after the so-called dead time T_d has expired). Under this condition:
mass inflow per time $= Tt_i \cdot v_i$
mass discharge per time $= Tt_o \cdot v_o$
with the both boundary conditions

$$v_o = v_i \tag{5.2}$$

$$Tt_o = Tt_i \cdot \exp(-p \cdot T_d) \tag{5.3}$$

Figure 5.2 shows a realisation of a thread (transport) dead time line. Equation 5.3 expresses the dead time relation between Tt_o and Tt_i in the

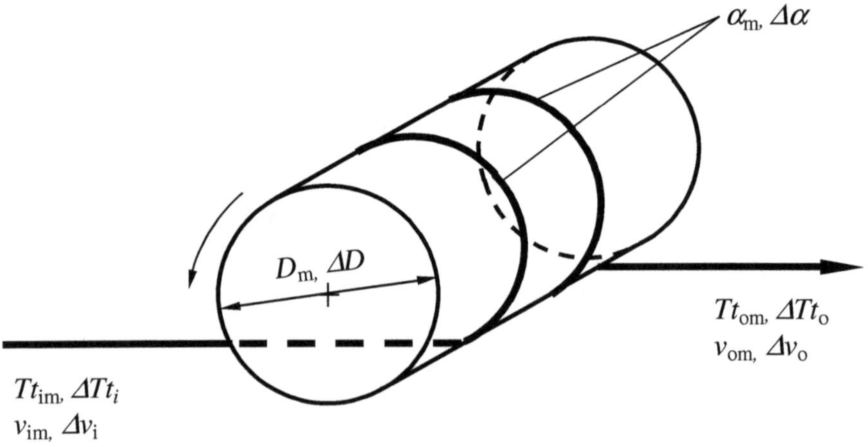

Fig. 5.2. Technological scheme of a dead time thread line

style of operator writing. It is the dynamic transfer equation for dead time elements known from the automatic control engineering (see for instance [10]). If the dead time T_{d} is realised by the thread wrap around a cylindrical thread transport element (for instance a wrapped godet which is a typical dead time transport element) with the diameter D, the angle of wrap α (in radian measure) and the circumference velocity v_{i} can then be written as:

$$T_{\mathrm{d}} = \frac{D \cdot \alpha}{2 \cdot v_{\mathrm{i}}} \tag{5.4}$$

One gets the dynamic model equation (first of all nonlinearly according to the variables Tt_{i}, Tt_{o}, v_{i}, D and α) for the dead time thread line by putting Eq. 5.4 into 5.3:

$$\Phi = Tt_{\mathrm{o}} - Tt_{\mathrm{i}} \cdot \exp\left(-p \cdot \frac{D \cdot \alpha}{2 \cdot v_{\mathrm{i}}}\right) = 0 \tag{5.5}$$

Linearisation by means of partial differentiation then results in the linearised final motion DEq. of the dead time thread line to:

$$\begin{aligned}
&\Delta Tt_{\mathrm{o}} - \exp\left(-p \cdot \frac{D_{\mathrm{m}} \cdot \alpha_{\mathrm{m}}}{2 \cdot v_{\mathrm{im}}}\right) \cdot \Delta Tt_{\mathrm{i}} \\
&-p \cdot \frac{Tt_{\mathrm{im}} \cdot D_{\mathrm{m}} \cdot \alpha_{\mathrm{m}}}{2 \cdot v_{\mathrm{im}}^2} \cdot \exp\left(-p \cdot \frac{D_{\mathrm{m}} \cdot \alpha_{\mathrm{m}}}{2 \cdot v_{\mathrm{im}}}\right) \cdot \Delta v_{\mathrm{i}} \\
&+p \cdot \frac{Tt_{\mathrm{im}} \cdot \alpha_{\mathrm{m}}}{2 \cdot v_{\mathrm{im}}} \cdot \exp\left(-p \cdot \frac{D_{\mathrm{m}} \cdot \alpha_{\mathrm{m}}}{2 \cdot v_{\mathrm{im}}}\right) \cdot \Delta D \\
&+p \cdot \frac{Tt_{\mathrm{im}} \cdot D_{\mathrm{m}}}{2 \cdot v_{\mathrm{im}}} \cdot \exp\left(-p \cdot \frac{D_{\mathrm{m}} \cdot \alpha_{\mathrm{m}}}{2 \cdot v_{\mathrm{im}}}\right) \cdot \Delta\alpha = 0
\end{aligned}$$

(5.6)

Equation 5.6 can be transmitted into the, in each case interesting, dynamic transfer function for determined concrete cause-effect-questions. From the latter the frequency, the amplitude and the phase frequency responses are calculable. The effect variable output thread fineness Tt_o will surely be of special interest in most cases. It is to be remarked that the calculation of the step response on the basis of the transfer function (by means of the residue theorem of the LAPLACE-transformation) is not possible here because the step response function of a dead time thread line is not a continuous function. The practical application of Eq. 5.6 will be demonstrated by means of an example in Sect. 5.1.5.

5.1.4 Dynamic Model to the Description of a Friction Thread Line

A *friction thread line* is in the present relationship a part of a transported, under a tensile force situated, thread with the length l which runs into this part with the velocity v_i and the fineness Tt_i and comes out of this part with the velocity v_o and the fineness Tt_o. Thus the following statement is valid: The tensile force of the thread increases permanently along the part according to the tensile friction law and the thread fineness decreases (a constant E-modulus of the thread material is presumed) permanently according to the same law (in a modified manner).
Friction lines appear practically on all fixed thread guides at the thread transport. They are as you know even consciously designed parts of the appropriate friction test apparatuses. The technological scheme of such a friction thread line is shown in Fig. 5.3. It is at first necessary to determine the thread mass on the friction element to the development of the continuity equation for the dynamic working case. The following is valid for the tensile force in the input and output thread (F_i and F_o):

$$F_o = F_i \cdot e^{\mu \cdot \alpha} \tag{5.7}$$

$e^{\mu \cdot \alpha}$ rope friction factor
μ coefficient of friction
α angle of wrap (in radian measure)

The tensile force within a thread (which possesses the fineness Tt_{zm} before the force influence on it and the fineness Tt_i during the force influence on it) can be written as:

$$F_i = A_{zm} \cdot Tt_{zm} \left[\frac{Tt_{zm}}{Tt_i} - 1 \right] \tag{5.8}$$

A_{zm} to the thread fineness related rise of the force-elongation-curve of the thread; dimension: force $\cdot$ fineness^{-1} $\cdot$ (relative length change)$^{-1}$

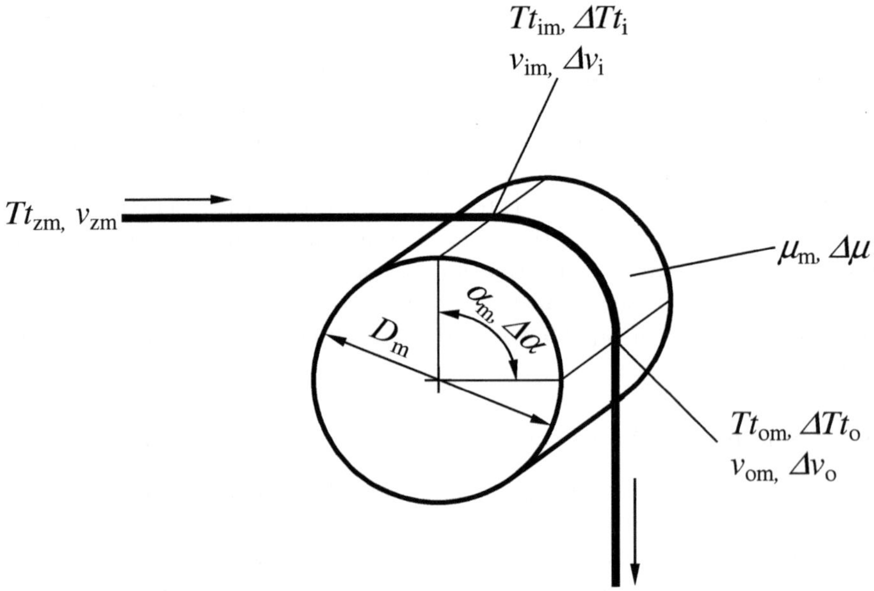

Fig. 5.3. Technological scheme of a friction thread line

Equation 5.8 put in Eq. 5.7 results:

$$A_{zm} \cdot Tt_{zm} \left[\frac{Tt_{zm}}{Tt_o} - 1 \right] = A_{zm} \cdot Tt_{zm} \left[\frac{Tt_{zm}}{Tt_i} - 1 \right] \cdot \exp(\mu \cdot \alpha) \tag{5.9}$$

Equation 5.9 results for Tt_o:

$$Tt_o = \frac{Tt_{zm}}{\left[\frac{Tt_{zm}}{Tt_i} - 1 \right] \cdot \exp(\mu \cdot \alpha) + 1} \tag{5.10}$$

The, on the friction element situated, thread mass M can now be determined by means of Eq. 5.10. The fineness Tt_o (dependent on the angle of wrap α) is to be multiplied by the circular arc $\alpha \cdot D_m/2$ to this (D_m = diameter of the friction element). Because T_o is a function of α it is to be integrated over the whole way of friction:

$$M = \frac{D_m}{2} \int_o^{\alpha} \frac{Tt_{zm}}{\left[\frac{Tt_{zm}}{Tt_i} - 1 \right] \cdot \exp(\mu \cdot \alpha) + 1} \cdot d\alpha \tag{5.11}$$

The solution of Eq. 5.11 results after substitution:

$$M = \frac{D_m \cdot Tt_{zm}}{2 \cdot \mu} \left\{ \mu \cdot \alpha + \ln \frac{\left[\frac{Tt_{zm}}{Tt_i} - 1 \right] + 1}{\left[\frac{Tt_{zm}}{Tt_i} - 1 \right] \cdot \exp(\mu \cdot \alpha) + 1} \right\} \tag{5.12}$$

If Eq. 5.10 is reduced to Tt_{i} and this value is inserted in Eq. 5.12 then one gets the thread mass on the friction element M in dependent upon the output thread fineness Tt_{o}:

$$M = \frac{D_{\mathrm{m}} \cdot Tt_{\mathrm{zm}}}{2 \cdot \mu} \left[\mu \cdot \alpha + \ln \frac{(Tt_{\mathrm{zm}} - Tt_{\mathrm{o}}) \cdot \exp(-\mu \cdot \alpha) + Tt_{\mathrm{o}}}{Tt_{\mathrm{zm}}} \right] \tag{5.13}$$

The time differential of Eq. 5.13 can be formed now because the value $\mathrm{d}m/\mathrm{d}t$ is nothing else than the changing thread mass on the friction element if Tt_{o} changes. This change element is easy to calculate:

$$\frac{\mathrm{d}M}{\mathrm{d}t} = \frac{D_{\mathrm{m}} \cdot Tt_{\mathrm{zm}}[1 - \exp(-\mu \cdot \alpha)]}{2 \cdot \mu(Tt_{\mathrm{zm}} - Tt_{\mathrm{o}}) \cdot \exp(-\mu \cdot \alpha) + Tt_{\mathrm{o}}} \cdot \dot{Tt}_{\mathrm{o}} \tag{5.14}$$

Equation 5.14 represents the change of stored mass. The nonlinear DEq. as the dynamic model equation for a friction thread line according to the basic Eq. 2.23 can now be written with the latter and both quantities

- mass inflow per time $= Tt_{\mathrm{i}} \cdot v_{\mathrm{i}}$ and
- mass discharge per time $= Tt_{\mathrm{o}} \cdot v_{\mathrm{o}}$

$$\Phi = v_{\mathrm{o}} \cdot Tt_{\mathrm{o}} + \frac{\mathrm{d}M}{\mathrm{d}t} = \frac{D_{\mathrm{m}} \cdot Tt_{\mathrm{zm}}[1 - \exp(-\mu \cdot \alpha)]}{2 \cdot \mu(Tt_{\mathrm{zm}} - Tt_{\mathrm{o}}) \cdot \exp(-\mu \cdot \alpha) + Tt_{\mathrm{o}}} \cdot Tt_{\mathrm{o}} \cdot p - v_{\mathrm{i}} \cdot Tt_{\mathrm{i}} = 0 \tag{5.15}$$

After partial derivation to all quantities which can change (v_{o}, v_{i}, Tt_{o}, Tt_{i}, μ, α) the following linearised motion-DEq. for a friction thread line results from Eq. 5.15:

$$\begin{aligned} & Tt_{\mathrm{om}} \cdot \Delta v_{\mathrm{o}} - Tt_{\mathrm{im}} \cdot \Delta v_{\mathrm{i}} - v_{\mathrm{im}} \cdot \Delta Tt_{\mathrm{i}} \\ & + \left\{ v_{\mathrm{om}} + p \frac{2Tt_{\mathrm{zm}}^2 D_{\mathrm{m}} \mu_{\mathrm{m}} \exp(-\mu_{\mathrm{m}} \alpha_{\mathrm{m}})[1 - \exp(-\mu_{\mathrm{m}} \alpha_{\mathrm{m}})]}{[2\mu_{\mathrm{m}}(Tt_{\mathrm{zm}} - Tt_{\mathrm{om}})\exp(-\mu_{\mathrm{m}} \alpha_{\mathrm{m}}) + Tt_{\mathrm{om}}]^2} \right\} \cdot \Delta Tt_{\mathrm{o}} \\ & + p \frac{2Tt_{\mathrm{zm}} Tt_{\mathrm{om}} D_{\mathrm{m}} \exp(-\mu_{\mathrm{m}} \alpha_{\mathrm{m}})[Tt_{\mathrm{om}} \alpha_{\mathrm{m}}/2 - (Tt_{\mathrm{zm}} - Tt_{\mathrm{om}})[\mu_{\mathrm{m}} \alpha_{\mathrm{m}} - \exp(-\mu_{\mathrm{m}} \alpha_{\mathrm{m}})]]}{[2\mu_{\mathrm{m}}(Tt_{\mathrm{zm}} - Tt_{\mathrm{om}})\exp(-\mu_{\mathrm{m}} \alpha_{\mathrm{m}}) + Tt_{\mathrm{om}}]^2} \cdot \Delta \mu \\ & + p \frac{2Tt_{\mathrm{zm}} Tt_{\mathrm{om}} D_{\mathrm{m}} \mu_{\mathrm{m}} \exp(-\mu_{\mathrm{m}} \alpha_{\mathrm{m}})[Tt_{\mathrm{om}}/2 + \mu_{\mathrm{m}}(Tt_{\mathrm{zm}} - Tt_{\mathrm{om}})]}{[2\mu_{\mathrm{m}}(Tt_{\mathrm{zm}} - Tt_{\mathrm{om}})\exp(-\mu_{\mathrm{m}} \alpha_{\mathrm{m}}) + Tt_{\mathrm{om}}]^2} \cdot \Delta \alpha = 0 \end{aligned} \tag{5.16}$$

The structure of Eq. 5.16 is a bit more complicated than the derivated equations until now. Nevertheless, it is the basis of the dynamic transfer functions and their derivated functions in the frequency range (frequency, amplitude, and phase frequency responses) which can answer to appointed cause-effect-questions. In the following subsection the use of the dynamic model Eqs. 5.1, 5.6 and 5.16 of the delay, dead time and friction lines thread will be represented by means of some practical examples. From these derived statements it will also be shown for the carrying out of the process.

5.1.5 Examples of Quantitative Investigations of Fibre Transport Processes

Fibre Influence in a Series Arrangement of Delay and Dead Time Thread Lines

The investigated delay thread line with a following dead time thread line is shown in Fig. 5.4. A practical realisation could be, for instance, that the thread runs from a supply bobbin into the manifold wrapped taking in godet of a drawing zone or the thread runs in an elongation zone realised between a taking in godet and a manifold wrapped taking out godet.

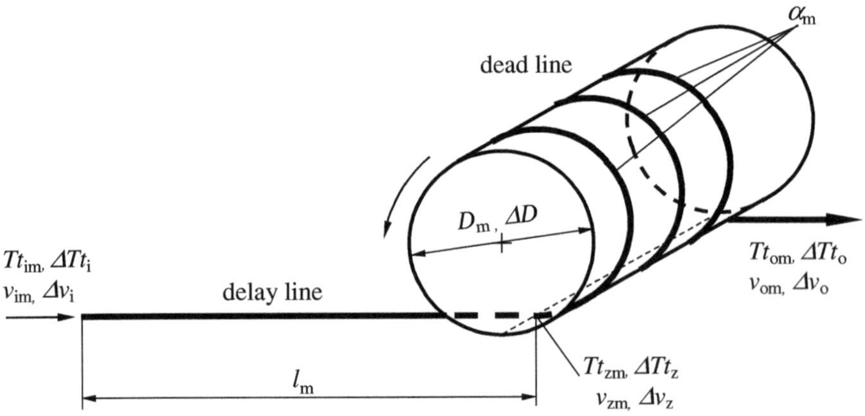

Fig. 5.4. Series arrangement of a delay and a dead time thread line

The used abbreviations mean:

Tt_{im}, v_{im}	mean value of the thread fineness or velocity at the input into the delay line
ΔTt_i, Δv_i	changes of the thread fineness or velocity around their mean values at the input into the delay line
Tt_{zm}, v_{zm}	mean value of the thread fineness or velocity at input into the dead time line (≡ output of the delay line)
ΔTt_z, Δv_z	changes of the thread fineness or velocity around their mean values at input into the dead time thread line (≡ output of the delay line)
Tt_{om}, v_{om}	mean value of the thread fineness or velocity at output of the dead time line
ΔTt_o, Δv_o	changes of the thread fineness or velocity around their mean values at output of the dead time line
l_m	length of the delay line
D_m	diameter of the godet
ΔD	change of the godet diameter
α_m	angle of wrap of the thread around the godet (in radian measure)

The dynamic model equations of a delay thread line (Eq. 5.1) and a dead time thread line (Eq. 5.6) – applied to the present case – are used for the calculation of the dynamic transfer function for a cause-effect relation whose aim is also to be defined. It should be assumed that ΔD (no eccentricity of the godet) as well as Δv_z and Δv_o (no speed changes of the transport godet) and ΔTt_i (no changes of the input thread fineness) are equal to zero. Moreover, if it is considered that $v_{zm} = v_{om}$ and $Tt_{zm} = Tt_{om}$ then the equations can be written as

delay line

$$(v_{om} + p \cdot l_m) \cdot \Delta Tt_z - Tt_{im} \cdot \Delta v_i = 0 \tag{5.17}$$

dead time line

$$\Delta Tt_o - \exp\left(-p \cdot \frac{D_m \cdot \alpha_m}{2 \cdot v_{om}}\right) \cdot \Delta Tt_z = 0 \tag{5.18}$$

It should now be investigated whether the reactions of the fineness at the output of the transport godet (ΔTt_o) are if the velocity is changing at the input of the whole transport line (Δv_i). Now the *dynamic transfer function* is to be formed

$$G(p) = \frac{\Delta Tt_o}{\Delta v_i}$$

which is to be calculated from Eqs. 5.17 and 5.18 as follows:

$$G(p) = \frac{\Delta Tt_o}{\Delta v_i} = \frac{Tt_{im}}{v_{om}} \cdot \frac{\exp\left(-p \cdot \dfrac{D_m \cdot \alpha_m}{2 \cdot v_{om}}\right)}{1 + p \cdot \dfrac{l_m}{v_{om}}} \tag{5.19}$$

amplitude frequency response

$$|G(jf)| = \left|\frac{\widetilde{\Delta Tt_o}}{\widetilde{\Delta v_i}}\right| = \frac{Tt_{im}}{\sqrt{v_{om}^2 + (2\pi f \cdot l_m)^2}} \tag{5.20}$$

phase frequency response

$$\varphi(f) = \arc\tan\left[-\frac{2\pi f \cdot l_m \cdot \cos\left(\dfrac{D_m \alpha_m \pi}{v_{om}} \cdot f\right) + v_{om} \cdot \sin\left(\dfrac{D_m \alpha_m \pi}{v_{om}} \cdot f\right)}{v_{om} \cdot \cos\left(\dfrac{D_m \alpha_m \pi}{v_{om}} \cdot f\right) - 2\pi f \cdot l_m \cdot \sin\left(\dfrac{D_m \alpha_m \pi}{v_{om}} \cdot f\right)}\right] \tag{5.21}$$

The phase shift angle $\varphi(f)$ (Eq. 5.21) indicates (as you know) in which angle φ (related to a full cycle duration of the disturbance oscillation) the effect-oscillation follows the cause-(disturbance-)oscillation. It is easily possible to conclude from this phase shift angle $\varphi(f)$ by means of the known output velocity v_{om} and of Eq. 2.50 to the thread length L_{d} which passes the transport system at its output before an input disturbance reaches this output. An independent representation of this delay thread length L_{d} from the velocity v_{om} can be given if it does not use the disturbance frequency f as an independent variable, but the wavelength λ_{f} of a full cycle duration of the disturbance in the thread. It is connected with the disturbance frequency f and the thread output velocity v_{om} according to the relation

$$f = \frac{v_{\mathrm{om}}}{\lambda_{\mathrm{f}}} \tag{5.22}$$

Comparable to this is also the equivalent Eq. 4.57.

The nomogram Fig. 5.5 should be inserted at this point. It allows the conversion of the single quantities into each other and it can be used because it is generally valid for other interests of the same kind which are presented for instance in the Sects. 4.3.3, 4.4.2, 4.4.3 and the Sect. 4.5.

The amplitude frequency response of Eq. 5.20 and the delay thread length L_{d} are represented in Figs. 5.6 and 5.7 depending upon the disturbance frequency f and of the correlated thread length (in connection with the thread output velocity), which represents a full disturbance oscillation. The corresponding dependences are epitomised presented for the following process and product variables (only for v_{im}=100 m/min) relating to the model arrangement in Fig. 5.4:

$$
\begin{aligned}
Tt_{\mathrm{im}} &= 3.4 \text{ tex}\\
Tt_{\mathrm{om}} &= 3.3 \text{ tex}\\
D_{\mathrm{m}} &= 0.1 \text{ m},\ 0.2 \text{ m}\\
\alpha_{\mathrm{m}} &= 0 \text{ (thread goes out the delay thread line directly)}\\
\alpha_{\mathrm{m}} &= 2\pi,\ 6\pi,\ 10\pi \ (\equiv 1, 3, 5 \text{ wraps round the godet})\\
l_{\mathrm{m}} &= 0.5 \text{ m},\ 1.0 \text{ m},\ 2.0 \text{ m}\\
v_{\mathrm{im}} &= 100 \text{ m/min},\ 500 \text{ m/min},\ 1000 \text{ m/min}\\
v_{\mathrm{om}} &= 103 \text{ m/min},\ 515 \text{ m/min},\ 1030 \text{ m/min}
\end{aligned}
$$

The following statements are to be derived also concerning the not presented results of the higher thread velocities:

a) The amplitude frequency responses (Fig. 5.6) show that the amplitude of the output fineness change $\widetilde{\Delta Tt_{\mathrm{o}}}$ (caused by an input disturbance of the velocity $\widetilde{\Delta v_{\mathrm{i}}}$) is independent on it, if a dead time thread line of any length follows to the delay thread line or not. The result means that the imprinted

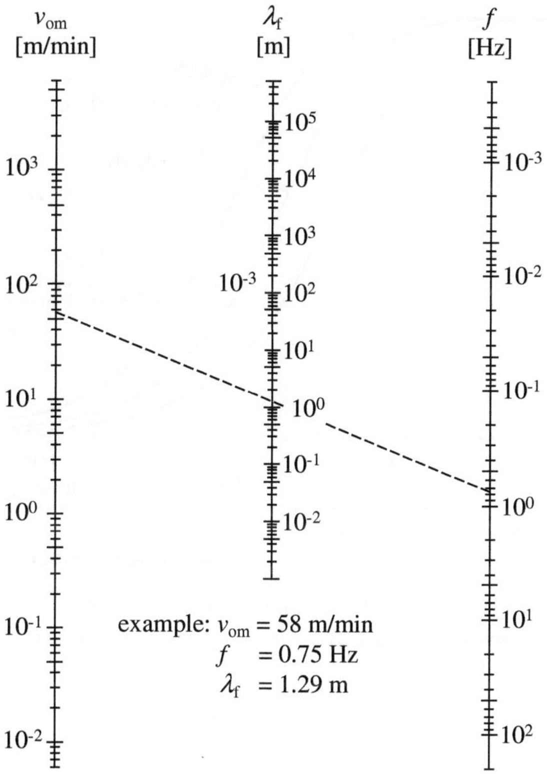

Fig. 5.5. Nomogram to the estimation of the relationship between the thread velocity v_{om}, the disturbance frequency f and the disturbance wavelength λ_{f}

thread fineness changes $\widetilde{\Delta Tt_{\mathrm{o}}}$ along a godet line with more or less wraps will neither decrease nor increase. They appear at the output unchanged and are only delayed by the pure transport time (see definition of the dead time in the Sect. 5.1.3).

b) Disturbances of the input velocity $\widetilde{\Delta v_{\mathrm{i}}}$ of $\geq$ 50 m/min (according to a frequency of the disturbance f of $\leq 3.3 \cdot 10^{-2}$ Hz at $v_{\mathrm{im}} = 100$ m/min or $\leq 3.3 \cdot 10^{-1}$ Hz at $v_{\mathrm{im}} = 1000$ m/min) will be transmitted practically undamped to the output thread fineness Tt_{o}.

c) A practical complete dampening is reached if the disturbance is $\leq$ 50 mm (according to a frequency of the disturbance f of $\geq$ 33 Hz at $v_{\mathrm{im}} = 100$ m/min or $\geq$ 330 Hz at $v_{\mathrm{im}} = 1000$ m/min). It is furthermore to read that the dampening effect is already put in at smaller frequencies the greater the length of the delay line.

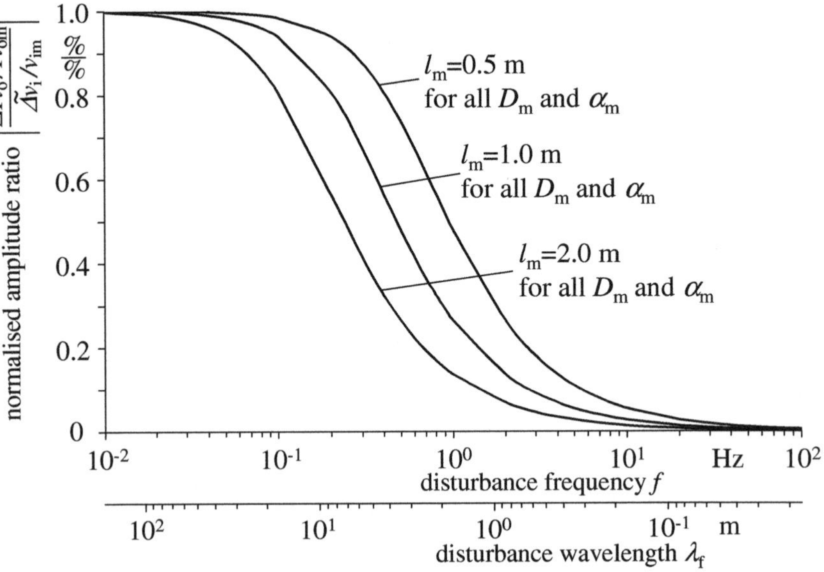

Fig. 5.6. Normalised amplitude frequency responses of fineness changes $\widetilde{\Delta Tt_o}$ caused by input velocity changes $\widetilde{\Delta v_i}$ after the thread line arrangement Fig. 5.4, input velocity $v_{im} = 100$ m/min

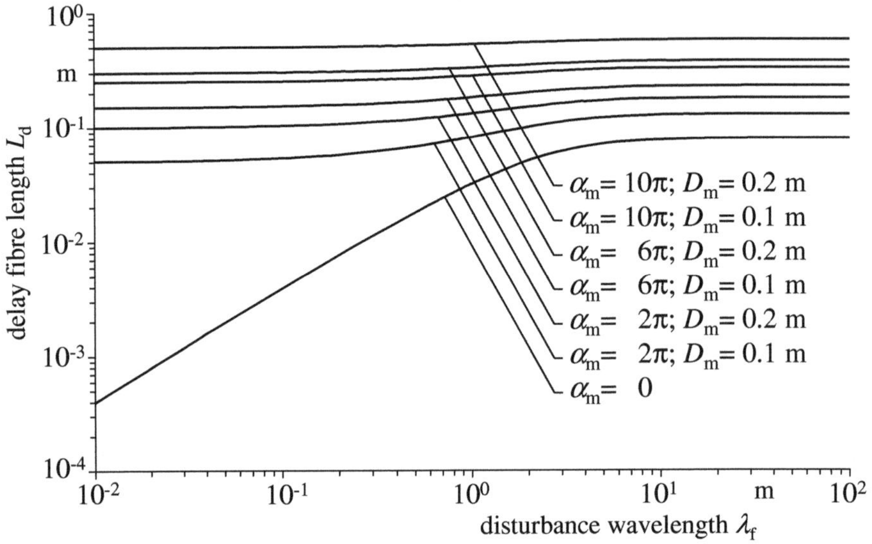

Fig. 5.7. Delay fibre length L_d, length of the delay thread line $l_m = 0.5$ m, thread line arrangement after Fig. 5.4

d) It is furthermore remarkable that a variation of the delay thread line length lowers or raises the critical frequencies in a similar ratio. From this it is derivable that the dampening of such input velocity disturbances is better the longer the line can be selected from the thread unwinding point to the machine input. Such disturbances happen for instance at the twisting, draw twisting or knitting in form of so-called "thread plucks" or also in longer periodic unwinding fluctuations by means of oil fluctuations on the thread. An extension of this line alone must also work for thread break behaviour, because the thread break is an extreme case of fineness change ($\Delta Tt = -Tt_{\mathrm{m}}$). Each measurement must effect diminishing, to the thread breaks, which fineness changes damp or dismantle, indifferent of which cause is produced.

e) The knowledge of the delay thread length L_{d} is also important for the cause research of unevennesses (Fig. 5.7). The dead times (caused by different roll wraps and diameters) are of decisive influence on L_{d} (contrary to the amplitude frequency response) specifically in the disturbance wavelength range ≤ 1 m. Whereas the delay thread length L_{d} with missing dead time ($\alpha_{\mathrm{m}} = 0$) line increases very strongly with increasing disturbance wavelength λ_{f} (according to decreasing disturbance frequency f, notice the double logarithm axes in Fig. 5.7) the dependence on the disturbance wavelength decreases with the increasing dead time line. The delay thread length L_{d} approaches a limit value versus the disturbance wavelength λ_{f} which is exactly equivalent to the length of the delay thread line l_{m}. This behaviour means practically that L_{d} is determined for $\lambda_{\mathrm{f}} \leq 1$ m (according to great disturbance frequencies f) almost fully by the dead time thread line, whereas the influence of the delay thread line (according to the length of this) will be relatively stronger for great disturbance wavelengths $\lambda_{\mathrm{f}} \geq 5$ m (according to small disturbance frequencies f).

f) The relationships show that specific short-time disturbances of high frequencies (with short and shortest wavelengths) in the thread (they influence the thread break behaviour especially drasticly) are very difficult to characterise only by their cause-effect relationship. The reason is that their amplitudes are probably strongly dampened (specifically for long delay lines) and because (specifically for short delay lines) the delay thread length L_{d} can amount to the multiple of the disturbance wavelength λ_{f}.

It is absolutely necessary that dynamic model investigations be carried out for such practical interesting cases to the design of unique relationships out of dynamic measurements. It can also be necessary that the interpretation of the measuring results be found by cross-correlation analysis of the disturbance time functions and their effects to the thread fineness (see to this Chap. 6).

Fibre Influence at Fixed Thread Guides (Friction Thread Lines)

The thread line at fixed thread guides (thread deflection elements, thread guides, friction brakes) has been defined as the friction line. The already used scheme of the thread guidance around a fixed friction element (Fig. 5.3) will be investigated quantitatively in the following more throughly. The used symbols of the dynamic model Eq. 5.16 and their meaning should be given as:

Tt_{zm}, v_{zm}	mean values of the thread fineness or velocity of the tensionless thread
Tt_{im}, v_{im}	mean values of the thread fineness or velocity at the input into the friction thread line
Tt_{om}, v_{om}	mean values of the thread fineness or velocity at the output of the friction thread line
D_{m}	diameter of the fixed (friction) thread guide
α_{m}	mean value of the angle of wrap of the thread around the fixed (friction) thread guide (in radian measure)
μ_{m}	mean value of the friction coefficient thread-thread guide
$\Delta Tt_{\mathrm{i}}, \Delta Tt_{\mathrm{o}}$, $\Delta v_{\mathrm{i}}, \Delta v_{\mathrm{o}}$, $\Delta\mu_{\mathrm{m}}, \Delta\alpha_{\mathrm{m}}, \Delta D$	changes of the adequate sizes around their mean values

It is assumed according to Eq. 5.16 that D, Tt_{z}, v_{z}, v_{o} and α are constant (the appropriate possible changes ΔD, ΔTt_{z}, Δv_{z}, Δv_{o} and $\Delta\alpha$ are assumed to be zero). Equation 5.16 is simplified appropriately then.

It is investigated in the following how the fineness change ΔTt_{o} acts if either the velocity at the input of the friction thread line v_{i} or the friction coefficient μ fluctuate. The fineness of the thread at the input should be constant ($\Delta Tt_{\mathrm{i}} = 0$).

The *amplitude frequency responses* for the named disturbances are:

$$|G(\mathrm{j}f)| = \left| \frac{\widetilde{\Delta Tt_{\mathrm{o}}}}{\widetilde{\Delta v_{\mathrm{i}}}} \right| = \frac{Tt_{\mathrm{im}} \cdot A_{\mathrm{a}}}{\sqrt{A_{\mathrm{a}}^2 \cdot B_{\mathrm{a}}^2 + (2\pi f \cdot C_{\mathrm{a}})^2}} \tag{5.23}$$

$$|G(\mathrm{j}f)| = \left| \frac{\widetilde{\Delta Tt_{\mathrm{o}}}}{\widetilde{\Delta \mu}} \right| = \frac{2\pi f \cdot E_{\mathrm{a}}}{\sqrt{A_{\mathrm{a}}^2 \cdot B_{\mathrm{a}}^2 + (2\pi f \cdot C_{\mathrm{a}})^2}} \tag{5.24}$$

with the abbreviations

$$A_{\mathrm{a}} = [2\mu_{\mathrm{m}}(Tt_{\mathrm{zm}} - Tt_{\mathrm{om}}) \cdot \exp(-\mu_{\mathrm{m}} \cdot \alpha_{\mathrm{m}}) + Tt_{\mathrm{om}}]^2$$

$$B_{\mathrm{a}} = v_{\mathrm{om}}$$

$$C_{\mathrm{a}} = 2Tt_{\mathrm{zm}}^2 \cdot D_{\mathrm{m}} \cdot \mu_{\mathrm{m}} \cdot \exp(-\mu_{\mathrm{m}} \cdot \alpha_{\mathrm{m}})[1 - \exp(-\mu_{\mathrm{m}} \cdot \alpha_{\mathrm{m}})]$$

$$E_{\mathrm{a}} = 2Tt_{\mathrm{zm}} \cdot Tt_{\mathrm{om}} \cdot D_{\mathrm{m}} \cdot \exp(-\mu_{\mathrm{m}} \cdot \alpha_{\mathrm{m}}) \cdot \{(Tt_{\mathrm{zm}} - Tt_{\mathrm{om}}) \cdot [1 - \mu_{\mathrm{m}}\alpha_{\mathrm{m}} - \exp(-\mu_{\mathrm{m}} \cdot \alpha_{\mathrm{m}})] - Tt_{\mathrm{om}} \cdot \alpha_{\mathrm{m}}/2\}$$

The correlated *phase frequency responses* for the disturbance $\widetilde{\Delta v_{\mathrm{i}}}$ are:

$$\varphi(f) = \arctan\left[-\frac{2\pi f \cdot C_{\mathrm{a}}}{A_{\mathrm{a}} \cdot B_{\mathrm{a}}}\right] \tag{5.25}$$

and for the disturbance $\widetilde{\Delta\mu}$:

$$\varphi(f) = \arctan\left[\frac{A_{\mathrm{a}} \cdot B_{\mathrm{a}}}{2\pi f \cdot C_{\mathrm{a}}}\right] \tag{5.26}$$

By means of Eqs.+5.25, 5.26 and 2.50 it is possible to give the delay thread lines L_{dv} (for a velocity disturbance Δv_{i}) or $L_{\mathrm{d}\mu}$ (for a friction coefficient disturbance $\Delta\mu$) as a quantity for the thread length which passes the transport system between the cause imprinting and the effect reaction of periodic disturbances. Considering Eq. 5.22 it is also possible to give here a result presentation versus the disturbance wavelength λ_{f} which is free of the output velocity v_{om}.

Before a quantitative analysis can be done it is necessary to calculate the sizes v_{im}, v_{zm}, Tt_{im} and Tt_{om} from the given sizes v_{om} and Tt_{zm}. If a preelongation of 0.3% is assumed in the input thread then $Tt_{\mathrm{im}} = Tt_{\mathrm{zm}}/1.003$ follows according to the continuity equation. Equation 5.10 was valid between input and output fineness of the thread along the friction line. With this it is also possible to calculate the velocities v_{im} and v_{zm} if the continuity equation $Tt_{\mathrm{im}} \cdot v_{\mathrm{im}} = \text{const.}$ is considered additionally.

Equations 5.23 to 5.26 have been analysed for the following combinations of process and product variables:

Tt_{zm}	= 3.4 tex
α_{m}	$= \pi/2$ (1/4 wrap)
	$= 2\pi$ (1 wrap)
Tt_{im}	= 3.3898 tex
Tt_{om}	= 3.384 tex (for $\alpha_{\mathrm{m}} = \pi/2$)
	= 3.334 tex (for $\alpha_{\mathrm{m}} = 2\pi$)
v_{om}	= 100 m/min, 500 m/min, 1000 m/min
$v_{\mathrm{zm}}(\alpha = \pi/2)$	= 99.54 m/min, 497.76 m/min, 995.72 m/min
$v_{\mathrm{zm}}(\alpha = 2\pi)$	= 98.10 m/min, 490.41 m/min, 980.82 m/min
$v_{\mathrm{im}}(\alpha = \pi/2)$	= 99.84 m/min, 499.26 m/min, 998.52 m/min
$v_{\mathrm{im}}(\alpha = 2\pi)$	= 98.40 m/min, 491.94 m/min, 983.82 m/min
D_{m}	= 2 mm, 5 mm
μ_{m}	= 0.3

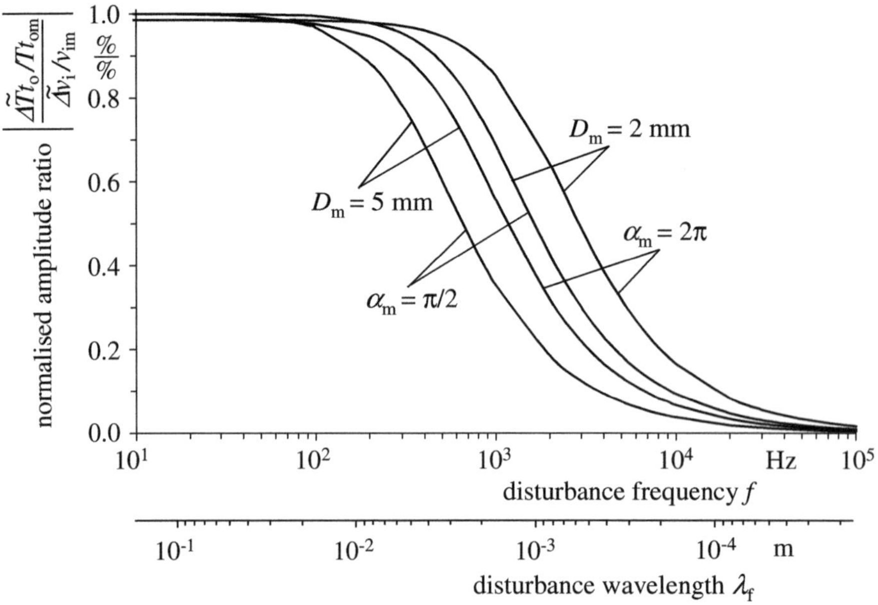

Fig. 5.8. Normalised amplitude frequency responses of fineness changes $\widetilde{\Delta Tt_o}$ caused by input velocity changes $\widetilde{\Delta v_i}$ of a friction thread line after Fig. 5.3, output velocity $v_{om} = 100$ m/min

The results of the quantitative calculations are partly shown (for $v_{om} = 100$ m/min only) in Figs. 5.8 to 5.11. The following statements are also to be derived concerning the not presented results of the higher thread velocities:

a) $\widetilde{\Delta Tt_o}$ will be smaller so the disturbance frequency f will be greater if a disturbance $\widetilde{\Delta v_i}$ effects. The reason for this is the dampening effect of the friction thread guide line (Fig. 5.8). This dampening effect starts at smaller frequencies the thicker the friction thread guide and the smaller the wrap angles are. It does not occur unless the disturbed thread lengths (disturbance wavelengths λ_f) are smaller than the wrap line on the thread guide.

b) A practically complete dampening of an input velocity disturbance occurs only at disturbance wavelengths in the thread of $\lambda_f < 0.15$ mm (according to influence times of the disturbance of < 0.1 ms!). Because all dynamic disturbances have as a rule a much longer influence time it is to be stated that all Δv_i disturbances at the input of a friction thread line will be transmitted greatly proportionally in the same ΔTt_o changes at the output (under the most suitable conditions of a constant output thread velocity).

c) The friction thread line effects (only for high frequency disturbances) a better disturbance decrease the greater its diameter and the smaller its

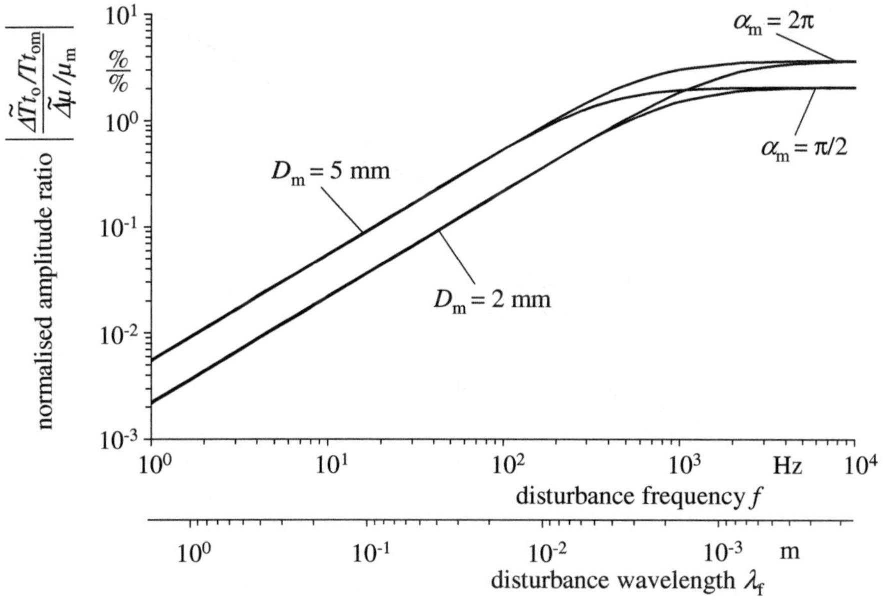

Fig. 5.9. Normalised amplitude frequency responses of fineness changes $\widetilde{\Delta Tt_o}$ caused by friction coefficient changes $\widetilde{\Delta\mu}$ of a friction thread line after Fig. 5.3, output velocity $v_{om} = 100$ m/min

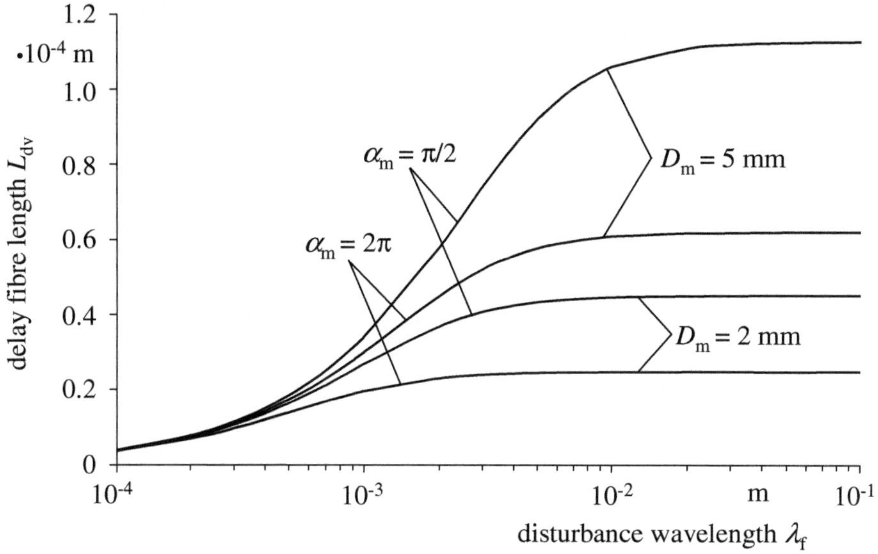

Fig. 5.10. Delay fibre length L_{dv} of a friction thread line after Fig. 5.3

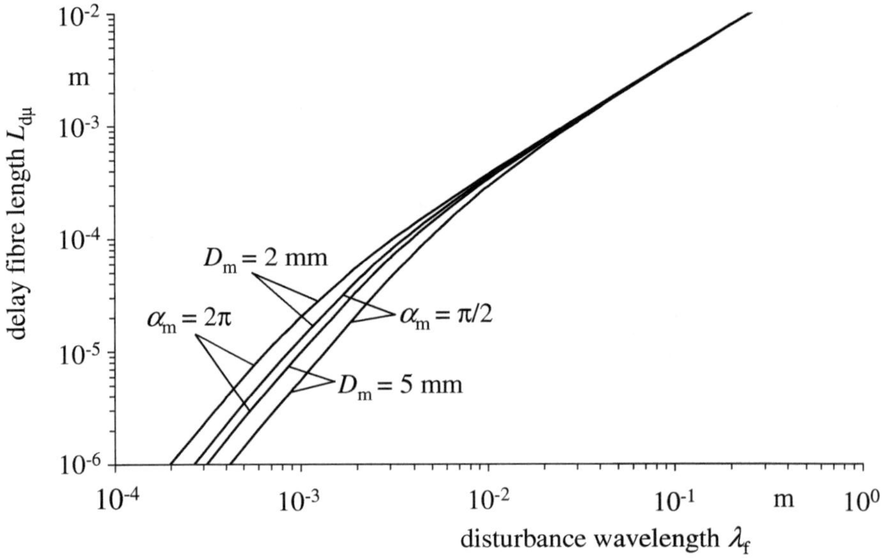

Fig. 5.11. Delay fibre length $L_{d\mu}$ of a friction thread line after Fig. 5.3

wrap angle is. This result suggests the following recommendation: To the realisation of an appointed mean thread tensile force at the input of a processing machine it is better to arrange serially several friction lines with large diameters and in each case small thread wrap angles (for instance a lattice brake) than only one friction thread line with a small diameter and a great wrap angle. If the use of only one friction thread line is possible then a great diameter is in its turn more favourable than a smaller on the same wrap angle.

d) The effect of friction coefficient changes or oscillations $\widetilde{\Delta\mu}$ to the fineness Tt_o are shown in Fig. 5.9. The relationships are valid for the case that the thread which runs into the friction thread line is braked with a constant brake force and it is constantly elongated (here 0.3%). Further it should be as valid as before: $v_o = v_{om} =$ const.
Disturbances with a frequency of < 1 Hz (consequently all quasi steady state changes of the friction coefficient too) will not be of influence on the fineness. The fineness changes reach only less than 0.01% in this range for instance if the friction coefficient changes by 1%. The effect of the disturbances with the same dimensions is here (in opposite to the case of a input velocity disturbance discussed before) the greater in the rest of the frequency range the thicker the friction thread guide and the greater the thread wrap angle are. The amplitude frequency response of the model variants approximated before always to zero for high disturbance frequencies. In the present case of a friction coefficient disturbance the amplitude frequency response reaches a constant value (dependent on the thread line velocity; Fig. 5.9 only shows the

output velocity $v_{\mathrm{om}} = 100$ m/min) for disturbance frequencies of $> 5 \cdot 10^3$ to $> 5 \cdot 10^4$ Hz. It is the question here of a differential action which follows from the amplitude frequency response (5.24). Friction coefficient changes of 1% effect in this range even 2% (for $\alpha = \pi/2$) to 3% (for $\alpha = 2\pi$) fineness changes. However, this range is not effective because the according disturbance wavelengths in the thread are < 1 mm, that means below the length of the wrap or friction line. This amounts to, for the selected model relations, between 1.6 mm (for $D_{\mathrm{m}} = 2$ mm and $\alpha = \pi/2$) and 15.7 mm (for $D_{\mathrm{m}} = 5$ mm and $\alpha = 2\pi$). A disturbance dampening occurs for such short disturbance wavelengths because a mean value of the friction coefficient can only effect along a thread part which just passes the friction line.

e) The delay thread line L_{dv} always amounts to < 0.12 mm in the entire interesting disturbance wavelength range > 1 mm (Fig. 5.10). These only insignificant delays (compared to the disturbance wavelengths) are of completely no account for dynamic measurements of threads which are running about fixed friction elements due to their smallness.

f) The relations are similar for disturbances of the friction coefficient (Fig. 5.11). The amplitude frequency response (see Fig. 5.9) is so small for disturbance wavelengths > 10 cm (the delay thread length $L_{\mathrm{d}\mu}$ could amount here to more than 1 to 3 cm) that a nearly complete disturbance dampening exists. The delay thread lengths $L_{\mathrm{d}\mu}$ are already insignificantly small again (< 0.3 mm) at disturbance wavelengths below 2 to 3 mm if the amplitude frequency response reaches its full value.

Possibilities to the Dampening of Tensile Force and Tensile Elongation Variations in Thread Input Lines

In Sect. 5.1.5 it has been hinted that a skillful designing of delay lines can effect dampening to input velocity disturbances and their effects to fineness unevennesses. Such delay lines are found in many machines of thread or fibre processing. The thread is unwound normally from fixed or rotating supported, but not actively driven, supply bobbins by means of the machine taking-in elements. The so-called axial “over end unwinding” is predominant here. But, the radial thread unwinding from rotating supported supply bobbins can also be found. Numerous examples can be given as to the spinning, drawing, twisting, winding, texturing, warping, sectional warping, and knitting. Such thread input lines (Fig. 5.12) are more or less significant (according to process step and material kind) under the view point of the cause research to process disturbances or thread and fabrics unevennesses. The reason for this is that each uncontrolled thread input (which is realised in the described manner) is connected to thread tensile force and thread elongation changes.

Such changes can be the cause for changes of appointed textile-physical quality characteristics of the thread which will only be formed in the following process steps or will be optically visible displaced to the fabric possibly. But, they can "only" be responsible for an increased thread break frequency. In the following, possibilities and their efficiency to an effective dampening of thread tensile force and thread elongation changes without the use of special devices (for instance automatic controlled thread brakes) in connection with [296] and [297] will be discussed. These investigations are independent of the causing factors for such thread tensile force and thread elongation changes for instance hooked and plastered up threads on the supply bobbins, periodic changes of the unwinding geometry at threads which are unwinded radially from rotating supported supply bobbins.

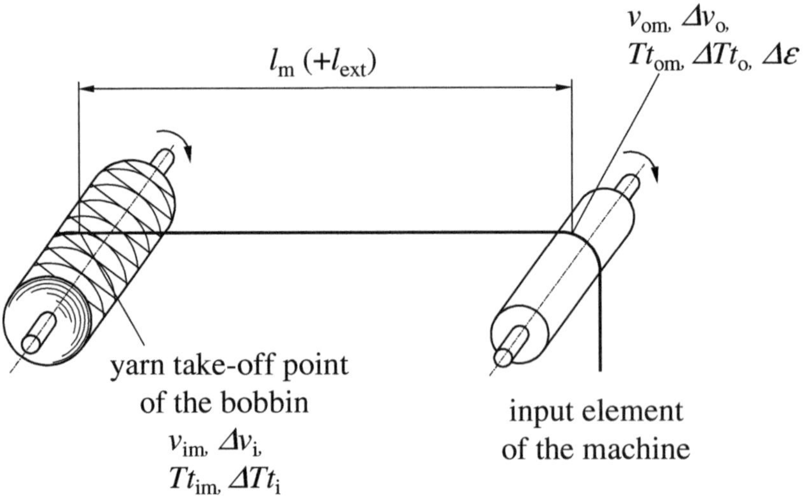

Fig. 5.12. Technological scheme of a thread input line

Input velocity changes Δv_i according to Fig. 5.12 are the most frequent cause for fineness changes ΔTt_o (and by this elongation and force changes $\Delta\varepsilon$ and ΔF, as will be demonstrated later in Sect. 6.2). Therefore the starting point of our views is the known *dynamic transfer function* of a delay line, which is derived from DEq. 5.1:

$$G(p) = \frac{\Delta Tt_o}{\Delta v_i} = \frac{Tt_{om}}{v_{im}} \cdot \frac{1}{1 + p \cdot \dfrac{l_m}{v_{om}}} \tag{5.27}$$

The elongation ε of a thread with the fineness Tt_i in the unloaded state and the fineness Tt_o in the elongated state is

$$\varepsilon = \frac{Tt_i}{Tt_o} - 1 \tag{5.28}$$

The partial differentiation of Eq. 5.28 to the three variables, which can be changeable, results in

$$\Delta\varepsilon = \frac{1}{Tt_\mathrm{o}} \cdot \Delta Tt_\mathrm{i} - \frac{Tt_\mathrm{im}}{Tt_\mathrm{om}^2} \cdot \Delta Tt_\mathrm{o} \tag{5.29}$$

We will not consider fineness changes of the input thread Tt_i in the present case ($\Delta Tt_\mathrm{i} = 0$). Equation 5.29 is simplified then to

$$\Delta\varepsilon = -\frac{Tt_\mathrm{im}}{Tt_\mathrm{om}^2} \cdot \Delta Tt_\mathrm{o} \tag{5.30}$$

ΔTt_o from Eq. 5.30 introduced into Eq. 5.27 results to the *amplitude frequency response*

$$|G(\mathrm{j}f)| = \left| \frac{\widetilde{\Delta\varepsilon}}{\widetilde{\Delta v_\mathrm{i}}/v_\mathrm{im}} \right| = (-)\frac{Tt_\mathrm{im}}{Tt_\mathrm{om}} \left[1 + \left(\frac{2\pi f \cdot l_\mathrm{m}}{v_\mathrm{om}} \right)^2 \right]^{-1/2} \tag{5.31}$$

The amplitude frequency response of Eq. 5.31 is shown versus the length of the input line l_m for different frequencies of the disturbance $\widetilde{v_\mathrm{i}}$ in Fig. 5.13. The diagram is related to an experimental investigation of a twister. The velocity v_om amounted to 58 m/min.

Thread input velocity oscillations $\widetilde{\Delta v_\mathrm{i}}$ should only effect small $\widetilde{\Delta\varepsilon}$ oscillations in a stable process. That means, the amplitude frequency response should be as small as possible or even zero. A relatively simple technological-constructive method insists on the extension of the thread input (delay) line as long as possible (compare Fig. 5.13). A further modified quantitative analysis of the solution equation for the dynamic mathematical model can be given as follows.

Dependent on

- the length l_m of any existing thread input line on a machine,
- the thread transport velocity v_om,
- the frequency f of the thread tensile force and the thread elongation changes

an exactly quantitative predetermined extension of the input line l_ext can be calculated, in which the existing amplitudes of the thread tensile force and thread elongation oscillations are reduced by a wished reduction factor R. This reduction factor R is defined as

$$R = 1 - \frac{|G(\mathrm{j}f)| \ \ (\text{input line } (l_\mathrm{m} + l_\mathrm{ext}))}{|G(\mathrm{j}f)| \ \ (\text{input line } l_\mathrm{m})} \tag{5.32}$$

Equation 5.32 inserted into Eq. 5.31 yields:

$$R = 1 - \sqrt{\frac{1 + \left(\dfrac{2\pi f \cdot l_\mathrm{m}}{v_\mathrm{om}} \right)^2}{1 + \left[\dfrac{2\pi f \cdot (l_\mathrm{m} + l_\mathrm{ext})}{v_\mathrm{om}} \right]^2}} \tag{5.33}$$

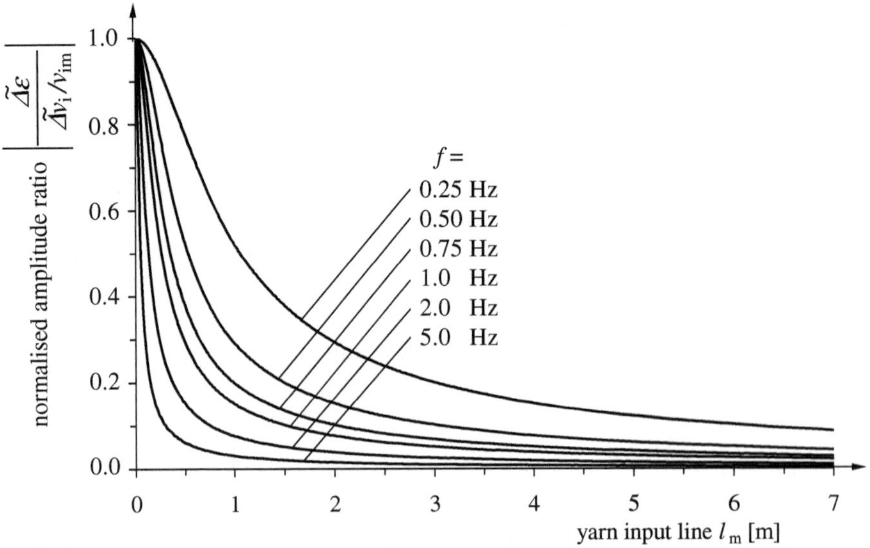

Fig. 5.13. Normalised amplitude transfer coefficients for changes of thread elongations $\widetilde{\Delta\varepsilon}$ caused by changes of the thread input velocity $\widetilde{\Delta v_i}$ of a thread input line

Equation 5.33 dissolved to l_{ext} results in the condition for the necessary extension of the thread input line at the wished reduction factor R

$$l_{ext} = \frac{\sqrt{R \cdot v_{om}^2 \cdot (2 - R) + (120\pi \cdot f \cdot l_m)^2}}{120\pi \cdot f \cdot (1 - R)} - l_m \tag{5.34}$$

The following dimensions are used:

l_m, l_{ext}	m
v_{om}	m/min
f	Hz
R	dimensionless, possible range from 0 (no reduction of disturbances) until 1 (full dampening of disturbances)

The necessary input line extension l_{ext} can also be appointed by means of a simplified approximation relation 5.35 (the basis for this is also Eq. 5.34), if a reduction of the thread tensile force and thread elongation changes have reached the half according to those by an unextended input line (reduction factor $R = 0.5$). This is much better to handle, but it requires somewhat greater l_{ext}-values for high thread line velocities ($v_{om} > 500$ m/min) and simultaneously small disturbance frequencies ($f < 5$ Hz). These l_{ext}-values are concerning the desired effect.

$$l_{ext} \approx 15 \cdot v_{om}/f + l_m \quad \text{(valid for } R{=}0.5\text{)} \tag{5.35}$$

The following dimensions are used:

l_m, l_{ext}	m
v_{om}	m/s
f	$\min^{-1}$

Quantitative data to the choice of input line extensions under concrete technological conditions can be taken away in Fig. 5.14.

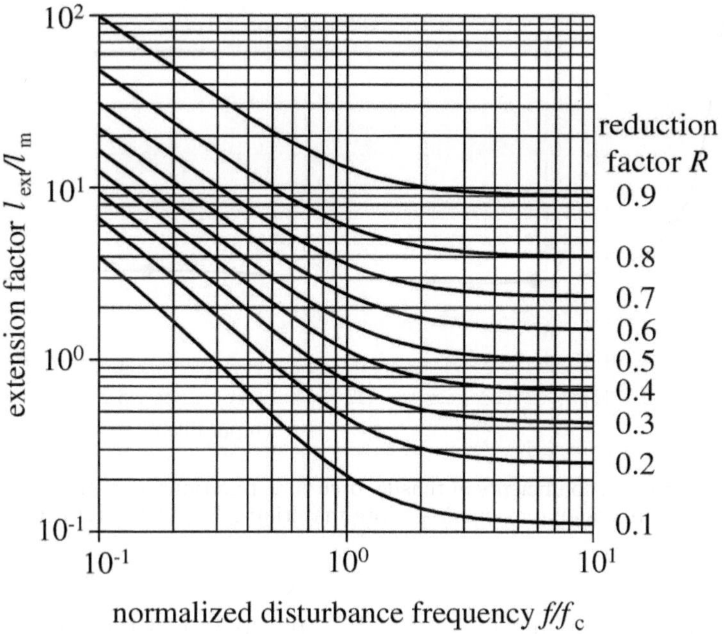

Fig. 5.14. Necessary extension factor l_{ext}/l_m dependent upon the disturbance frequency ratio f/f_c and on the reduction factor R

The use is the following:
One appoints the critical frequency f_c of the input line by means of Eq. 2.49 or Fig. 4.7 on the basis of the already present input line length l_m and winding up thread or input velocity into the machine v_{om}. It is also possible to find out the extension factor l_{ext}/l_m according to the already present input line l_m, which is to be realised for a dampening of the changes with the appointed reduction factor R.
If the necessary input line extension cannot be realised in a straight line on the machine then the whole length can be realised by means of thread deflection elements. However, the latter must be thread guide elements with light running rolls to reach the full dampening effect [297].

The efficiency of an input line extension (calculated on the basis of the developed theoretical fundamentals) should be demonstrated finally using variation coefficients derived from the experimentally measured thread tensile force time functions in the input line of a ring twister (Fig. 5.15).
The normal input line amounted to $l_m = 0.42$ m, and the thread input ve-

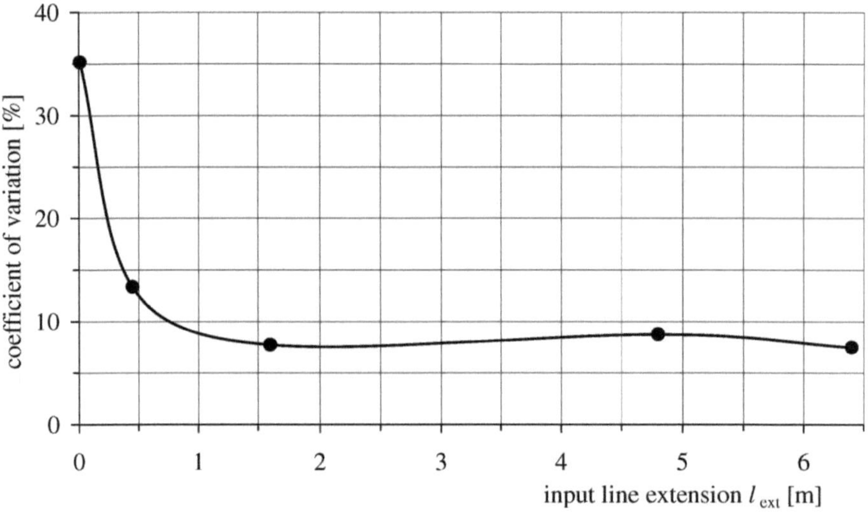

Fig. 5.15. Variation coefficients of tensile force time functions on a *Mouliné*-twister for different input line extensions l_{ext}, fibre material: PET 11 tex (20), texturised, $v_{om} = 58$ m/min

locity to $v_{om} = 58$ m/min ($f_c = 0.37$ Hz). It was to observed through this condition that a main frequency of the thread tensile force changes (and with this also of the thread elongation) of 0.75 Hz caused by a radially unround thread backing-off from a rotating supported supply bobbin. The tensile force changes amount nevertheless to ±50% around the mean value of 75 mN and result in a variation coefficient of 35.15% ($l_{ext} = 0$). These led to the induced elongation changes of the same relative quantity to distinct shade disturbances of the manufactured plied threads which consisted of two such supply threads of different fundamental colours. Input line extensions of $l_{ext} = 0.45$ m led to a variation coefficient of 13.34% and the disturbed quality deficiencies were removed at the input line extensions $l_{ext} \geq 1.6$ m. The variation coefficients of the tensile force time functions decrease then to $< 9\%$ (see also Sect. 6.4.2 and Fig. 5.15 again).
This technological example shows that an undisturbed thread run can be realised throughout without greater additional expense, if the dynamic transmission regularities are strictly used.

Fibre Influence by Means of Dynamics of Thread Traverse Motion at Winders

A further interesting application of the dynamic model of a delay thread line (applied to a concrete technological situation) is the analysis of the fineness influence caused by the periodic thread traverse motion at the winders. The characteristics of the appropriate extensive investigations [298] and some results are presented as follows. First, a few remarks to the motive of such investigations:

a) The traverse motion system permanently imprints the thread periodic property changes along its length axis which can be recorded as elongation and fineness changes. Such changes can evoke considerable molecular structure changes of the threads, particularly at the spinning machines for man-made fibres, because these changes are imprinted during the decisive phase of the structure development [299].

b) Already periodic elongation and fineness changes of 1% (!) can be the cause for visible dyeing defects in special products of man-made fibres [300].

c) A well-defined relationship exists between the quantity of the thread breaks at special processing processes and the fineness fluctuation amplitudes imprinted by the thread traverse motion process at the man-made fibre spinning machine.

d) Elongation and thread tensile force fluctuations caused by the traverse motion can involve deviations of the reel body from the desired cylindrical shape which leads to local different mill work particularly at the friction roll drives as a consequence of different contact pressures.

The thread line between the delivery godet and the wind-up bobbin can be taken in as a delay thread line (Fig. 5.16) which also includes the common case of an asymmetric traverse motion triangle (Fig. 5.17).

The basis for the derivation of the dynamic model is therefore the common linearised DEq. for a delay line. The cause variable of the fineness disturbances ΔTt_{o} of the winded thread are the periodically enforced changes of the length Δl of the delay line. The short friction lines on the thread guides at the top of the traverse motion triangle and of the traverse motion are itself neglected in the following. The dynamic model DEq. derives from Eq. 4.7 and is changed to:

$$(v_{\mathrm{om}} + p \cdot l_{\mathrm{m}}) \cdot \Delta Tt_{\mathrm{o}} + p \cdot Tt_{\mathrm{om}} \cdot \Delta l = 0 \tag{5.30}$$

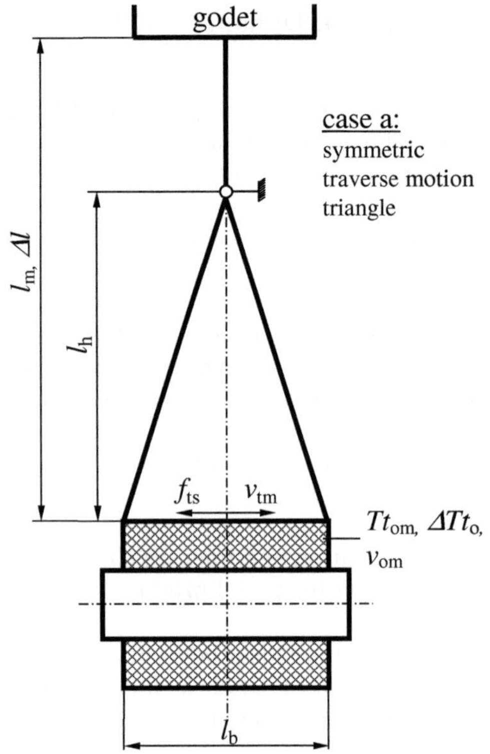

Fig. 5.16. Technological scheme of the thread traverse motion, case a

The symbols in Figs. 5.16, 5.17 and Eq. 5.36 mean:

- Tt_{om} mean value of the fineness of thread at the bobbin wind-up point approaching from the traverse motion triangle
- l_m mean value of the thread length between godet and wind-up point of the bobbin
- v_{om} mean value of the velocity of thread which the bobbin winds up (output or wind-up velocity)
- l_h height of the traverse motion triangle
- v_{tm} mean value of the linear velocity of traverse motion thread guide
- l_b length of the bobbin
- f_{ts} frequency of the traverse motion thread guide for the operation of one twice stroke, that means its motion from the left bobbin edge to the right and back
- e asymmetry parameter
- p differential operator $\mathrm{d}/\mathrm{d}t$

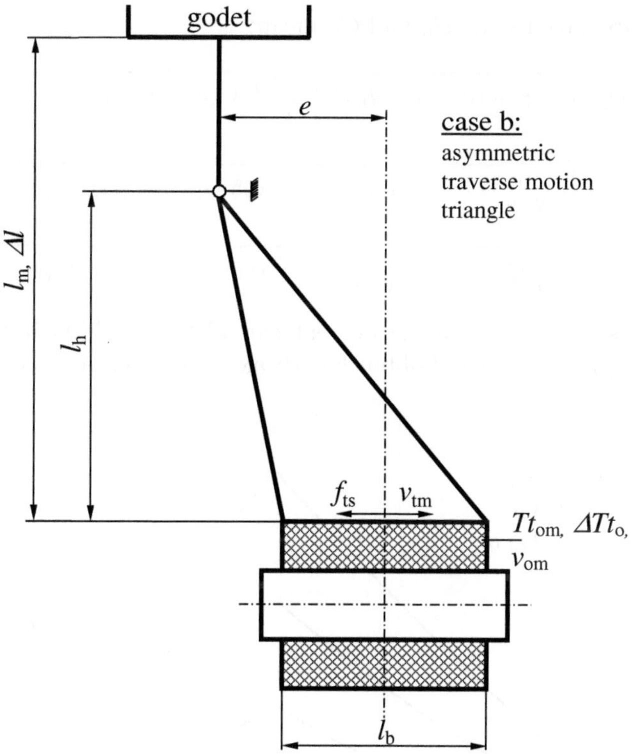

Fig. 5.17. Technological scheme of the thread traverse motion, case b

The resolution of the DEq. 5.36 for impulse-like Δl-disturbances (it is equivalent to the first derivation of the step response Eq. 4.36) is

$$\Delta Tt_{\mathrm{o}} \perp \Delta l = \Delta l \cdot \frac{Tt_{\mathrm{om}} \cdot v_{\mathrm{om}}}{l_{\mathrm{m}}^2} \cdot \exp\left(-\frac{v_{\mathrm{om}}}{l_{\mathrm{m}}} \cdot t\right) \tag{5.37}$$

If the change Δl caused by the traverse motion is impulse-like then the effected fineness change can already be calculated by means of Eq. 5.37. However, the time function of the change Δl is to be derived from the root laws of the rectangular triangle (see Figs. 5.16, 5.17). This can be derived after several intermediate steps (not further described here) and simplifications:
Forward motion (thread guide goes from the left bobbin edge to the right with the linear velocity $v_{\mathrm{tm}} = 2 \cdot l_{\mathrm{b}} \cdot f_{\mathrm{ts}}$)

$$\Delta l(t) = A_{\mathrm{b}} \cdot t^2 + B_{\mathrm{b}} \cdot t \tag{5.38}$$

Backward motion (thread guide goes from the right bobbin edge to the left with the linear velocity $v_{\mathrm{tm}} = 2 \cdot l_{\mathrm{b}} \cdot f_{\mathrm{ts}}$)

$$\Delta l(t) = A_{\mathrm{b}} \cdot t^2 + C_{\mathrm{b}} \cdot t \tag{5.39}$$

The abbreviations $A_{\mathrm{b}}, B_{\mathrm{b}}$ and C_{b} mean:

$$A_{\mathrm{b}} = 8f_{\mathrm{ts}}^2 \cdot \left[\sqrt{l_{\mathrm{h}}^2 + (e + 0.5 \cdot l_{\mathrm{b}})^2} + \sqrt{l_{\mathrm{h}}^2 + (e - 0.5 \cdot l_{\mathrm{b}})^2} - 2\sqrt{l_{\mathrm{h}}^2 + e^2}\right]$$

$$B_{\mathrm{b}} = 2f_{\mathrm{ts}} \cdot \left[4\sqrt{l_{\mathrm{h}}^2 + e^2} - 3\sqrt{l_{\mathrm{h}}^2 + (e - 0.5 \cdot l_{\mathrm{b}})^2} - \sqrt{l_{\mathrm{h}}^2 + (e + 0.5 \cdot l_{\mathrm{b}})^2}\right]$$

$$C_{\mathrm{b}} = 2f_{\mathrm{ts}} \cdot \left[4\sqrt{l_{\mathrm{h}}^2 + e^2} - 3\sqrt{l_{\mathrm{h}}^2 + (e + 0.5 \cdot l_{\mathrm{b}})^2} - \sqrt{l_{\mathrm{h}}^2 + (e - 0.5 \cdot l_{\mathrm{b}})^2}\right]$$

Figure 5.18 shows the resulting percent thread length shifts in the traverse motion triangle versus the bobbin length axis for different geometrical conditions.

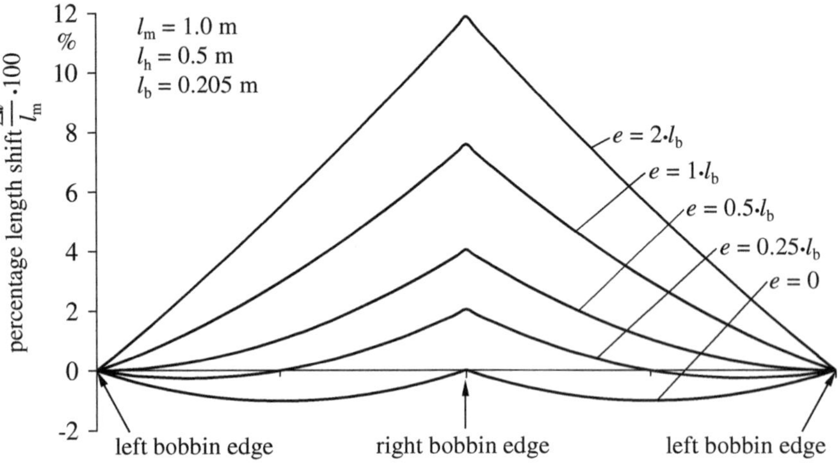

Fig. 5.18. Per cent thread length shifts in the traverse motion triangle caused by the traverse motion

In the next step, the resolution of the DEq. 5.36 for the time courses of the disturbances Δl defined by Eqs. 5.38 and 5.39 has to be found. The LAPLACE-transformation makes a resolution algorithm available for any time course by means of the convolution integral (see for instance [13]). The integral applied to the present case can be read as:

$$\underline{\Delta Tt_{\mathrm{o}}/\overline{\Delta l}} = \int_0^t \Delta l(\tau) \cdot \frac{Tt_{\mathrm{om}} \cdot v_{\mathrm{om}}}{l_{\mathrm{m}}^2} \cdot \exp\left(-\frac{t-\tau}{l} \cdot v_{\mathrm{om}}\right) d\tau \qquad (5.40)$$

The integrand consists of the product of the disturbance time function (here $\Delta l(\tau)$, the time variable t is to be substituted by the integration variable τ – Eqs. 5.38 and 5.39 are to be inserted practically) and the impulse response function of the wanted goal variable (here ΔTt_{o}) for the same cause variable

(here Δl, that means Eq. 5.37). The time variable is to be substituted finally by shifting the term $t-\tau$.
After the integrations, one gets the following related time functions of the percent fineness changes caused by the traverse motion thread guide:
Forward motion

$$\frac{\Delta Tt_o}{Tt_{om}}=\frac{100}{v_{om}}\left\langle\left(\frac{2A_b\cdot l_m}{v_{om}}-B_b\right)\left[1-\exp\left(-\frac{v_{om}}{l_m}\cdot t\right)\right]-2A_b\cdot t\right\rangle \quad (5.41)$$

Backward motion

$$\frac{\Delta Tt_o}{Tt_{om}}=\frac{100}{v_{om}}\left\langle\left(\frac{2A_b\cdot l_m}{v_{om}}-C_b\right)\left[1-\exp\left(-\frac{v_{om}}{l_m}\cdot t\right)\right]-2A_b\cdot t\right\rangle \quad (5.42)$$

The mathematical basis is now given for quantitative investigations by means of Eqs. 5.41 and 5.42. The changes ΔTt_o are completely calculable only for a single stroke of the traverse motion thread guide, because the time functions $\Delta l(t)$ are also not informable close too (the derivations are not defined at the turning back points of the bobbin edges and should be taken from one equation to the other). The destination of the steady state oscillation state is only possible by repeatedly modified joinings of Eqs. 5.41 and 5.42. It should be observed when the change time functions of two successive double strokes of the traverse motion thread guide do not distinguish themselves. It is also only possible to calculate the steady state oscillation state iteratively as a sequence of mathematical completely calculable single time courses of the thread fineness changes which partly depend on each other. Closed resolutions have become known (these develop the periodic, not continuous $\Delta l(t)$-course into a *Fourier*-row and only use the first element for further calculations [299]), but only with greater mistakes at the turning back points. However, the $\Delta l(t)$-course possesses the greatest gradation at only this points.

Extensive quantitative investigations have been carried out which included variations of the wind-up velocity v_{om} (1000...6000 m/min), the height of the traverse motion triangle l_h (0.3...1 m), the length of the bobbin l_b (0.135...0.205 m), the twice stroke frequency of the traverse motion thread guide f_{ts} (500...1500 min^{-1}), and the asymmetry of the top of the traverse motion triangle e ($0...2\cdot l_b$). Some selected result diagrams are shown in Figs. 5.19 (transient oscillations of the percent fineness shifts), 5.20 (percent maximal spans of the fineness shifts), and 5.21 (percent fineness shifts on the bobbin edges). The valid technological conditions should be learned from the signatures of the figures in each case.

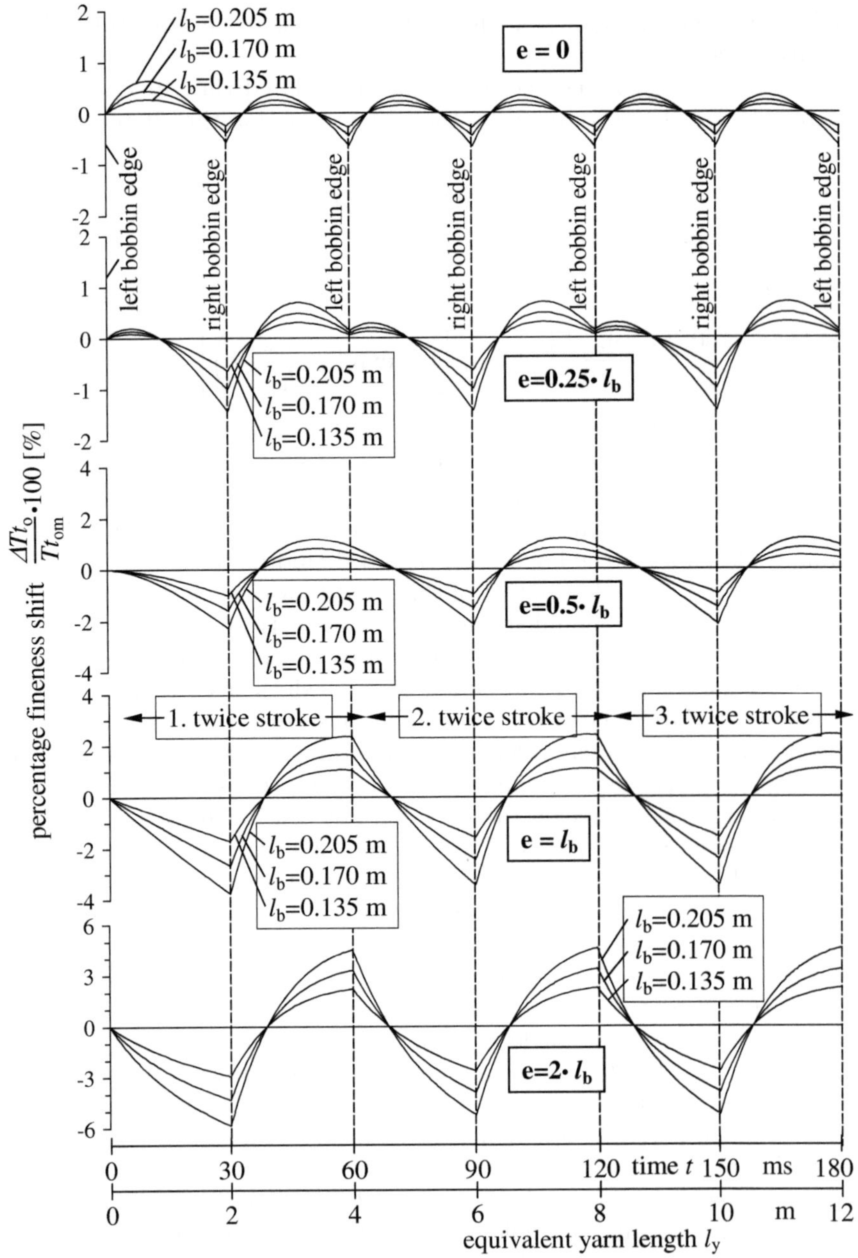

Fig. 5.19. Transient oscillations of percent fineness shift caused by the thread traverse motion; $l_m = 1.0$ m, $l_h = 0.5$ m, $v_{om} = 4000$ m/min, $f_{ts} = 1000$ min^{-1}

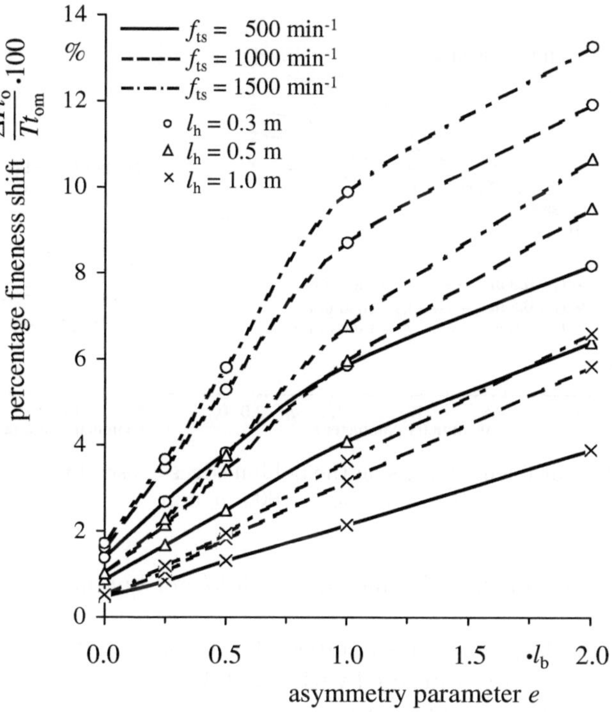

Fig. 5.20. Percent maximal span of the fineness shifts caused by the thread traverse motion; $l_m = 1.0$ m, $l_b = 0.205$ m, $v_{om} = 4000$ m/min

The following statements and conclusions can be given on the basis of all calculation results (also those which are not shown in the figures):

a) The fineness change course shows (for $< 0.5 \cdot l_b$) twice the frequency compared to systems with $> 0.5 \cdot l_b$ (see Fig. 5.19) at symmetric and little asymmetric traverse motion triangles.

b) The amplitudes of the shifts are smaller than $\pm 1\%$ for symmetric and slightly asymmetric traverse motion triangles and the so-called technological operating point ranges. They increase more for large asymmetries of the traverse motion triangle. If e increases from 0 to $0.5 \cdot l_b$ (triangle top is situated above the bobbin edge) then the fineness shift will nearly quadruple (see Fig. 5.20).

c) A strong ascent of the fineness shifts can be observed if the top of the traverse motion triangle moves nearer to the bobbin. The fineness shift is nearly doubled in the investigated operation range if l_h is halved (see Fig. 5.20).

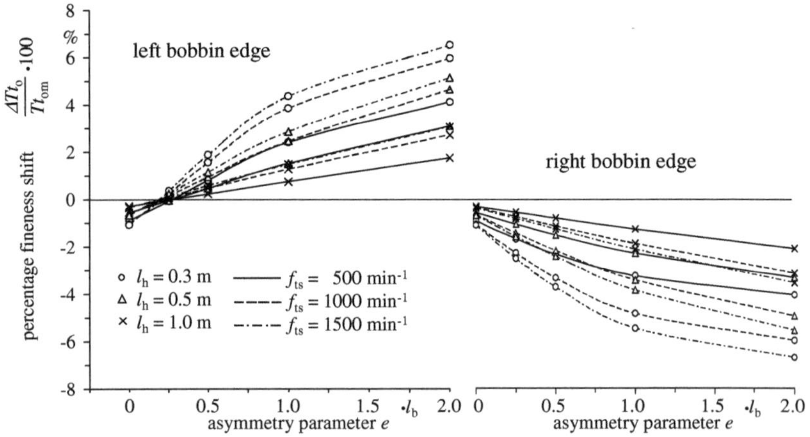

Fig. 5.21. Percent fineness shifts on the bobbin edges caused by the thread traverse motion; $l_m = 1.0$ m, $l_b = 0.205$ m, $v_{om} = 4000$ m/min

d) The fineness shifts also increase with longer bobbins (see Fig. 5.19).

e) High wind-up velocities dampen the fineness shifts. This is then the special case if they are connected with small double stroke frequencies of the traverse motion thread guide, because the thread will be transported faster out of the traverse motion field for these cases the fineness shift is then imprinted by the Δl-lengthening or -shortening.

f) The influence of the traverse motion frequency is quantitatively different. In principle high frequencies effect greater fineness shifts, but their influence is strongly dependent upon the selected wind-up velocity. The influence is weakly stamped for small wind-up velocities and strongly stamped for high wind-up velocities (see Fig. 5.20).

g) Symmetric traverse motion systems produce a symmetric bobbin structure in which fineness minima appear at the bobbin edges. These are double the size of the fineness maxima in the middle of the bobbin (not presented here). Bale-shaped bobbins with reduced bobbin edges will arise always with this (see Fig. 5.21).

h) Asymmetric traverse motion triangles produce in principle greater fineness shifts. But, the bobbin structure is also seemingly unsymmetric for such systems: The one bobbin edge will be built up stronger and corresponds to the mean value of the fineness, the other will then be provided with the maximum negative fineness shifts. It does not appear bale-shaped or barrel-shaped, but leads to conical bobbins (see Fig. 5.21).

i) The following constructive and technological conditions can be proposed which allow for the smallest fineness shifts (caused by the traverse motion):
- symmetric structure of the traverse motion system ($e = 0$)
- greatest possible height of the traverse motion triangle ($l_{\mathrm{h}} > 0.5$ m)
- as short as possible bobbins (manifold wind-up technology is ideal if each single bobbin can get its own symmetric traverse motion triangle!)
- lowest possible traverse motion velocity, that means small double stroke frequency.

The fineness shifts or changes (caused by the traverse motion system) can be held constant under these conditions at smaller than $\pm 0.2\%$ without special resources (for instance stress compensation rolls).

5.2 Dynamics of the Twist Transfer at the False Twist Texturing

5.2.1 Task

The majority of the PET and PA fine yarns manufactured today is texturised. The preferred process is the false twist (or FT-) texturing which realises more than 90% of all texturised fine yarns.
A twisting element gives the yarn torsional twists which are strongly imprinted (fixed) on this by means of consecutive heating and cooling. The imprinted so-called false twists will be dissolved after passing the twisting element and the texturised (equipped with a strong fixed curling) yarn is wound up. One distinguishes the magnetic spindle and the friction disk principles according to the kind of twist generation by means of the different false twist spindles. Both principles should generate torsional, or false twists, to the yarn on an only limited, usually short distance of the yarn length axis by means of quickly rotating machine elements. The torsion twist is to be generated by an intensive, as slip poor, as possible friction transmission through the yarn surface. It is not the task of this section to discuss the different advantages and disadvantages of the different basic principles of the false twist texturing. A lot of special literature is available concerning this.
However, some dynamic cause-effect relations of this process should be explained which regard the effect or goal variable twist density T_{D} of the texturised yarn in the texturing and setting zone. The expert knows from a-priori knowledge (one should once again recall to Sect. 2.5.1 - proposed steps for the working out the dynamic model) that texturing mistakes (that means, changes of the curling intensity along the texturised yarn) are caused mainly by dynamic changes of the product variable twist density T_{D} which only exists during the running process.

It is therefore obvious to elaborate a dynamic model of this product variable which theoretically founds the relationship of the different process and product variables. This should help to clear up the process analytical relationships between the disturbances of the variable T_D and the product quality. In the past, investigations have been done for either the steady state process only or the experimental results of measured twist densities and (also partly) for their changes (see [301] until [310]).

Nevertheless, it is to be remarked that simplifications and approximations (more than in other sections of the book) have been necessary for the following demonstrated investigations because of the complicated structure of the exact physical-analytical relationships. This necessity aims at presenting explicit resolution equations which should be usable without any further special knowledge. This should allow for qualitatively correct tendency statements in their proportions between each other on the one hand, but on the other is only restricted quantitative exactness connected with this.

5.2.2 Mathematic-Dynamic Model

Differential Equation of the Twist Density and its Solutions

The development of the DEq. for the twist density of the yarn in the texturing and setting zone T_D enforces some previous considerations and results which were informed in Sect. 5.1 (as a-priori knowledges to be declared now). The earlier recommended registration and sorting of all process and product variables, which are involveded in the process and are probably connected with each other via cause-effect relations (see Sect. 2.5.1) should be carried out here only verbally. The basis is the Fig. 5.22 which shows the necessary process and product variables of the thread course of a false twist texturing process with magnetic spindle twist element(magnetic false twist spindle) in a strongly simplified form.

It is irrelevant, by the way, whether the twists are generated by a friction spindle element or by a magnetic spindle. One can imagine that the several, short in series operating twists transmitting yarn touch points of a friction spindle are collected only in one integral effectual point. This one point then carries out a sum of effectual friction on the yarn. An explanation of this will be given at the end of this section.

The twist density T_D, as the quantity of twists referred to the thread length, (measured in the twisted state) obviously depends on

- the speed of the false twist spindle n_s,
- the present velocity of the twisted thread in the texturing and setting zone v_i, and
- the thread fineness Tt.

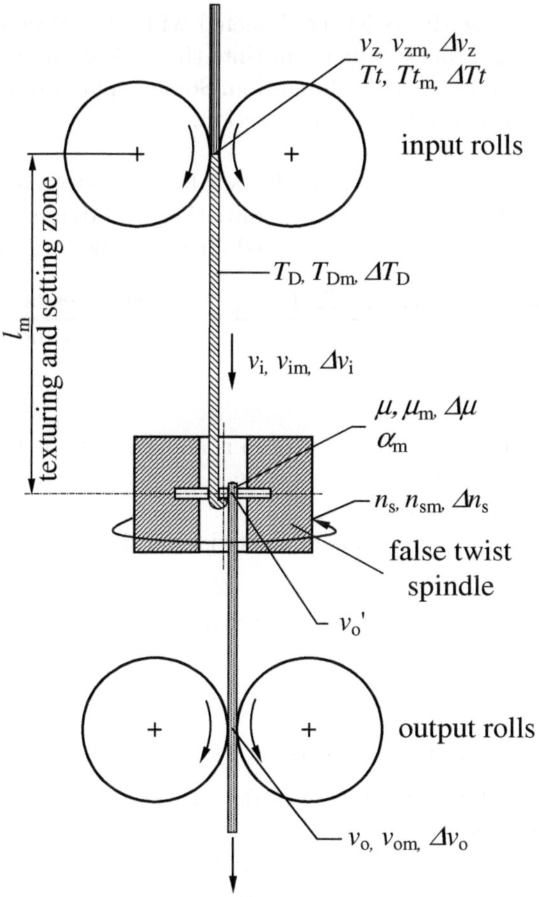

Fig. 5.22. Technological scheme of the false twist texturing process with magnetic twist spindle

The velocity of the twisted thread v_i (simultaneously equal to the input velocity of the twisted thread into the false twist spindle) finally turns up dependent upon

- the input velocity of the untwisted thread into the input rolls v_z,
- the thread output velocity v_o,
- the rope friction factor $e^{\mu\alpha_m}$,
- the thread fineness Tt, and
- the speed of the false twist spindle n_s.

Changes of the thermic process and product variables (for instance thread temperatures) and their effects on the thread properties (for instance changes of the E-moduli and the following changes of the thread tensile force relations before and after the twisting element) will be neglected, simultaneous to a draw texturing process which imprints the thread additionally a greater plastic longitudinal orientation elongation. Some explanatory remarks will be given to the latter at the end of this section.

If the texturing and setting zone l_m is taken as a store and exchange line of the yarn torsional twists per time unit, then the dynamic basic Eq. 2.23 (applied to the present case in a modified form) can be formulated as follows:

$$\frac{\text{twists inflow}}{\text{time}} = \frac{\text{twists discharge}}{\text{time}} + \frac{\text{changes of twists}}{\text{time}} \tag{5.43}$$

It is to be put in as:

twists inflow/time $= n_s$ (where no slip is assumed at the twisting element)
twists discharge/time $= T_D \cdot v_i$
changes of twists/time $= \dot{T}_D \cdot l_m$
(in which the same mean value of the twist density is assumed through the whole line l_m)

The DEq. of the twist density can be written then as:

$$n_s = T_D \cdot v_i + p \cdot T_D \cdot l_m \tag{5.44}$$

where $p = \dfrac{d}{dt}$ is the well-known LAPLACE-operator.
The steady state relation for the twist density $T_D = n_s/v_i$ follows for $\dot{T}_D = 0$ of course unrestricted.
Equation 5.44 contains the independent not-freely-adjustable process variable v_i which must be expressed by means of technological independent process and product variables. The shortening factor K_s which represents the ratio of the twisted length l_t to the stretched, untwisted length l_u of one twist of the same in each case unloaded yarn piece must first of all be defined:

$$K_s = l_t/l_u \tag{5.45}$$

Obviously it is also valid:

$$K_s = v_i/v_z \tag{5.46}$$

The thread length of one twist l_t is

$$l_t = l_m/z \tag{5.47}$$

where z is the number of the total twists in the zone l_m, that means:

$$z = \frac{n_s \cdot l_m}{v_i}$$

consequently

$$l_t = \frac{v_i}{n_s} = \frac{K_s \cdot v_z}{n_s} \tag{5.48}$$

The stretched, untwisted length l_u, of the same yarn piece, is calculated from the periphery of the cross sectional area of the multifilament (twisted) yarn which we will formally introduce as $\pi \cdot D_y$ (D_y is in this aspect the diameter of the multifilament (twisted) yarn).
The stretched, untwisted length of one yarn twist finally results as the length of the line of a screw with the diameter D_y and with the ascent height l_u to:

$$l_u = \sqrt{(\pi \cdot D_y)^2 + \frac{v_i^2}{n_s^2}} \tag{5.49}$$

v_i is determined in the real process by v_o (caused by the output rolls) which effects the friction body of the twisting element into the process line l_m.

We assume that the yarn along the whole friction thread line will be twisted on the twisting element and the false twists will be immediately dissolved in turn, if the twisted yarn leaves the friction body down to the output. In this peel-off moment in the twisted state, the yarn should possess the velocity v'_o and, immediately afterwards (after dissolving of the twists), the output velocity v_o. This is valid as:

$$v_o = \frac{v'_o}{K_s} \tag{5.50}$$

The already deviated relationships of the thread transport along a friction thread line (see Sect. 5.1.4) must be validated to the estimation of the velocity transition from v_i to v'_o. The desired relation can be derived from Eq. 5.10 considering the continuity equation of a transported yarn without losses $Tt_i \cdot v_i = \text{const}$. It is read as, using the symbols of our problem:

$$\frac{v'_o}{v'_z} = \left(\frac{v_i}{v'_z} - 1\right) \cdot \exp(\mu \cdot \alpha_m) + 1 \tag{5.51}$$

The independent given yarn input velocity v_i is to be inserted into Eq. 5.51 in its reduced form

$$v'_z = v_z \cdot K_s \tag{5.52}$$

because the twisted yarn is this aspect on the friction body.
Finally v_i is to be now written by using the Eqs. 5.45 and 5.48 to 5.52

$$v_i = \sqrt{[(v_o - v_z) \cdot \exp(-\mu\alpha_m) + v_z]^2 - (\pi \cdot D_y \cdot n_s)^2} \tag{5.53}$$

Before the final DEq. of the twist density T_D will be formulated, the yarn diameter D_y shall be expressed depending on the yarn fineness Tt in the following form:

$$D_y = K_p \cdot \sqrt{Tt} \tag{5.54}$$

K_p is a polymer-specific constant. The yarn diameter D_y will be reached in the unit of m, if Tt is set in tex and for K_p the following values are set at:

$K_p = 33.3 \cdot 10^{-6}$ for the fibre material PA (density $= 1.12\ \mathrm{g/cm^3}$)
$K_p = 30 \cdot 10^{-6}$ for the fibre material PET (density $= 1.38\ \mathrm{g/cm^3}$)

The final mathematic-dynamic model equation of the twist density T_D can be calculated now on the basis of DEq. 5.44 and under the consideration of the Eqs. 5.53 and 5.54 as follows:

$$\Phi = n_s - T_D \cdot \sqrt{[(v_o - v_z) \cdot \exp(-\mu\alpha_m) + v_z]^2 - Tt(\pi K_p \cdot n_s)^2} - p \cdot T_D \cdot l_m = 0 \tag{5.55}$$

Differential equation 5.55 can be partially derived (for the purpose of its linearisation) from the potentially changeable variables T_D, n_s, v_z, v_o, Tt and μ according to the already often demonstrated manner:

$$\frac{\partial \Phi}{\partial T_D} \cdot \Delta T_D + \frac{\partial \Phi}{\partial n_s} \cdot \Delta n_s + \frac{\partial \Phi}{\partial v_z} \cdot \Delta v_z + \frac{\partial \Phi}{\partial v_o} \cdot \Delta v_o + \frac{\partial \Phi}{\partial Tt} \cdot \Delta Tt + \frac{\partial \Phi}{\partial \mu} \cdot \Delta \mu = 0$$

If the partial differentiations are carried out then one gets the linearised DEq. of the twist density T_D as follows. It should be considered through this that $T_{Dm} = n_{Dm}/W_a$

$$\begin{aligned}
&\left[\frac{W_a^2}{n_{sm}} + Tt_m \cdot n_{sm}(\pi K_p)^2\right] \cdot \Delta n_s - \langle N_a[1 - \exp(-\mu_m\alpha_m)]\rangle \cdot \Delta v_z \\
&- N_a \cdot \exp(-\mu_m\alpha_m) \cdot \Delta v_o + 0.5 \cdot (\pi K_p \cdot n_{sm})^2 \cdot \Delta Tt \\
&+ [\alpha_m \cdot N_a(v_{om} - v_{zm}) \cdot \exp(-\mu_m\alpha_m)] \cdot \Delta\mu - \frac{W_a^2}{n_{sm}}[W_a + p \cdot l_m] \cdot \Delta T_D = 0
\end{aligned} \tag{5.56}$$

The abbreviations W_a and N_a mean:

$$W_a = \sqrt{[(v_{om} - v_{zm}) \cdot \exp(-\mu_m\alpha_m) + v_{zm}]^2 - Tt_m(\pi K_p \cdot n_{sm})^2} \tag{5.57}$$

$$N_a = (v_{om} - v_{zm}) \cdot \exp(-\mu_m\alpha_m) + v_{zm} \tag{5.58}$$

Equation 5.56 can be used now to calculate the *dynamic transfer functions*, which inform about the influence of Δn_s-, Δv_z-, Δv_o-, ΔTt-, or $\Delta\mu$- changes to changes of the twist density ΔT_D:

$$G_1(p) = \frac{\Delta T_\mathrm{D}}{\Delta n_\mathrm{s}} = \frac{\frac{W_\mathrm{a}^2}{n_\mathrm{sm}} + Tt_\mathrm{m} \cdot n_\mathrm{sm}(\pi K_\mathrm{p})^2}{\frac{W_\mathrm{a}^2}{n_\mathrm{sm}}(W_\mathrm{a} + p \cdot l_\mathrm{m})} \tag{5.59}$$

$$G_2(p) = \frac{\Delta T_\mathrm{D}}{\Delta v_\mathrm{z}} = -\frac{N_\mathrm{a}[1 - \exp(-\mu_\mathrm{m}\alpha_\mathrm{m})]}{\frac{W_\mathrm{a}^2}{n_\mathrm{sm}}(W_\mathrm{a} + p \cdot l_\mathrm{m})} \tag{5.60}$$

$$G_3(p) = \frac{\Delta T_\mathrm{D}}{\Delta v_\mathrm{o}} = -\frac{N_\mathrm{a} \cdot \exp(-\mu_\mathrm{m}\alpha_\mathrm{m})}{\frac{W_\mathrm{a}^2}{n_\mathrm{sm}}(W_\mathrm{a} + p \cdot l_\mathrm{m})} \tag{5.61}$$

$$G_4(p) = \frac{\Delta T_\mathrm{D}}{\Delta Tt} = \frac{0.5(\pi K_\mathrm{p} \cdot n_\mathrm{sm})^2}{\frac{W_\mathrm{a}^2}{n_\mathrm{sm}}(W_\mathrm{a} + p \cdot l_\mathrm{m})} \tag{5.62}$$

$$G_5(p) = \frac{\Delta T_\mathrm{D}}{\Delta \mu} = \frac{\alpha_\mathrm{m} \cdot N_\mathrm{a}(v_\mathrm{om} - v_\mathrm{zm}) \cdot \exp(-\mu_\mathrm{m}\alpha_\mathrm{m})}{\frac{W_\mathrm{a}^2}{n_\mathrm{sm}}(W_\mathrm{a} + p \cdot l_\mathrm{m})} \tag{5.63}$$

The corresponding normalised *amplitude frequency responses* of these transfer functions (all of them show proportional action with delay of first order) are the following:

$$|G_1(\mathrm{j}f)| = \left|\frac{\widetilde{\Delta T_\mathrm{D}}/T_\mathrm{Dm}}{\widetilde{\Delta n_\mathrm{s}}/n_\mathrm{sm}}\right| = \frac{W_\mathrm{a}^2 + Tt_\mathrm{m}(\pi K_\mathrm{p} \cdot n_\mathrm{sm})^2}{W_\mathrm{a}\sqrt{W_\mathrm{a}^2 + (2\pi f \cdot l_\mathrm{m})^2}} \tag{5.64}$$

$$|G_2(\mathrm{j}f)| = \left|\frac{\widetilde{\Delta T_\mathrm{D}}/T_\mathrm{Dm}}{\widetilde{\Delta v_\mathrm{z}}/v_\mathrm{zm}}\right| = (-)\frac{v_\mathrm{zm} \cdot N_\mathrm{a}[1 - \exp(-\mu_\mathrm{m}\alpha_\mathrm{m})]}{W_\mathrm{a}\sqrt{W_\mathrm{a}^2 + (2\pi f \cdot l_\mathrm{m})^2}} \tag{5.65}$$

$$|G_3(\mathrm{j}f)| = \left|\frac{\widetilde{\Delta T_\mathrm{D}}/T_\mathrm{Dm}}{\widetilde{\Delta v_\mathrm{o}}/v_\mathrm{om}}\right| = (-)\frac{v_\mathrm{om} \cdot N_\mathrm{a} \cdot \exp(-\mu_\mathrm{m}\alpha_\mathrm{m})}{W_\mathrm{a}\sqrt{W_\mathrm{a}^2 + (2\pi f \cdot l_\mathrm{m})^2}} \tag{5.66}$$

$$|G_4(\mathrm{j}f)| = \left|\frac{\widetilde{\Delta T_\mathrm{D}}/T_\mathrm{Dm}}{\widetilde{\Delta Tt}/Tt_\mathrm{m}}\right| = \frac{0.5 \cdot Tt_\mathrm{m}(\pi K_\mathrm{p} \cdot n_\mathrm{sm})^2}{W_\mathrm{a}\sqrt{W_\mathrm{a}^2 + (2\pi f \cdot l_\mathrm{m})^2}} \tag{5.67}$$

$$|G_5(\mathrm{j}f)| = \left|\frac{\widetilde{\Delta T_\mathrm{D}}/T_\mathrm{Dm}}{\widetilde{\Delta \mu}/\mu_\mathrm{m}}\right| = \frac{\mu_\mathrm{m} \cdot \alpha_\mathrm{m} \cdot N_\mathrm{a}(v_\mathrm{om} - v_\mathrm{zm}) \cdot \exp(-\mu_\mathrm{m}\alpha_\mathrm{m})}{W_\mathrm{a}\sqrt{W_\mathrm{a}^2 + (2\pi f \cdot l_\mathrm{m})^2}} \tag{5.68}$$

The corresponding *phase frequency response* of the *complex frequency responses* $G_1(\mathrm{j}f)$, $G_4(\mathrm{j}f)$ and $G_5(\mathrm{j}f)$ are uniformly:

$$\varphi(f) = \mathrm{arc\ tan}\left[-\frac{2\pi f \cdot l_\mathrm{m}}{W_\mathrm{a}}\right] \tag{5.69}$$

The corresponding *phase frequency response* of the *complex frequency responses* $G_2(\mathrm{j}f)$ and $G_3(\mathrm{j}f)$ are:

$$\varphi(f) = \text{arc tan}\left[-\frac{2\pi f \cdot l_\mathrm{m}}{W_\mathrm{a}}\right] - \pi \tag{5.70}$$

An incidental evaluation of the achieved equations should be undertaken at this point before we can discuss a quantitative result presentation of a technological process example:

a) In each case, the amplitude transmission factors of Eqs. 5.64 to 5.68 do not only depend on the disturbance frequency f but also, strongly, on the mean values of the independent process and product variables which define the viewed technological operation point. This is an expression for the declared complexity of the investigated general problem in spite of the simplifications and neglects made in the statements.

b) The time constant T_c and the critical frequency f_c of the process line are qualitatively the same for all kinds of viewed disturbances and depend likewise on the mean values of all independent process and product variables. They can be best written from the phase frequency response Eq. 5.70 (see to this Sect. 4.2.1):

$$\frac{1}{2\pi f_\mathrm{c}} = T_\mathrm{c} = \frac{l_\mathrm{m}}{W_\mathrm{a}} \tag{5.71}$$

The abbreviation W_a (see Eq. 5.57) is identical with the yarn velocity in the texturing and setting zone.

c) The parameter W_a allows the data of a limit condition. This is not allowed to be exceeded if a proper thread line (that really means the carrying out of the process) should be reached. This deals with the yarn tensile stress, yarn velocity and yarn twist density relations. This limited condition can be defined as maximum eligible speed of the false twist spindle n_sm dependent upon the other free eligible process and product variables. They can be derived from the trivial condition that the radicand of the root (Eq. 5.57) is not allowed to be negative. It must always be fulfilled as:

$$n_\mathrm{sm} \leq \frac{(v_\mathrm{om} - v_\mathrm{zm}) \cdot \exp(-\mu_\mathrm{m}\alpha_\mathrm{m}) + v_\mathrm{zm}}{\pi \cdot K_\mathrm{p} \cdot \sqrt{Tt_\mathrm{m}}} \tag{5.72}$$

The following dimensions are to be inserted in Eqs. 5.53 to 5.72 for quantitative calculations:

v_{om}, v_{zm}	numerical values to put in m/min
n_{sm}	numerical values to put in min^{-1}
l_m	numerical values to put in m
Tt_m	numerical values to put in tex
f	numerical values to put in min^{-1} or 60Hz
K_p	numerical values to put in as described in Eq. 5.54
α_m	numerical values to put in the radian measure (dimensionless)
μ_m	coefficient of friction (dimensionless)

Technological Application and Analysis

At first Fig. 5.23 shows the limit curves of the eligible revolutions per minute of the false twist spindle n_{sm} for the fine yarn range (2 until 20 tex) and dependent upon different velocity and angle of wrap levels for the fibre materials PA and PET. These correspond to the further investigated technological variants which are collected in Table 5.1.

The curves confirm the practical experiences that the eligible (and also necessary to reach an appointed twist density T_{Dm}) speed of the false twist spindle n_{sm} increases if the fineness Tt_m decreases and the yarn take-up velocity v_{om} increases. It is to be remarked that the results have the correct size. The latter is moreover an indirect proof for the exactness for the principle of the selected model idea.

Relating to practice, variants of the process realisation have been selected for detailed analysis by means of the developed model equations. They have been collected in Table 5.1 considering the preliminary investigations. The last two columns contain the time constants T_c and the critical frequencies f_c of the texturing and setting zone which roughly characterise the dynamic transmission behaviour.

The time constants T_c amount to about 2.50 s for the velocity level 100 m/min and about 1.65 s for the velocity level 200 m/min (indeed with an assumed increased length of l_m = 4 m). It can be concluded from this that aperiodic disturbances of the viewed process and product variables will effect undefined bulkiness properties along a thread length of about 12 m in the first case (v_{om} = 100 m/min) and of about 17 m in the second case (v_{om} = 200 m/min).

The small critical frequencies f_c of about 0.065 Hz at v_{om} = 100 m/min and of about 0.1 Hz at v_{om} = 200 m/min mean that periodic disturbances of only relatively small frequency will have a considerable influence on the twist density $\widetilde{\Delta T_D}$. Frequencies > 0.5 Hz will already have a sufficient dampening in each case. The following figures also show this.

The normalised amplitude frequency responses $|G_1(jf)|$ to $|G_5(jf)|$ are presented in Figs. 5.24 to 5.28 for all technological variants according to Table 5.1.

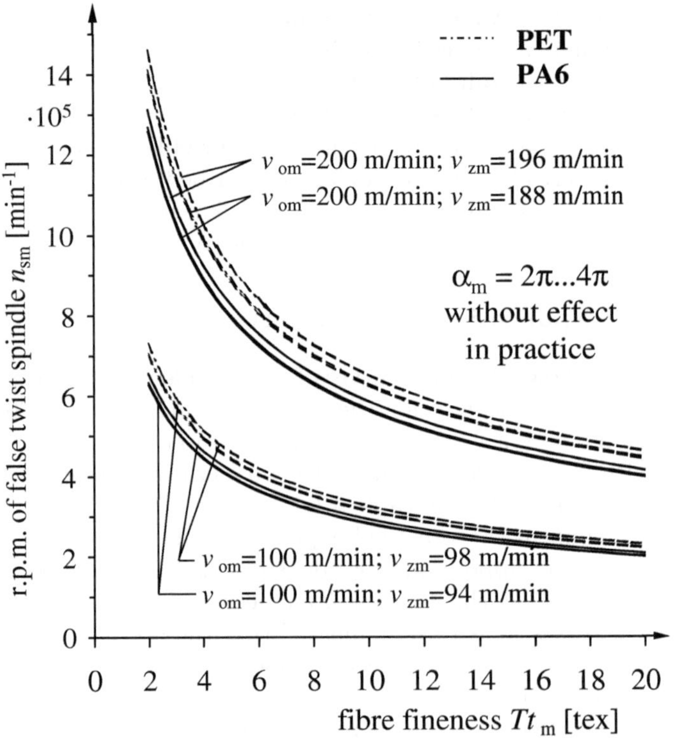

Fig. 5.23. Limit curves of eligible revolutions per minute of the false twist spindle n_{sm} dependent upon different technological conditions according to Eq. 5.72. Further data see Table 5.1

Comparing the study of the figures leads to the following general conclusions for the disturbance estimation:

a) Speed changes of the twisting element $\widetilde{\Delta n_s}$ have the greatest effect on the twist density T_D ($|G_1(jf)|$, Fig. 5.24). Relative amplification factors of 1.0 to 2.0 will be reached in the disturbance frequency range $f < f_c$. This means that $\widetilde{\Delta n_s}$-changes of 1% can effect $\widetilde{\Delta T_D}$-changes from 1 to 2%. The finenesses of the texturised thread Tt_m are of only small influence, in which the coarser thread will be influenced more (specifically at the quasi steady state disturbances – f nearly 0 Hz) than the finer. The usual variations of the thread input velocity v_z relative to the thread output velocity v_o (present lags are assumed from –2% to –6%) and number of wraps around the pin of the twisting element (assumed 1 or 2 wraps) practically have no influence on the start height of the curves and their further courses. On the other hand, a PA-thread will be influenced more strongly than a PET-thread of the same fineness.

Table 5.1. Technological variants of the investigated FT-texturing process

Process variant	v_{om} in m/min	v_{zm} in m/min	T_{Dm} in m^{-1}	n_{sm} in m^{-1}	α_m	μ_m	K_p	Tt_m in tex	l_m in m	T_c in s	f_c in Hz
							PA-fibres				
1	100	98	4300	325348	2π	0.3	$33.3 \cdot 10^{-6}$	3.4	3.0	2.38	0.067
2	100	94	4300	314120	2π	0.3	$33.3 \cdot 10^{-6}$	3.4	3.0	2.46	0.065
3	100	98	4300	324496	4π	0.3	$33.3 \cdot 10^{-6}$	3.4	3.0	2.39	0.067
4	100	94	4300	311562	4π	0.3	$33.3 \cdot 10^{-6}$	3.4	3.0	2.48	0.064
							PA-fibres				
5	100	98	3000	221776	2π	0.3	$33.3 \cdot 10^{-6}$	7.8	3.0	2.43	0.065
6	100	94	3000	214122	2π	0.3	$33.3 \cdot 10^{-6}$	7.8	3.0	2.52	0.063
7	100	98	3000	221195	4π	0.3	$33.3 \cdot 10^{-6}$	7.8	3.0	2.44	0.065
8	100	94	3000	212379	4π	0.3	$33.3 \cdot 10^{-6}$	7.8	3.0	2.54	0.063
							PET-fibres				
9	100	98	3000	232598	2π	0.3	$30.0 \cdot 10^{-6}$	7.6	3.0	2.32	0.069
10	100	94	3000	224570	2π	0.3	$30.0 \cdot 10^{-6}$	7.6	3.0	2.40	0.066
11	100	98	3000	231988	4π	0.3	$30.0 \cdot 10^{-6}$	7.6	3.0	2.33	0.068
12	100	94	3000	222742	4π	0.3	$30.0 \cdot 10^{-6}$	7.6	3.0	2.42	0.066
							PET-fibres				
13	100	98	2300	169244	2π	0.3	$30.0 \cdot 10^{-6}$	16.7	3.0	2.45	0.065
14	100	94	2300	163403	2π	0.3	$30.0 \cdot 10^{-6}$	16.7	3.0	2.53	0.063
15	100	98	2300	168800	4π	0.3	$30.0 \cdot 10^{-6}$	16.7	3.0	2.45	0.065
16	100	94	2300	162072	4π	0.3	$30.0 \cdot 10^{-6}$	16.7	3.0	2.55	0.062
							PA-fibres				
17	200	196	4300	650696	2π	0.3	$33.3 \cdot 10^{-6}$	3.4	4.0	1.59	0.100
18	200	188	4300	628240	2π	0.3	$33.3 \cdot 10^{-6}$	3.4	4.0	1.64	0.097
19	200	196	4300	648992	4π	0.3	$33.3 \cdot 10^{-6}$	3.4	4.0	1.59	0.100
20	200	188	4300	623125	4π	0.3	$33.3 \cdot 10^{-6}$	3.4	4.0	1.66	0.096
							PA-fibres				
21	200	196	3000	443552	2π	0.3	$33.3 \cdot 10^{-6}$	7.8	4.0	1.62	0.098
22	200	188	3000	428244	2π	0.3	$33.3 \cdot 10^{-6}$	7.8	4.0	1.68	0.095
23	200	196	3000	442390	4π	0.3	$33.3 \cdot 10^{-6}$	7.8	4.0	1.63	0.098
24	200	188	3000	424757	4π	0.3	$33.3 \cdot 10^{-6}$	7.8	4.0	1.70	0.094
							PET-fibres				
25	200	196	3000	465196	2π	0.3	$30.0 \cdot 10^{-6}$	7.6	4.0	1.55	0.103
26	200	188	3000	449141	2π	0.3	$30.0 \cdot 10^{-6}$	7.6	4.0	1.60	0.099
27	200	196	3000	463977	4π	0.3	$30.0 \cdot 10^{-6}$	7.6	4.0	1.55	0.103
28	200	188	3000	445484	4π	0.3	$30.0 \cdot 10^{-6}$	7.6	4.0	1.62	0.098
							PET-fibres				
29	200	196	2300	338487	2π	0.3	$30.0 \cdot 10^{-6}$	16.7	4.0	1.63	0.098
30	200	188	2300	326806	2π	0.3	$30.0 \cdot 10^{-6}$	16.7	4.0	1.69	0.094
31	200	196	2300	337600	4π	0.3	$30.0 \cdot 10^{-6}$	16.7	4.0	1.64	0.097
32	200	188	2300	324145	4π	0.3	$30.0 \cdot 10^{-6}$	16.7	4.0	1.70	0.093

b) Changes $\widetilde{\Delta v_z}$ effect changes $\widetilde{\Delta T_D}$ quantitatively in the same manner ($|G_2(jf)|$, Fig. 5.25). Amplification factors of 1.0 to 1.9 are also to be observed in the frequency range $f < f_c$ for such disturbances.

Smaller differences are given as follows: The coarser thread and the thread with the greater number of wraps around the pin of the twisting element will be disturbed relatively stronger than the finer thread and the thread with only one wrap around the pin ($\alpha_m = 2\pi$). PA-threads are more disturbance endangered than PET-threads. The influence of the lag of v_{zm} is relatively

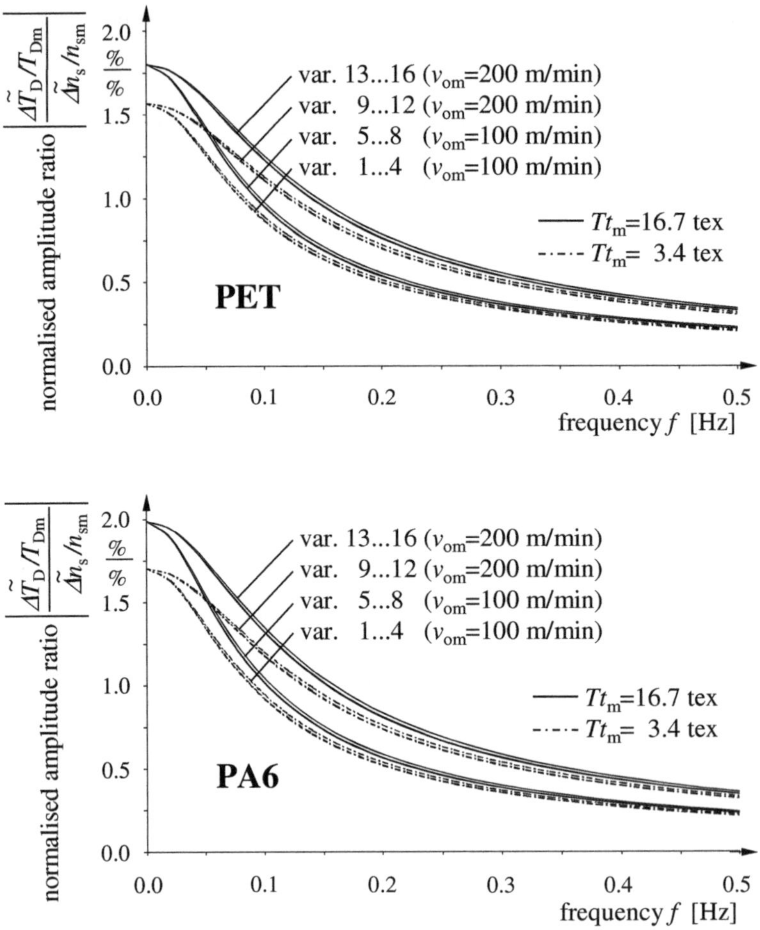

Fig. 5.24. Normalised amplitude frequency responses of twist density changes $\widetilde{\Delta T_D}$ caused by changes of the false twist spindle speed $\widetilde{\Delta n_s}$ at the FT-texturing process. Further data see Table 5.1

insignificant; the variants with –6% lag have only somewhat lower amplitude frequency response curves than those with –2% lag.

c) The output velocity v_o clearly influences the goal variable $\widetilde{\Delta T_D}$ less than the process variables n_s and v_z, named under a) and b) ($|G_3(jf)|$, Fig. 5.26). Changes $\widetilde{\Delta v_o}$ reach relative amplitude amplification factors of < 0.32, that means a more than 3% periodic change of $\widetilde{\Delta v_o}$ around its mean value v_{om} will effect a 1% periodic change of the twist density $\widetilde{\Delta T_D}$ in an appropriately small frequency range. A clear separation of the curves dependent on the angle of wrap α_m appears as well. The amplification factors nearly reach 0.32 (PA 6)

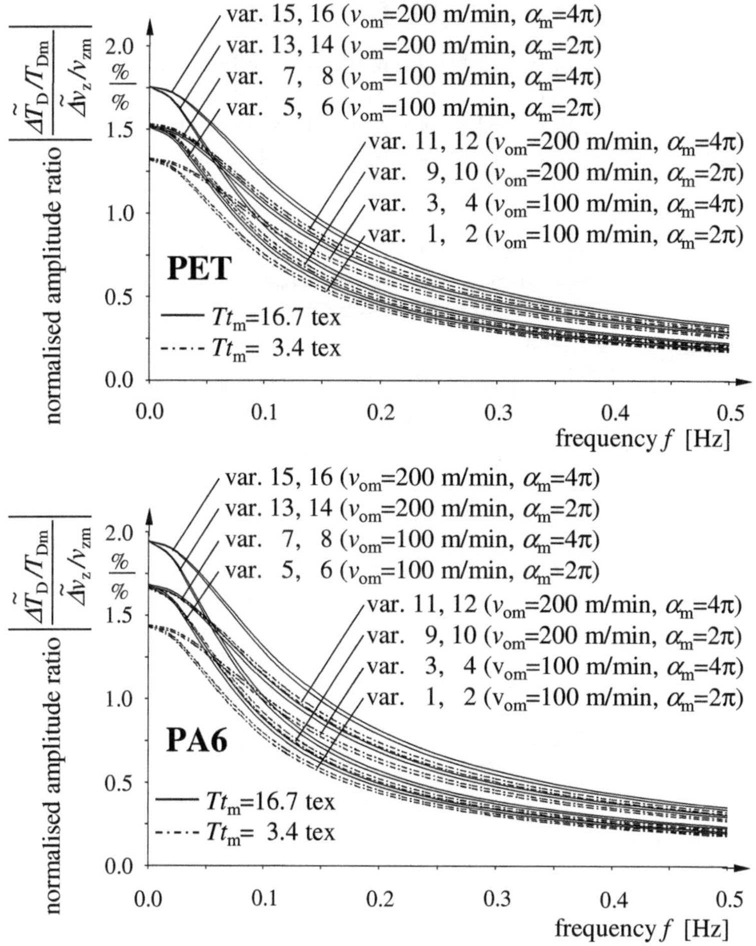

Fig. 5.25. Normalised amplitude frequency responses of twist density changes $\widetilde{\Delta T_D}$ caused by changes of the input velocity $\widetilde{\Delta v_z}$ at the FT-texturing process. Further data see Table 5.1

for 1 wrap around the pin of the twisting element and they only reach insignificant values of 0.05 for 2 wraps. Smaller insignificant differences can be observed as follows: The process with the greater lag of v_{zm} opposite v_{om} and the coarser threads show somewhat greater disturbance transmission factors. PA is to be assessed again more unfavourably than PET.
This total result is also physical-obviously plausible so far, as output velocity changes must be transmitted into the texturing and setting zone of the friction wrap line at first which effects dampening in this case. Tho output velocity changes are effective only indirectly to the change $\widetilde{\Delta T_D}$ of the influence of the real velocity pertaining to the twisted thread v_i (see Eq. 5.57,

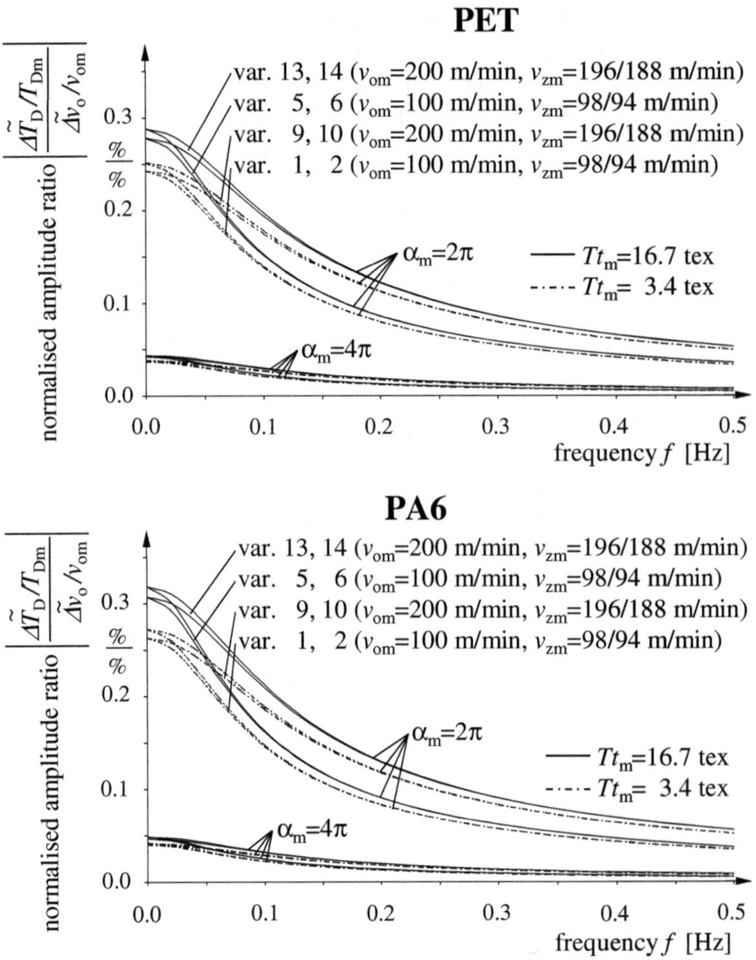

Fig. 5.26. Normalised amplitude frequency responses of twist density changes $\widetilde{\Delta T_D}$ caused by changes of the output velocity $\widetilde{\Delta v_o}$ at the FT-texturing process. Further data see Table 5.1

abbreviation W_a is identical with v_{im}). It is intelligible that a lengthening of this dampening friction thread line additionally degrades the influence of $\widetilde{\Delta v_o}$ to $\widetilde{\Delta T_D}$ drastically.

d) Also the fineness of the texturising thread Tt and its changes $\widetilde{\Delta Tt}$ effect changes of the twist density $\widetilde{\Delta T_D}$ only with a maximum amplitude transmission factor of 0.3 to 0.5 ($|G_4(jf)|$, Fig. 5.27). The curves obviously show the larger, more unfavourable, values for coarser threads and (frequently pointed out previously) for PA compared to PET.

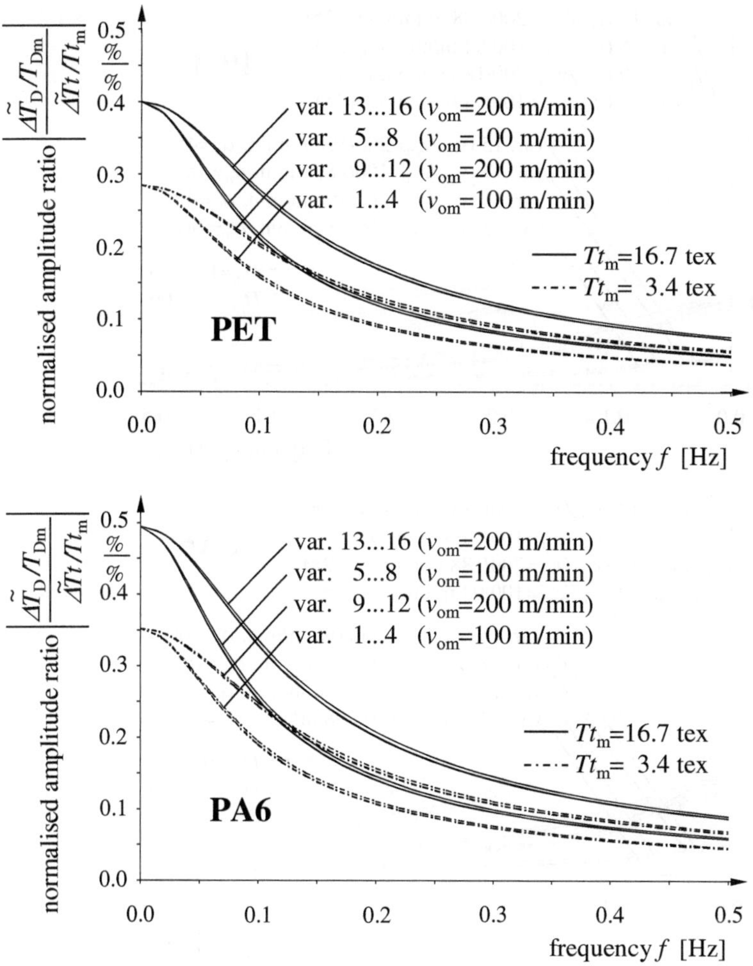

Fig. 5.27. Normalised amplitude frequency responses of twist density changes $\widetilde{\Delta T_{\mathrm{D}}}$ caused by changes of the thread fineness $\widetilde{\Delta Tt}$ at the FT-texturing process. Further data see Table 5.1

e) The system behaviour regarding to changes $\widetilde{\Delta\mu}$ and their effect on changes $\widetilde{\Delta T_{\mathrm{D}}}$ ($|G_5(\mathrm{j}f)|$) is demonstrated in Fig. 5.28. The amplitude transmission factors are extremely small with ≤ 0.04 for which a great angle of wrap and/or a small v_{z}-lag causes a further drastic decrease. Changes $\widetilde{\Delta\mu}$ of about 30 to 40% would be necessary to effect a 1% change $\widetilde{\Delta T_{\mathrm{D}}}$. This is to be practically excluded. But, it is to be remarked additionelly, that the present model comprises changes $\widetilde{\Delta\mu}$ which only effect twist density changes by means of velocity changes Δv_{i} of the texturised thread. Effects are not included, on the other hand, which twist slip appearences at the pin of the twisting ele-

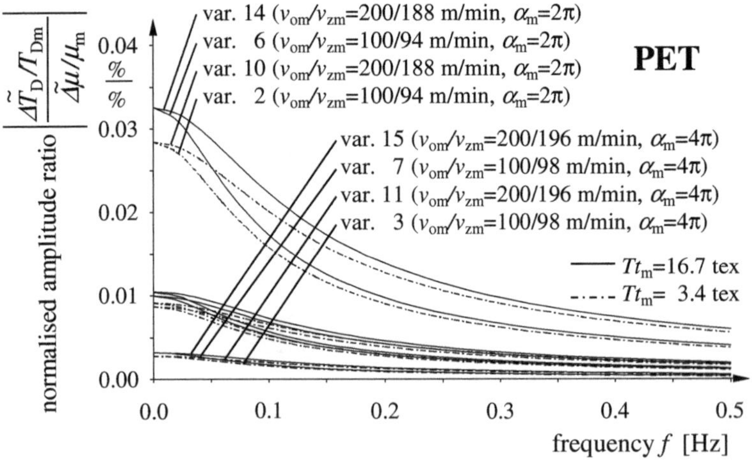

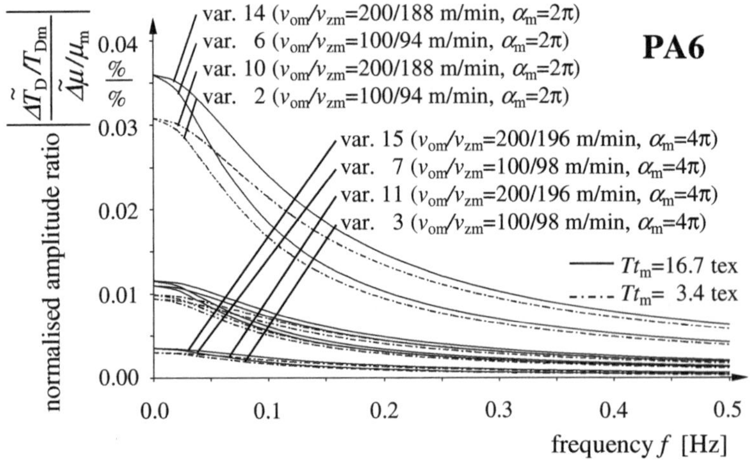

Fig. 5.28. Normalised amplitude frequency responses of twist density changes $\widetilde{\Delta T_D}$ caused by changes of the friction coefficient $\widetilde{\Delta\mu}$ at the FT-texturing process. Further data see Table 5.1

ment (missing friction adhesion) evoked by means of great changes to $\widetilde{\Delta\mu}$. Of course, these twist slip appearences possibly effect much greater twist density changes but they are not covered by the chosen model.

The common valid phase frequency response curves for the dynamic transmission functions $G_1(p)$ to $G_5(p)$ (Eqs. 5.69 and 5.69) are shown in Fig. 5.29 for all technological variants according to Table 5.1. The delay thread length L_d can be calculated by means of these because this has been demonstrated in Sect. 5.1.5 (see also Eq. 2.50). The length L_d represented in the present example which thread length leaves the texturing and setting zone before a

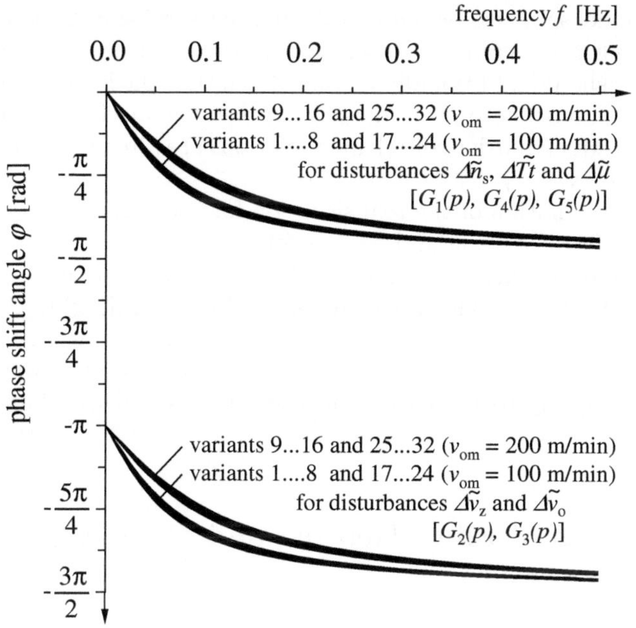

Fig. 5.29. Phase frequency responses of twist density changes $\widetilde{\Delta T_D}$ caused by changes of all regarded disturbance quantities at the FT-texturing process. Further data see Table 5.1

change $\widetilde{\Delta n_s}$, $\widetilde{\Delta v_o}$, $\widetilde{\Delta v_z}$, $\widetilde{\Delta Tt}$ or $\widetilde{\Delta \mu}$ will or can be visible in the output thread as changes $\widetilde{\Delta T_D}$. It has already been mentioned before that certain correlations exist here with the time constant T_c or the critical frequency f_c.

Two remarks should be given at the end of this subsection to the transmission of the presented model to other process realisations of the FT-texturing:

1. All model equations should also be transmittable to the friction texturing process considering the developed imaginations and derivations in Sects. 5.1.4 and 5.1.5, if

a) the angle of wrap α_m is interpreted as the sum of single angles which are imprinted repeatedly to the thread by means of the friction disk. The friction unit is assumed to be a collected effect at this point.

b) the efficiency of the twist transmission from the twisting element to the thread[1] is considered in the DEqs. 5.44 or 5.56 to be in the form of $0.7 \cdot T_D$, $0.7 \cdot T_{Dm}$ or $0.7 \cdot \Delta T_D$. The normalised results of Eqs. 5.64 to 5.68 are then applicable in the same manner.

[1] It is usually realised for the magnetic spindle principle with 100%; the efficiency is only $\leq 70\%$ for the friction principle

2. The model seems to be more problematic to transmit to a simultaneous draw-texturing process because twist generation and drawing are superimposed here. It should additionally be considered that the heating of the thread (not considered until here) well-known decreases the mechanical draw energy in the thread. Much harder conceivable relations are already given under pure steady state view.
A coarse approximation of the real situation for the dynamic model could be that one modifies Eqs. 5.53 and 5.57 as follows:

a) The input velocity of the partial else to drawing thread v_z is to be corrected by multiplying it with the draw ratio which is realised in the texturing and setting zone. This product, named v_{zc}, is then (instead of v_z) the basis for the calculation of the velocity v_i which appoints the twist density.

b) It is generally to be inserted in the fineness and the diameter of the drawn thread only.

5.3 Dynamics of Fibre Heating and Cooling

5.3.1 Task

The mathematical description of lapses of thermic proceedings is of importance in the modelling of appointed fibre formation and processing processes. This becomes true specifically for process stages of manufacturing and processing processes of polymer threads. The fibre formation of melts, the draw process (if thermic energy is induced), and the texturing are examples of this. Heating as well as cooling processes are to be described for the modelling of the temperature-time-courses in the thread.

Thread heating and cooling processes are effected by means of *heat spreading processes* which are due to three different basic principles:

a) *Heat conduction* which is to be described as heat spreading in solid, resting fluid and resting gaseous bodies from points of higher temperature to points of lower temperature.

b) *Heat convection* or *heat transfer* which is to be described as heat transport by means of flowing fluids or flowing gases which is to be distinguished between the *enforced* convection (flow separately generated) and the *free* convection (flow arises from itself by means of density or pressure differences).

c) *Heat radiation* which is to be described as heat transfer between bodies by means of electro-magnetic waves of the infrared spectral range without the co-operation of a transmission medium.

All three principles are effective for the thread heating depending on the process stage, for the thread cooling essentially only the principles a) and b).

The *convection* or the *heat transfer* is of *essential importance* to the calculation – and with this the modelling – of heating and cooling processes of running threads in fibre formation lines of melt spinning processes, around the heated or unheated godet systems and in texturing heaters.

Questions relating to such dynamic models can be for instance:

a) At which distance from the spinneret and under which conditions a melt spun thread will be solidified?
b) How many thread wraps around a heated godet or a system of such godets are necessary (for appointed boundary conditions, for instance thread fineness, thread velocity, godet geometry), that a maximum possible thread temperature can be reached?
c) How high is the reached mean thread temperature (under given boundary conditions) in relation to the godet temperature?
d) How large are the fluctuations of the thread temperature of the mean value which the thread suffers at one wrap around the godet system.

The answer to those questions is necessary for an optimal process operation as well as to the design of machine elements. In the following a simplified description of the heat transfer between a thread and its surrounding which enables process applied calculations with sufficient exactness as a rule will be given. The heat transfer between threads and metallic surfaces and between threads and air is theoretically explained in more detail in the additional literature [311] to [313] considering heat transfer and heat equalise proceedings in a monofilament thread on the one hand and multifilament threads on the other. However, the study shows that appropriate results and methods are not applicable enough for engineers in practice and are also difficult to handle.

5.3.2 Differential Equation for the Description of Heat Transfer at Fibres

Starting point of the view is the common DEq. of the heat exchange processes on the surface of a cylindrical body, and without consideration of the heat conduction inside the body. The last assumption means an equal temperature across the whole thread cross section[2]. In other words, there is no temperature

[2] This assumption is fulfilled not in any practical case because an utter heating or cooling of the thread (general body) assumes a heat flow (being due to the heat conduction principle) from the surface to the core or vice versa. Nevertheless, temperature differences are just the impulse for this. But, these differences amount only to a few *Kelvin* for threads with their big surface-volume-ratios (for coarse threads more unfavourable than for fine) and they are very small compared to the normally imprinted mean temperature changes. Therefore this assumption does not involve aggravating mistakes.

difference between core and surface of the radial symmetric thread cross section. If the heat balance is viewed on a small disk of a cylindrical thread piece according to Fig. 5.30 then for heating [314] is valid

$$c \cdot \mathrm{d}m \cdot \mathrm{d}T_\mathrm{y} = \alpha \cdot \mathrm{d}S_\mathrm{y} \cdot (T_\mathrm{h} - T_\mathrm{y}) \cdot \mathrm{d}t \tag{5.73}$$

the left side of Eq. 5.73: absorbed heat quantity
the right side of Eq. 5.73: about the surface transmitted heat quantity

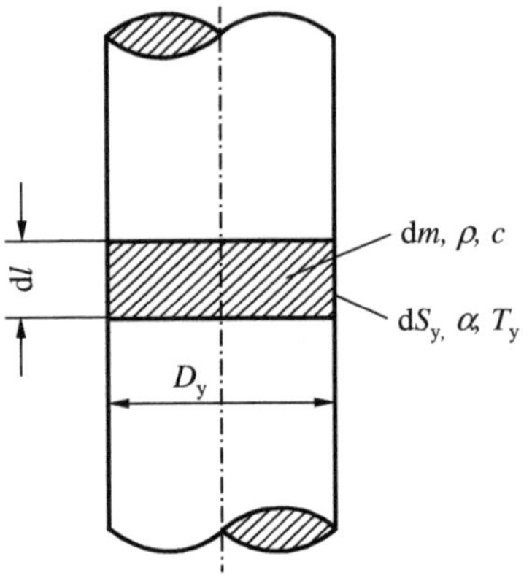

Fig. 5.30. Cylindrical thread piece

In Fig. 5.30 and Eq. 5.73 the symbols mean:

- $\mathrm{d}S_\mathrm{y}$ surface of the disk-shaped thread piece which takes part in the heat exchange
- D_y diameter of the thread piece
- $\mathrm{d}l$ length of the thread piece
- $\mathrm{d}m$ mass of the thread piece
- ϱ density of the thread material
- c specific heat capacity of the thread material
- α coefficient of the heat transfer
- T_y temperature of the thread piece
- T_h temperature of the heat medium
- $\mathrm{d}t$ time interval for the heat exchange
- $\mathrm{d}T_\mathrm{y}$ temperature change of the thread piece

The following mass-surface-ratio is valid for the cylindrical- or disk-shaped thread piece

$$\frac{\mathrm{d}m}{\mathrm{d}S_\mathrm{y}} = \frac{\pi \cdot D_\mathrm{y}^2 \cdot \varrho \cdot \mathrm{d}l}{4 \cdot \pi \cdot D_\mathrm{y} \cdot \mathrm{d}l} = \frac{D_\mathrm{y} \cdot \varrho}{4} \tag{5.74}$$

The DEq. 5.73 can be simply integrated considering Eq. 5.74

$$\int_{T_b}^{T_y} \frac{dT_y}{T_h - T_y} = \int_0^t \frac{4 \cdot \alpha}{c \cdot \varrho \cdot D_y} \cdot dt \tag{5.75}$$

The equation for the *fibre heating* results from Eq. 5.75 to

$$T_y = T_h - (T_h - T_b) \cdot \exp\left(-\frac{4 \cdot \alpha}{c \cdot \varrho \cdot D_y} \cdot t\right) \tag{5.76}$$

T_b is the thread temperature at the beginning of the heat exchange, that means to the time $t = 0$.

The DEq. 5.73 can be similarly solved for the thread cooling from the temperature at the beginning of the heat exchange T_b in a surrounding cool medium with the temperature T_r.

The equation for the *fibre cooling* can be read as:

$$T_y = T_r + (T_b - T_r) \cdot \exp\left(-\frac{4 \cdot \alpha}{c \cdot \varrho \cdot D_y} \cdot t\right) \tag{5.77}$$

Eqations 5.76 and 5.77 describe the thread heating and cooling as a dynamic transfer process (in the form of the step response function) in a approximated, but simplified manner. The practical use and the application of these equations to monofilament as well as multifilament man-made yarns are shown in the following: The thread diameter D_y of a multifilament yarn is to be calculated from its fineness (not from the fineness of the single filament!) as the equivalent diameter which would have a monofilament yarn of the same fineness [315]. This equivalent diameter (which has been proven correct for the heat transfer) is:

$$D_y = \sqrt{\frac{4 \cdot Tt}{\varrho \cdot \pi \cdot 10^9}} \tag{5.78}$$

The following dimensions are to be inputted:

fineness of the whole thread Tt in tex
density of the fibre material ϱ in g/cm^3

The equivalent diameter of the whole thread D_y will be obtained in m.

Equation 5.78 has already been used in Sect. 5.2.2 as a numerical value equation. All numerical constants for the fibre materials PA and PET (including the densities ϱ) have been collected to the numerical constants K_p (see Eq. 5.54).

For user friendly style of writing of the Eqs. 5.76 and 5.77 it is recommended here that the material specific constants c and ϱ, as well as all numerical constants in the exponent of the e-functions (which remain after the substitution of the diameter D_y by the fineness Tt according to Eq. 5.78), are collected to a material specific constant for the heat transfer K_h.

The equations can now be read as follows:

For the *fibre heating*:

$$T_y = T_h - (T_h - T_b) \cdot \exp\left(-\frac{K_h}{\sqrt{Tt}} \cdot \alpha \cdot t\right) \tag{5.79}$$

For the *fibre cooling*:

$$T_y = T_r + (T_b - T_r) \cdot \exp\left(-\frac{K_h}{\sqrt{Tt}} \cdot \alpha \cdot t\right) \tag{5.80}$$

The following dimensions are to be inputted:

T_y, T_h, T_r, T_b	in K or °C
Tt	in tex
α	in $W/(m^2 \cdot K)$
t	in s
K_h	as numerical value $5.9 \cdot 10^{-2}$ for PA
	as numerical value $7.6 \cdot 10^{-2}$ for PET

The value of $1.12\,g/cm^3$ (for PA) resp. $1.38\,g/cm^3$ (for PET) for the material density ϱ, and the value of $1.884\,kJ/(kg{\cdot}K)$ (for PA) resp. $1.256\,kJ/(kg{\cdot}K)$ (for PET) for the specific heat capacity c have been used for the calculation of K_h. The latter is almost valid in the temperature range of $40\,°C \leq T_y \leq 100\,°C$.

The data of the heat transfer coefficient α presents a certain problem for practical calculations. This depends strongly on the surrounding medium which participates in the heat transfer (for threads normally resting or moving air or metallic surfaces), on the surface quality of the thread and the metallic contact areas, on the thread velocity, on the thread fineness and at last also on the mean temperature level at which the heat transfer takes place. An (but not without any difficulties) experimental estimation is indispensable for more detailed investigations of special process stages [315].

From the literature and on the basis of own investigations it is possible to use the following ranges which are valid for PA and PET threads for calculations with a good exactness:

For the heat transfer in air:

$\alpha = 50...200 \text{ W}/(\text{m}^2 \cdot \text{K})$

For the heat transfer on metallic surfaces (for instance godets):

$\alpha = 450...600 \text{ W}/(\text{m}^2 \cdot \text{K})$

Greater values are valid in the tendency for finer threads and for higher velocities.

The velocity dependence of the heat transfer coefficient for the heat exchange between thread and air can also be given a relationship which is presented in [316]. This has been investigated experimentally as "air streams alongside of even, rough metallic surfaces" which is valid for velocities of $v > 5$ m/s and which also gives useful results for threads in an air stream too. The relationship has been used also in the technological example for Sect. 5.3.3 and to the calculation of the time constants and critical frequencies of Table 5.2. It is read as (converted into SI-units):

$$\alpha = 7.52 \cdot v^{0.78} \tag{5.81}$$

If v is inserted in m/s then α will be obtained in $\text{W}/(\text{m}^2 \cdot \text{K})$.

The time constant T_{ch} and the critical frequency f_{ch} which describe the heat transfer dynamic of a thread can be read from the exponent of the exponential function in Eqs. 5.79 and 5.80:

$$T_{\text{ch}} = \frac{\sqrt{Tt}}{\alpha \cdot K_{\text{h}}} \tag{5.82}$$

$$f_{\text{ch}} = \frac{\alpha \cdot K_{\text{h}}}{2\pi\sqrt{Tt}} \tag{5.83}$$

The appropriate numerical values are collected for the used PA- and PET-threads of the technological application example in Sect. 5.3.3. The heat transfer coefficient for the heat transfer thread-metallic surface was taken as a basis of $\alpha = 530 \text{ W}/(\text{m}^2 \cdot \text{K})$.

The heat transfer coefficient α for the heat transfer moved thread-air has been selected depending on the velocity according to Eq. 5.81 and K_{h} for PA and PET according to the data of Eqs. 5.79 and 5.80.

Qualitative conclusions to the dynamic transmission behaviour are possible by aid of the general explanations in Sect. 4.2.1 (specifically statements dealing with Eq. 4.41). Each heat area- or air-contact with another temperature than T_{y} means an imprinting step for the thread. The explanations to the dynamic transfer functions with proportional action and delay of first order are analogously valid here.

Table 5.2. Time constants T_{ch} and critical frequencies f_{ch} of PA- and PET-threads corresponding to Eqs. 5.82 and 5.83

Material	Fineness Tt [tex]	Velocity v [m/min]	Time constant T_{ch} [s]	Critical frequency f_{ch} [Hz]
Heating or cooling of the thread in contact with a metallic surface:				
PA	5		$7.53 \cdot 10^{-2}$	2.11
PA	10		$1.07 \cdot 10^{-1}$	1.49
PA	20		$1.51 \cdot 10^{-1}$	1.06
PET	5		$5.55 \cdot 10^{-2}$	2.87
PET	10		$7.85 \cdot 10^{-2}$	2.03
PET	20		$1.11 \cdot 10^{-1}$	1.43
Heating or cooling of the thread in contact with surrounded air:				
PA	5	500	1.02	$1.57 \cdot 10^{-1}$
PA	5	1000	$5.92 \cdot 10^{-1}$	$2.69 \cdot 10^{-1}$
PA	5	2000	$3.45 \cdot 10^{-1}$	$4.62 \cdot 10^{-1}$
PA	5	3000	$2.51 \cdot 10^{-1}$	$6.34 \cdot 10^{-1}$
PA	5	4000	$2.01 \cdot 10^{-1}$	$7.93 \cdot 10^{-1}$
PA	10	500	1.44	$1.11 \cdot 10^{-1}$
PA	10	1000	$8.37 \cdot 10^{-1}$	$1.90 \cdot 10^{-1}$
PA	10	2000	$4.87 \cdot 10^{-1}$	$3.27 \cdot 10^{-1}$
PA	10	3000	$3.55 \cdot 10^{-1}$	$4.48 \cdot 10^{-1}$
PA	10	4000	$2.84 \cdot 10^{-1}$	$5.61 \cdot 10^{-1}$
PA	20	500	2.03	$7.83 \cdot 10^{-2}$
PA	20	1000	1.18	$1.35 \cdot 10^{-1}$
PA	20	2000	$6.89 \cdot 10^{-1}$	$2.31 \cdot 10^{-1}$
PA	20	3000	$5.02 \cdot 10^{-1}$	$3.17 \cdot 10^{-1}$
PA	20	4000	$4.01 \cdot 10^{-1}$	$3.97 \cdot 10^{-1}$
PET	5	500	$7.49 \cdot 10^{-1}$	$2.13 \cdot 10^{-1}$
PET	5	1000	$4.36 \cdot 10^{-1}$	$3.65 \cdot 10^{-1}$
PET	5	2000	$2.54 \cdot 10^{-1}$	$6.27 \cdot 10^{-1}$
PET	5	3000	$1.85 \cdot 10^{-1}$	$8.60 \cdot 10^{-1}$
PET	5	4000	$1.48 \cdot 10^{-1}$	1.08
PET	10	500	1.06	$1.50 \cdot 10^{-1}$
PET	10	1000	$6.16 \cdot 10^{-1}$	$2.58 \cdot 10^{-1}$
PET	10	2000	$3.59 \cdot 10^{-1}$	$4.43 \cdot 10^{-1}$
PET	10	3000	$2.62 \cdot 10^{-1}$	$6.08 \cdot 10^{-1}$
PET	10	4000	$2.09 \cdot 10^{-1}$	$7.61 \cdot 10^{-1}$
PET	20	500	1.50	$1.06 \cdot 10^{-1}$
PET	20	1000	$8.72 \cdot 10^{-1}$	$1.83 \cdot 10^{-1}$
PET	20	2000	$5.08 \cdot 10^{-1}$	$3.13 \cdot 10^{-1}$
PET	20	3000	$3.70 \cdot 10^{-1}$	$4.30 \cdot 10^{-1}$
PET	20	4000	$2.96 \cdot 10^{-1}$	$5.38 \cdot 10^{-1}$

5.3.3 Technological Application Examples

Dynamics of Fibre Heating of Man-Made Fibres at Heated Godet Systems

The presented technological schemes of a heated godet roll duo (arrangement I, following abbreviated with A I) and a heated godet roll with unheated temple pulley (arrangement II, following abbreviated with A II) are taken as the basis for the following investigations (see Fig. 5.31). These arrangements could be realised for instance at the spin-draw-winding process (SDW-process). The example rests essentially upon already published results of a formerly revised paper ([317]).

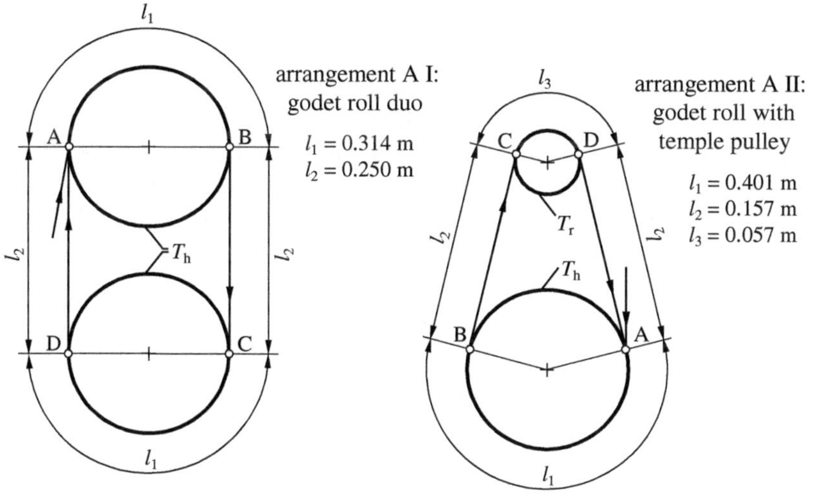

Fig. 5.31. Arrangements of heated godets

The following short signs have been chosen corresponding to Sect. 5.3.2:

T_b	thread temperature at the input into the godet arrangement
T_r	temperature of the cool medium, that means both the surrounding air temperature and the temperature of the unheated temple pulley corresponding to A II
T_h	temperature of the heated godet surface
$T_1, T_2, ...T_n$	thread temperatures
T_y	after passing the single heating and cooling lines
α_1	heat transfer coefficient between metallic surface and thread (for heating and cooling assumed of the same quantity)
α_2	heat transfer coefficient between air and thread
l_1	contact line of the thread with the heated godet surface

	(for A I and A II differently long corresponding to the different geometry of the thread way around the godet arrangements, see Fig. 5.31)
l_2	air line of the thread (for A I and A II differently long corresponding to the different geometry of the thread way around the godet
l_3	contact line of the thread with the surface of the unheated temple pulley
Tt	thread fineness
K_{h}	thread material specific constant for the heat transfer (definition see Sect. 5.3.2)
v	thread velocity
z	number of thread wraps around the geometric arrangement (one wrap = length of the way $\overline{\mathrm{ABCDA}}$)

The basis for the calculation is the common DEq. for the heat transfer 5.73 and its solutions 5.79 and 5.80 (step responses) for heating and cooling. The equation system reads for a full thread wrap around the godet arrangement A I (heated godet roll duo) corresponding to these derived basic equations ($l_1 ... l_3$ are to be put in m, v is to be put in m/s, the other quantities are to be put in as upper defined):

$$T_1 = T_{\mathrm{h}} - (T_{\mathrm{h}} - T_{\mathrm{b}}) \cdot \exp\left(-\frac{K_{\mathrm{h}} \cdot l_1 \cdot \alpha_1}{v \cdot \sqrt{Tt}}\right) \tag{5.84}$$

$$T_2 = T_{\mathrm{r}} + (T_1 - T_{\mathrm{r}}) \cdot \exp\left(-\frac{K_{\mathrm{h}} \cdot l_2 \cdot \alpha_2}{v \cdot \sqrt{Tt}}\right) \tag{5.85}$$

$$T_3 = T_{\mathrm{h}} - (T_{\mathrm{h}} - T_2) \cdot \exp\left(-\frac{K_{\mathrm{h}} \cdot l_1 \cdot \alpha_1}{v \cdot \sqrt{Tt}}\right) \tag{5.86}$$

$$T_4 = T_{\mathrm{r}} + (T_3 - T_{\mathrm{r}}) \cdot \exp\left(-\frac{K_{\mathrm{h}} \cdot l_2 \cdot \alpha_2}{v \cdot \sqrt{Tt}}\right) \tag{5.87}$$

The equation system reads for a full thread wrap around the godet arrangement A II (heated godet with unheated temple pulley) as:

$$T_1 = T_{\mathrm{h}} - (T_{\mathrm{h}} - T_{\mathrm{b}}) \cdot \exp\left(-\frac{K_{\mathrm{h}} \cdot l_1 \cdot \alpha_1}{v \cdot \sqrt{Tt}}\right) \tag{5.88}$$

$$T_2 = T_{\mathrm{r}} + (T_1 - T_{\mathrm{r}}) \cdot \exp\left(-\frac{K_{\mathrm{h}} \cdot l_2 \cdot \alpha_2}{v \cdot \sqrt{Tt}}\right) \tag{5.89}$$

$$T_3 = T_{\mathrm{r}} + (T_2 - T_{\mathrm{r}}) \cdot \exp\left(-\frac{K_{\mathrm{h}} \cdot l_3 \cdot \alpha_1}{v \cdot \sqrt{Tt}}\right) \tag{5.90}$$

$$T_4 = T_{\mathrm{r}} + (T_3 - T_{\mathrm{r}}) \cdot \exp\left(-\frac{K_{\mathrm{h}} \cdot l_2 \cdot \alpha_2}{v \cdot \sqrt{Tt}}\right) \tag{5.91}$$

The heating process similarly pursues the 2nd, 3rd, ... , n^{th}-wrap. The, in each case, actual thread temperatures T_1, T_2, ..., T_n are only to be taken over in the next following exponential equations which describe the next heating or cooling line of the thread.

In principle Fig. 5.32 shows the course of the heating process for both arrangements A I and A II. The end values of the e-functions (which the thread temperature T_y passes through the single thread lines) are connected simply by straight lines. The thread way l_y corresponds to the sizes of the geometrical godet arrangements which were the basis for the investigations. This thread way l_y is of course proportional to the running time t if the thread velocity is constant.

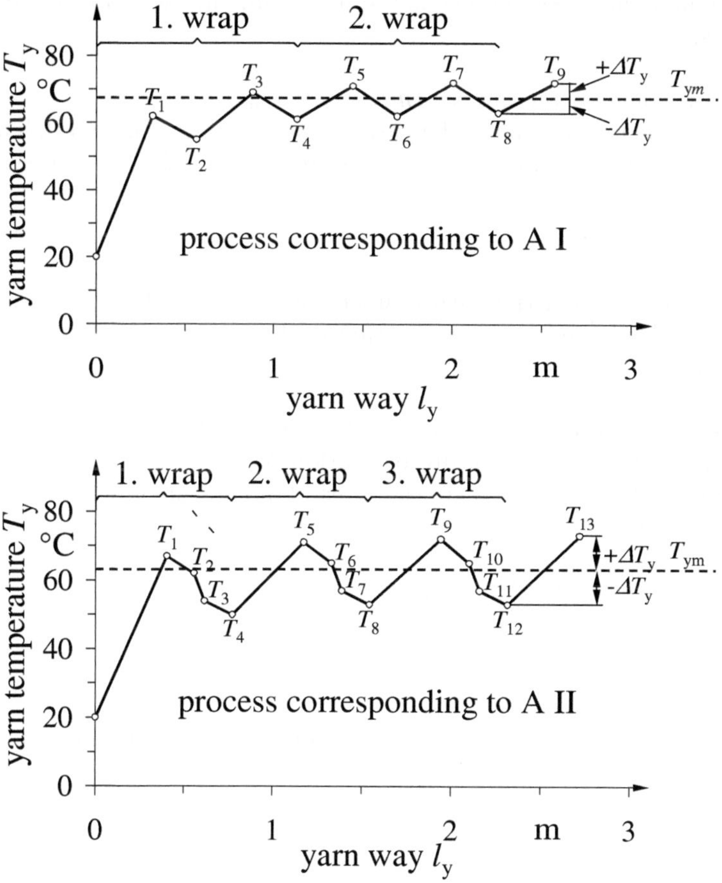

Fig. 5.32. Course of principle of thread heating on heated godet arrangements corresponding to Fig. 5.31

It can be seen by using the Eqs. 5.84 to 5.91 that the total heating process is only reproducible by means of a system of inputting each other and dependent on each other e-functions. It appears that the system behaviour can be described sufficiently by means of the following four aim quantities:

a) minimum number of thread wraps z
The quantity z is the minimum number of thread wraps which is necessary to reach the steady state condition of the heating and cooling process at a determined technological variant (Fig. 5.33). This has been defined in the calculation program by means of a breaking off criterion. The steady state condition is reached accordingly if the difference of the yarn temperatures at the end of both last calculated heating sections falls below which is a limit value, occurring for the first time, and which is given odds at your convenience. One K has been chosen as the limit value in the following numerical examples.

b) mean value of the thread temperature T_{ym}
T_{ym} means the mean value of the yarn temperature T_{y} (Fig. 5.34). The latter oscillates around T_{ym} after the steady state condition has been reached:

$$T_{\mathrm{ym}} = \frac{T_{\mathrm{n}} + T_{\mathrm{n}-1}}{2} \tag{5.92}$$

T_{n} is on this occasion the yarn temperature at the end of the last heating section at which the breaking off criterion has been reached, occurring for the first time.

c) relative thread temperature oscillations $\Delta T_{\mathrm{y}}/T_{\mathrm{ym}}$
The relative thread temperature oscillation $\Delta T_{\mathrm{y}}/T_{\mathrm{ym}}$ is the percent oscillation range (as the $\pm$ quantity) of the thread temperature referring to its mean value T_{ym} (Fig. 5.35).

$$\frac{\Delta T_{\mathrm{y}}}{T_{\mathrm{ym}}} \cdot 100\% = \frac{T_{\mathrm{n}} - T_{\mathrm{n}-1}}{T_{\mathrm{n}} + T_{\mathrm{n}+1}} \cdot 100\% \tag{5.93}$$

d) heating yield η_{h}
η_{h} represents the ratio of the thread temperature mean value T_{ym} to the temperature of the godet surface T_{h} (Fig. 5.36).

$$\eta_{\mathrm{h}} = \frac{T_{\mathrm{ym}}}{T_{\mathrm{h}}} \cdot 100\% \tag{5.94}$$

η_{h} is also a measurement for the approximation of the yarn temperature to the theoretical maximum possible quantity T_{h}. The latter could only be reached by a sufficiently large number of yarn wraps if the cooling lines would be reduced to zero. η_h is not in this respect a yield in the energy sense. Furthermore it is of course to be remarked that this quantity possesses only relative comparing character in case the temperatures are given (as usually) in °C.

Calculation results are presented following for the aim sizes previously defined in a choice of diagrams (Figs. 5.33 to 5.36) for the yarn materials PA and PET with the finenesses $Tt = 5$, 10 and 20 tex. Temperatures of the heated godets in each case of $T_h = 70\,°C$ and 100 °C are assumed for both geometrical arrangements A I and A II according to Fig. 5.31. Heat transfer coefficients are taken as the basis as follows:

yarn - metallic surface: $\alpha_1 = 530\ W/(m^2 \cdot K)$,
yarn - air: according to Eq. 5.81.

The systems have been investigated in the yarn velocity range of 500 to 4000 m/min (time constants and critical frequencies of these cases regarding heating and cooling see Table 5.2).

The following fundamental statements can be learned from the Figs. 5.33 to 5.36:

a) The minimum number of thread wraps z (which is necessary to reach the steady state yarn temperature state T_{ym}) is shown in Fig. 5.33. It must be higher, the higher the yarn velocity v, the coarser the heating yarn and the higher the godet temperature T_h. The arrangement A II enforces roughly 1.5 times more wraps than the arrangement A I, where PA enforces, on the average, a somewhat higher number of wraps than PET. It is remarkable that more than 10 wraps are necessary to reach the mean yarn temperature T_{ym} if coarser yarns are to be heated at higher velocities.

b) The mean yarn temperature T_{ym} (see Fig. 5.34) depends on the yarn velocity. T_{ym} decreases according to the expectation of higher velocity. This decrease amounts in the simulated range (dependent on the yarn fineness and the godet temperature) 15 ... 25 °C! The reached temperatures at smaller velocities ($v < 1000$ m/min for PA, $v < 1500$ m/min for PET) are significantly higher at the arrangement A I (godet roll duo) than at the arrangement A II. This tendency is obliterated at higher velocities ($v > 2000$ m/min) in so-far as (dependent upon the yarn fineness) the A II (godet roll with a temple pulley) can produce the somewhat higher mean yarn temperatures. The fineness influence comes forth clearly at higher velocities for both arrangements: Coarser yarns can be heated to a lower mean temperature compared to finer yarns. The difference can amount to $v = 4000$ m/min at 10 °C (5 tex compared with 20 tex). It is also remarkable that PET-yarns can be heated to higher temperature (especially at higher yarn velocities) on the average than PA-yarns under the same technological boundary conditions and despite

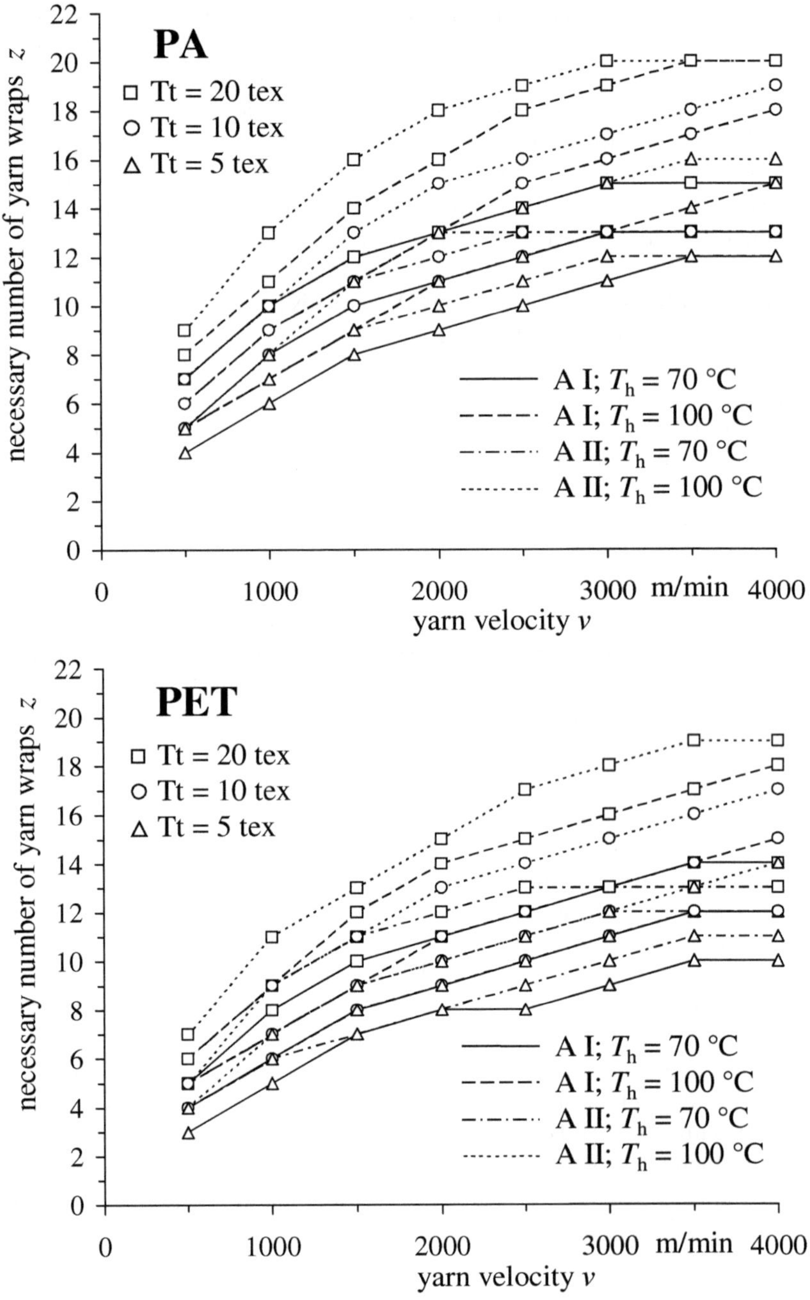

Fig. 5.33. Necessary numbers of thread wraps z around heated godet arrangements A I and A II, corresponding to Fig. 5.32

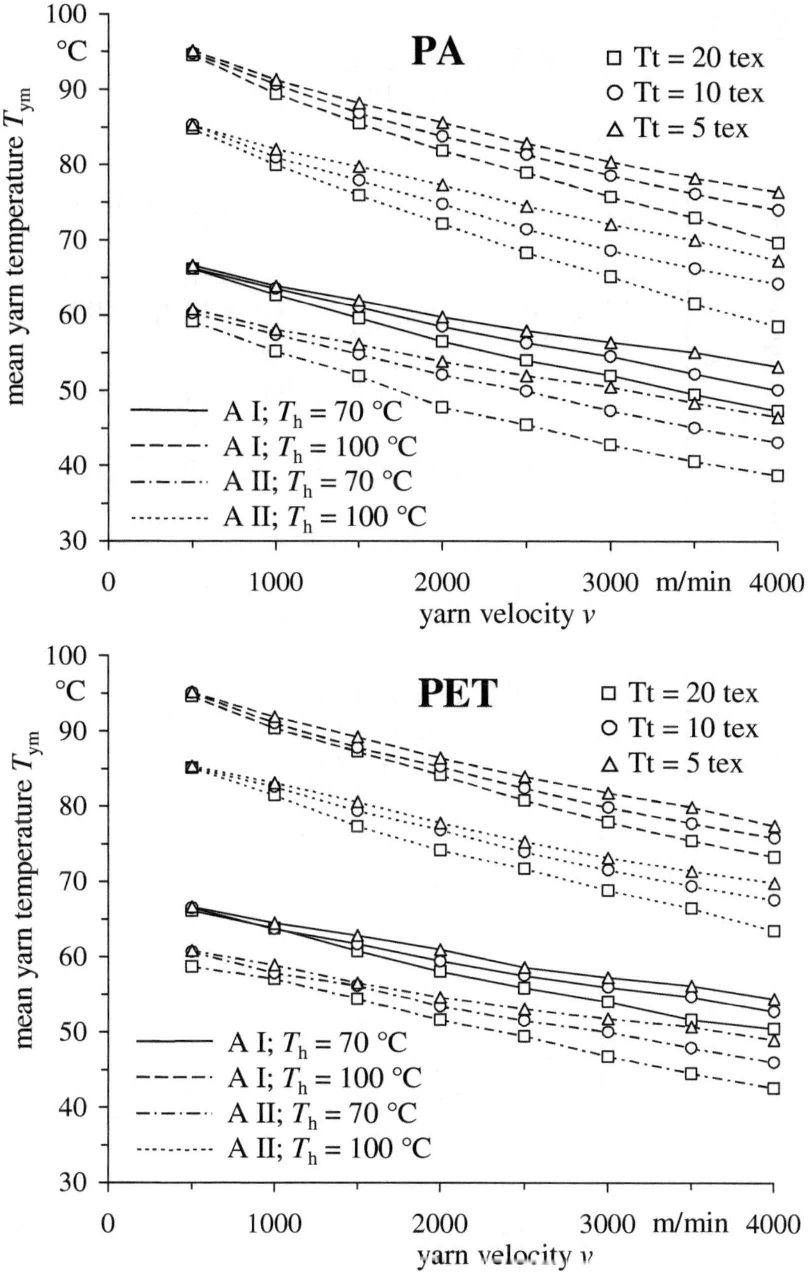

Fig. 5.34. Mean values of the thread temperatures T_{ym} at the thread running around heated godet arrangements A I and A II, corresponding to Fig. 5.32

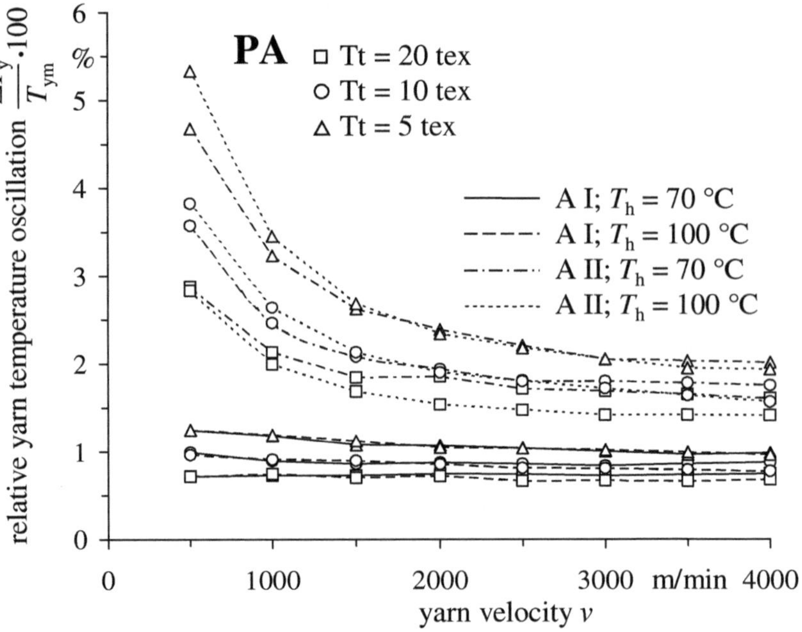

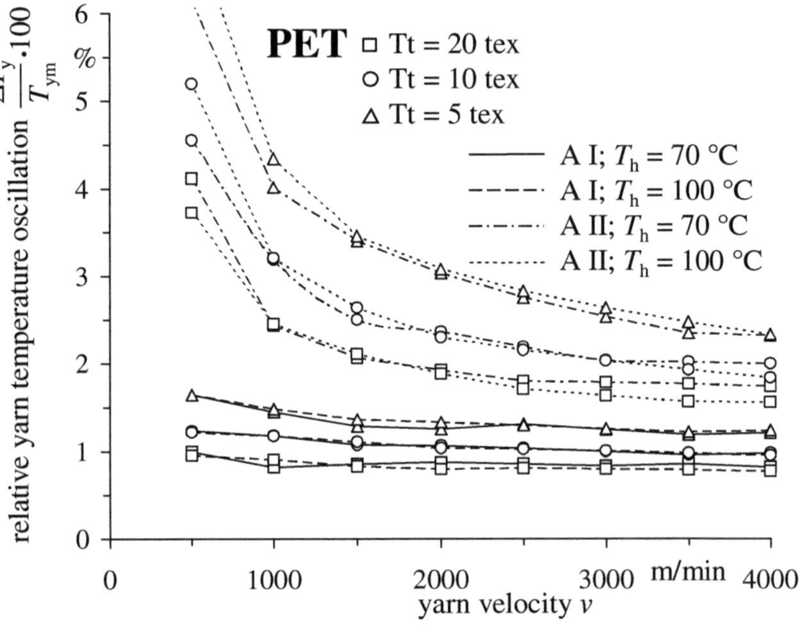

Fig. 5.35. Relative thread temperature oscillations $\Delta T_y/T_{ym}$ at the running around heated godet arrangements A I and A II, corresponding to Fig. 5.32

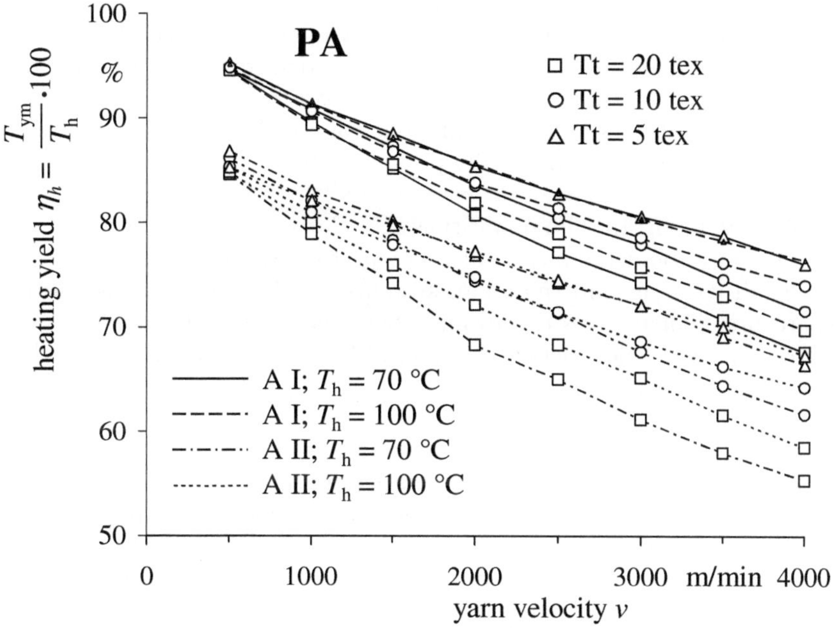

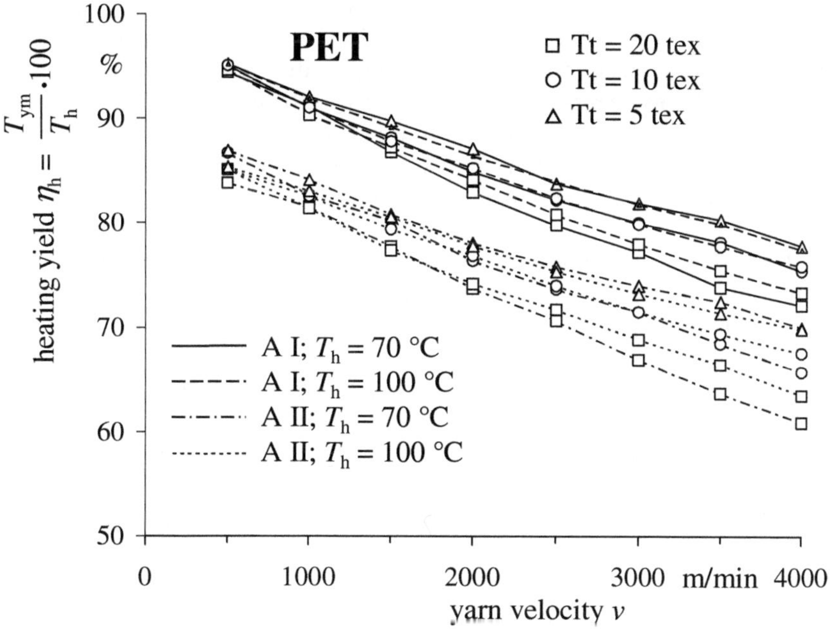

Fig. 5.36. Heating yield η_h at the thread running around heated godet arrangements A I and A II, corresponding to Fig. 5.32

the same godet temperatures. Nevertheless, the relative dependences are the same for both yarn materials.

It is quantitatively shown with these calculations that in principle the mean yarn temperatures cannot reach the temperature of the heated godet surface at such godet arrangements: The used godet temperatures will fail in the yarn (at $v = 4000$ m/min) for instance between 20 and 40 °C for PA and between 15 and 30 °C for PET. A difference of 5 to 15 °C can be seen even at $v = 500$ m/min to the theoretically possible value T_h.

c) A very important aim quantity is the relative yarn temperature oscillations $\Delta T_y / T_{ym}$ about the mean yarn temperature T_{ym} which the yarn suffers at each full wrap around the godet systems (see Fig. 5.35). This enormous temperature change stress is imprinted onto the yarn in the investigated velocity range nevertheless between 15 and 110 times/s, also with a disturbance frequency of 15 to 110 Hz!

The following tendencies can be read as:
The temperature change stress decreases with increasing velocity, coarser yarns and decreasing temperature of the heated godets. The arrangement A II strains the yarn more than the arrangement A I. Change stresses up to ±5 °C can be awaited specifically in the range of small velocities (< 1000 m/min) for finer yarns at the higher godet temperature ($T_h = 100$ °C). The dampening effect (which is correlated with the critical frequency of the system "to heating or to cooling thread") is effective in higher degrees ($f > f_{ch}$) for the higher velocities, this means for higher disturbance frequencies. The temperature change stress of the yarn decreases to uncritical values of < 2% (see also time constants and critical frequencies in Table 5.2). The yarn material PET is exposed to a somewhat greater temperature change stress than the yarn mateial PA under the same technological conditions.

d) If one looks at the heating yield η_h (see Fig. 5.36) then a clear decrease is to be noticed above all with increasing yarn velocity. The tendencies are confirmed which are named under a): A I is more favourable, on the average, according to the heat transfer (two heated godets are necessary, compared to only one at A II!). But, this is no longer valid for coarser yarns (which generally have the worse heat transfer conditions) at high yarn velocities. PET-yarns can be heated with a higher heating yield η_h than PA-yarns.

Fibre Cooling in the Melt Spinning of Polymers (simplified)

Extensive investigations to the modelling and the complex proceedings in the fibre formation distance have been described in detail in Sects. 3.1. and 3.2. One of the most important product variables is the temperature of the melt stream T_f which withdraws from the spinning die and solidifies along the spinning way l_s in the last amount of time to the filament with the fineness

Tt_f. This problem is not able to be solved by means of the developed equations in Sect. 5.3.2, but it is also pronounced as a problem of the thread cooling dynamic. In the following it should be made clear that calculation results describe the cooling process in a modified manner. These are based on the application of a simple to handle basic equation which has been developed in a formerly restricted paper the modelling of the fibre formation in melt spinning [280]. It is taken into account at this occasion that the heat transfer conditions and the heat capacity c of the solidifying polymer melt is subjected to greater temperature depending changes. The following result (based on the concentration of experimental data) is explained in [280]: The quotient of the specific heat capacity c and the NUSSELT-number[3] is roughly constant along the whole spinning way l_s. An analytical equation can be derived on the basis of this knowledge (after some intermediate steps which are not performed here) for the temperature course of the formated filament is as follows:

$$T_f = T_r + (T_s - T_r) \cdot \exp(l_s/x_0) \tag{5.95}$$

The length measurement x_0 can be calculated as follows:

$$x_0 = 1.5 \cdot K_{W1} \cdot (q_o)^{0.79} \cdot (v_s)^{-0.05} \tag{5.96}$$

or

$$x_0 = 1.5 \cdot K_{W2} \cdot (Tt_f)^{0.79} \cdot (v_s)^{0.74} \tag{5.97}$$

The single sizes and their dimensions in Eqs. 5.95 to 5.97 mean:

T_f	filament temperature in °C or K
T_r	temperature of the cooling medium (surrounding air) in °C or K
T_s	temperature of the spinneret = melt temperature at the spinneret output in °C or K
l_s	distance from the spinneret = spinning way in m
x_0	length measurement corresponding to Eqs. 5.96 or 5.97, dimension reads in m
q_o	throughput of one orifice in the spinneret = filament throughput in g/min
Tt_f	fineness of the filament
v_s	spinning velocity in m/min
K_{W1}	polymer-specific constant in Eq. 5.96 K_{W1}=0.81 for PA K_{W1}=0.57 for PET K_{W1}=0.91 for PP

[3] The NUSSELT-number is a dimensionless quantity to the description of the heat transfer which arises from the heat transfer coefficient α, the heat conductivity λ of the surrounding medium and a characteristic length measurement d as follows: $Nu = \alpha \cdot d/\lambda$.

K_{W2} polymer-specific constant in Eq. 5.97
$K_{W2} = 3.44 \cdot 10^{-3}$ for PA
$K_{W2} = 2.42 \cdot 10^{-3}$ for PET
$K_{W2} = 3.89 \cdot 10^{-3}$ for PP

Equation 5.95 also describes the filament temperature course depending on the distance from the spinneret, that means on the spinning way l_s. The knowledge of this dependence is specifically important for the best design of the spinning tube length or for the placing of filament treatment elements in the filament line (for instance preparation disk or preparation finger).

Some typical filament temperature courses are shown in Fig. 5.37 for the most important spinning materials PA and PET. Spinning velocities v_s of 1000, 4000 and 6000 m/min and finenesses of the spun filaments Tt_f of 1 and 0.3 tex for both materials have been chosen. The spinneret temperature T_s has been set at 280°C for PA and 300°C for PET. The temperature of the cooling medium (surrounding air) $T_r = 20$°C and the diameter of the orifice $d_o = 0.25$ mm (which is not of any effect) have been constantly chosen.

Quantitative tendencies of the cooling behaviour and conclusions in different directions can be easily learned using the single curves from this diagram. It can be seen as a quantitative completion of Sect. 3.1.7.

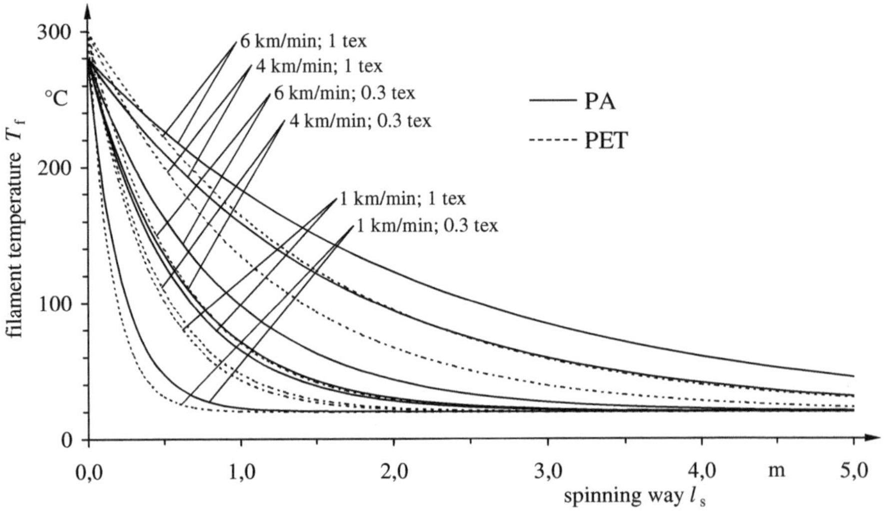

Fig. 5.37. Filament temperature courses T_f dependent upon the spinning way l_s in the melt spinning of polymers

6. Dynamics of the Tensile Force and its Importance for Process Stability

6.1 Task

Since the textile fibre formation and fibre processing processes have been carried out the special attention has been aimed at the reaction force of the yarn to an induced longitudinal deformation or elongation. The number of papers on the problem of yarn tensile force has increased meanwhile to probably more than a thousand. At first the steady state or dynamic measurements of the yarn tensile force (its mean value or the statistical description of its fluctuation parts) were, in the textile test lab, a sign of quality and use fitness of the threads. Yarn tensile force measurements have been carried out after the thread line on most different production machines in various manners. The measured signal courses have been analyzed by means of simple or also pretentious algorithms of the signal analysis or signal concentration. It is not the task of this chapter to sort and evaluate this specific textile-technological literature. Rather a row of universal regularities and hints to the measuring-methodical and analytic-technical practice which are worth mentioning should be given. These come from the theory of dynamic measurements as well as from own experimental results and they are not collected (to the knowledge of the authors) in this manner anywhere else.

The following specifics and problems are to be mentioned in connection with the product variable yarn tensile force:

a) Each fibre formation and yarn transport process is connected to a yarn elongation (see Sect. 5.1 - dynamics of the fibre transport). The so-called yarn tensile force therefore appears because each elongated yarn counteracts the elongation, a reaction force which is generally (as time function of the yarn tensile force) an expression for dynamic process reactions.
The mechanism of this appearance comes from the force-elongation diagram of the yarn (Fig. 6.1); each elongation ε (for instance ε_1, ε_2) is correlated with a yarn tensile force F (for instance F_1, F_2).

b) The yarn tensile force is a product variable which only exists during the process run. It provides information unlike any other process or product variable (measurable directly during the process) to the following points:

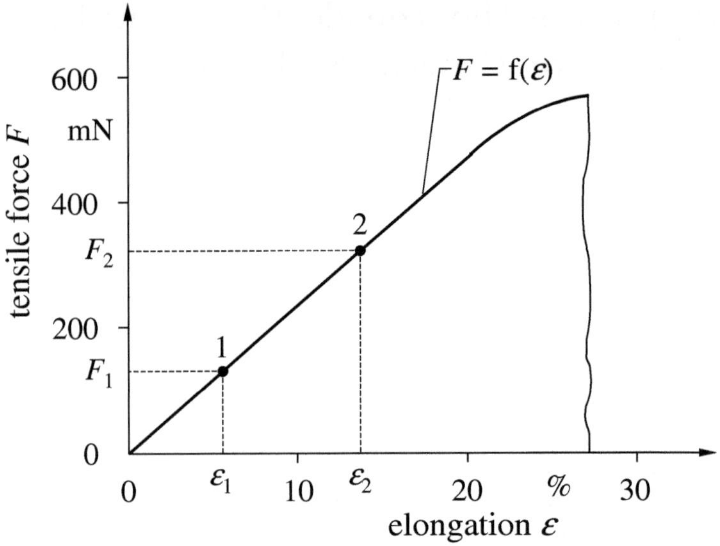

Fig. 6.1. Force-elongation diagram of a yarn (in principle)

- unevennesses of the force-elongation-behaviour of the threads (problem of the "inner unevennesses").
- unevennesses of the yarn transport (irregularities at the passing of the yarn transport lines which are necessary for carrying out the fibre formation and yarn processing processes).

c) The yarn tensile force is a complex effect quantity. Most of the different causes are reflected to it in the strongest superposed manner.

d) As a rule, only a visual analysis of recorded yarn tensile force signals does not allow an unambiguous research of disturbance causes.

e) The yarn tensile force is the cause of effected structural changes in man-made fibres (for instance orientation changes of the macromolecules). The yarn tensile force is because of that an essential cause quantity to the stress strain properties.

f) The time function of the yarn tensile force can be seen as an indicator of the process stability. High mean values of the yarn tensile force near the breakage limit lead similarly to increased yarn breakages and decrease of the productivity as great changes of the yarn tensile force at its low mean value. It is noticed that the breakage limit itself is not a constant value (see Fig. 6.2).

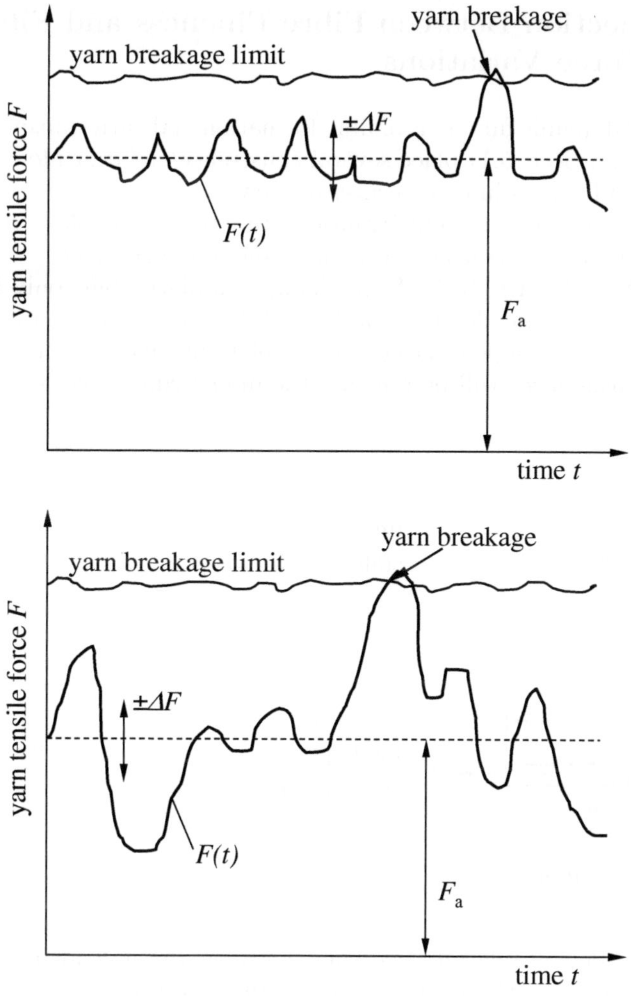

Fig. 6.2. Oscillations of the yarn tensile force – cause for the yarn breakage; *above*: great average F_a, small oscillations $\pm\Delta F$; *below*: small average F_a, great oscillations $\pm\Delta F$

g) The measurement of the yarn tensile force is in most cases without contact. Reactions to the thread line are therefore not excluded.

h) yarn tensile force measuring sensors must have a sufficiently high critical frequency because the time function of the yarn tensile force reflects the dynamics of most of the different disturbance causes (see Sect. 6.3).

6.2 Connection Between Fibre Fineness and Fibre Tensile Force Variations

The derived dynamic functions of fibre formation and yarn transport lines (see Sect. 4.2.1 and Sect. 5.1) place several cause-effect relations for the product variable yarn fineness into the center of view.

All derived dynamic time transient functions, complex frequency responses, amplitude frequency responses and phase frequency response are valid for effected yarn tensile force changes similarly, but, only if the same disturbance causes (as described in Sect. 4.2.1 and Sect. 5.1) take place and if the following developed mathematical relationships between yarn fineness and yarn tensile force will be put into the model equations.

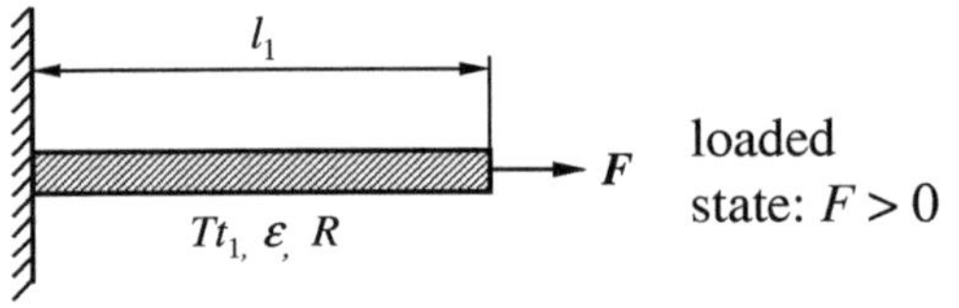

Fig. 6.3. Elongation

The well-known relations are valid between the unloaded (Tt_0, l_0) and the yarn (Tt_1, l_1, ε, R_y) loaded by means of the tensile force F according to Fig. 6.3:

$$R_y = E_y \cdot \varepsilon \tag{6.1}$$

R_y tensile stress $= F/Tt_1$
ε yarn elongation
E_y elastic modulus

Equation 6.1 can be written as:

$$F = E_y \cdot Tt_1 \cdot \varepsilon \tag{6.2}$$

The fineness of the loaded yarn Tt_1 is inserted into Eqs. 6.1 and 6.2 only because this mediates the approach to the tensile stress (fineness related tensile force) which the material will suffer from physically. This differs from the practice of yarn testing in a lab, where for the tensile stress (fineness related tensile force), the fineness of the unloaded yarn is in general use.

The elongation ε can be read as:

$$\varepsilon = \frac{l_1 - l_0}{l_0} = \frac{Tt_0 - Tt_1}{Tt_1} \tag{6.3}$$

Eq. 6.3 put into Eq. 6.2 results in:

$$F = E_y \cdot (Tt_0 - Tt_1) \tag{6.4}$$

If we take F and Tt_1 as variable sizes which consist of a mean value part F_m and Tt_{1m} as well as a change part ΔF and ΔTt then Eq. 6.4 can also be written as:

$$F_m + \Delta F = E_y \cdot [Tt_{0m} - (Tt_{1m} + \Delta Tt_1)] \tag{6.5}$$

The fineness of the unloaded yarn Tt_0 should apply to this constantly and designated (formal following the same agreement) with Tt_{0m}. Equation 6.5 now represents the conversion relations between the yarn fineness under elongation load and the appearing yarn tensile force:

$$F_m = E_y \cdot (Tt_{0m} - Tt_{1m}) \tag{6.6}$$

The change sizes are obtained through the relationship:

$$\Delta F = -E_y \cdot \Delta Tt_1 \tag{6.7}$$

It is also valid for the interesting output fineness (Tt_o, Tt_{om}, ΔTt_o) of the yarn which leaves the process line according to the generally used relationship in Sect. 4.2.1 and Sect. 5.1:

$$F_m = E_y \cdot (Tt_{0m} - Tt_{om}) \tag{6.8}$$

$$\Delta F = -E_y \cdot \Delta Tt_o \tag{6.9}$$

All dynamic functions which the yarn fineness change contains as effect variable can be converted immediately into the effect variable yarn tensile force by use of the conversion relationship of Eq. 6.9.
The formal procedure should be demonstrated using the dynamic transfer function (see Eq. 4.9) which the relationship represented between a change of the output velocity Δv_o and yarn fineness change ΔTt_o effected by this. Equation 4.9 in Sect. 4.2 can be read as:

$$G(p) = \frac{\Delta Tt_s}{\Delta v_s} = -\frac{Tt_{sm}}{v_{sm}} \cdot \frac{1}{1 + p \cdot \dfrac{l_m}{v_{sm}}}$$

The effect variable ΔTt_o is to be substituted by the new effect variable ΔF according to Eq. 6.9:

$$G(p) = \frac{\Delta F}{\Delta v_s} = -\frac{E_y \cdot Tt_{sm}}{v_{sm}} \cdot \frac{1}{1 + p \cdot \frac{l_m}{v_{sm}}} \tag{6.10}$$

The step response of the yarn tensile force change ΔF owing to a step of the output velocity Δv_o also comes from the step response of the yarn fineness change (Eq. 4.13) similarly:

$$\underline{\Delta F}|\overline{\Delta v_s} = \Delta v_s \cdot E_y \cdot \frac{Tt_{sm}}{v_{sm}} \left[1 - \exp\left(-\frac{v_{sm}}{l_m} \cdot t\right)\right] \tag{6.11}$$

The elastic modulus E_y of the elongation-loaded transported yarn also appears as the conversion factor with the dimension mN/tex. All diagram representations (for instance Figs. 2.13, 2.14 to 2.16, 4.4 to 4.6, 5.6, 5.8, 5.9, 5.20, 5.21) are usable in principle if the presented conversion relations are used. The appropriate ordinate measure is to be converted from ΔTt_o to ΔF according to the conversion relation. The normalised representation of ordinate measures (Figs. 4.4 to 4.6) do not cause again normalised (relative effect change referred to relative disturbance change) representations after the conversion from ΔTt_o to ΔF! The normalised amplitude ratio $\frac{\widetilde{\Delta Tt_o / Tt_{om}}}{\widetilde{\Delta v_o / v_{om}}}$ will be converted into $-\frac{\widetilde{\Delta F / (E_y \cdot Tt_{om})}}{\widetilde{\Delta v_o / v_{om}}}$ and ***not*** to $-\frac{\widetilde{\Delta F / F_m}}{\widetilde{\Delta v_o / v_{om}}}$ because of the simple conversion relation $F_m = -E_y \cdot Tt_{om}$ is ***not*** valid according to Eq. 6.8.

6.3 Dynamic Properties of Tensile Force Measuring Sensors and its Importance for Experimental Process Analytical Investigations

It is useful to remember some general theoretical knowledge of oscillation science and the measuring of dynamics before the evaluation of time functions from the yarn tensile force can be described.

Yarn tensile force measuring sensors are oscillatable elements related to the tools which are turned toward the yarn (usually bend-stiff elastic, one-sided chucked steel tongues are used for force transmission). Therefore, the yarn tensile force changes can only be transmitted up to an appointed upper frequency (the well-known critical frequency) without an amplitude falsification of the measured tensile force change course. Tensile force fluctuations above the critical frequency of the measuring sensor will be reflected damped, that means the amplitudes will either be reflected on a small scale or not even noticeable.

The demand for a measuring quantity transmission without any mistakes is:

critical frequency of the measuring sensor ≥ the greatest occurring oscillation frequency in the time function of the concerned process or product variable (6.12)

However, the relation 6.12 is only one necessary prerequisite. A further important prerequisite is an optimum dampening of the force transmitting, oscillatable bending tongue. Only then is it possible to fully treat the critical frequency f_c of a measuring sensor. Optimum dampening means that the amplitude frequency response of the (sinusoidal) bending tongue shift $\widetilde{\Delta w}$ is almost constant in the frequency range $0 \leq f \leq f_c$ by means of a sinusoidal imprinted tensile force at the free end of the bending tongue which is equipped with a yarn guiding element. The optimum dampening is realised by means of special design arrangements either as pure air dampening or as oil dampening (for instance with an oil filling of suitable viscosity between the bending tongue and the fixed sensor case).

Figure 6.4 shows an example of a one-sided chucked measurement sensor bending tongue which's bending way $\widetilde{\Delta w}$ is measured by means of two inductive way sensors (capacitive measurement is likewise possible). The necessary dampening is reached by oil which is filled into the slit between the inductive way sensors and the bending tongue. Such a design solution (here only outlined schematically) is described in [318].

In principle the dynamic transfer properties and the possible amplitude frequency responses are shown in the diagram in Fig. 6.4 for such a system with differently strong dampening of the bending tongue.

The curves are based upon the generally known amplitude frequency response equation of a mass-spring-dampening system which can be found for instance in [5]. For the case of our system:

$$|G(\mathrm{j}f)| = \left|\frac{\widetilde{\Delta w}}{\widetilde{\Delta F}}\right| = K_\mathrm{N} \cdot \left[\left(1 - \frac{f^2}{f_\mathrm{n}^2}\right)^2 + 4d_\mathrm{f}^2 \frac{f^2}{f_\mathrm{n}^2}\right]^{-1/2} \qquad (6.13)$$

The symbols mean:

- $\widetilde{\Delta F}$ amplitude of the induced sinus-like force oscillations at the bending tongue
- $\widetilde{\Delta w}$ amplitude of the sinus-like shifts at the bending tongue
- K_N normalising factor
- f frequency of the induced sinus-like force oscillations
- f_n natural frequency of the bending tongue
- d_f dampening factor

A dampening factor that is too small ($d_\mathrm{f} < 0.6$) as well as a dampening factor that is too large ($d_\mathrm{f} > 0.6$) is unsuitable for a measuring signal trans-

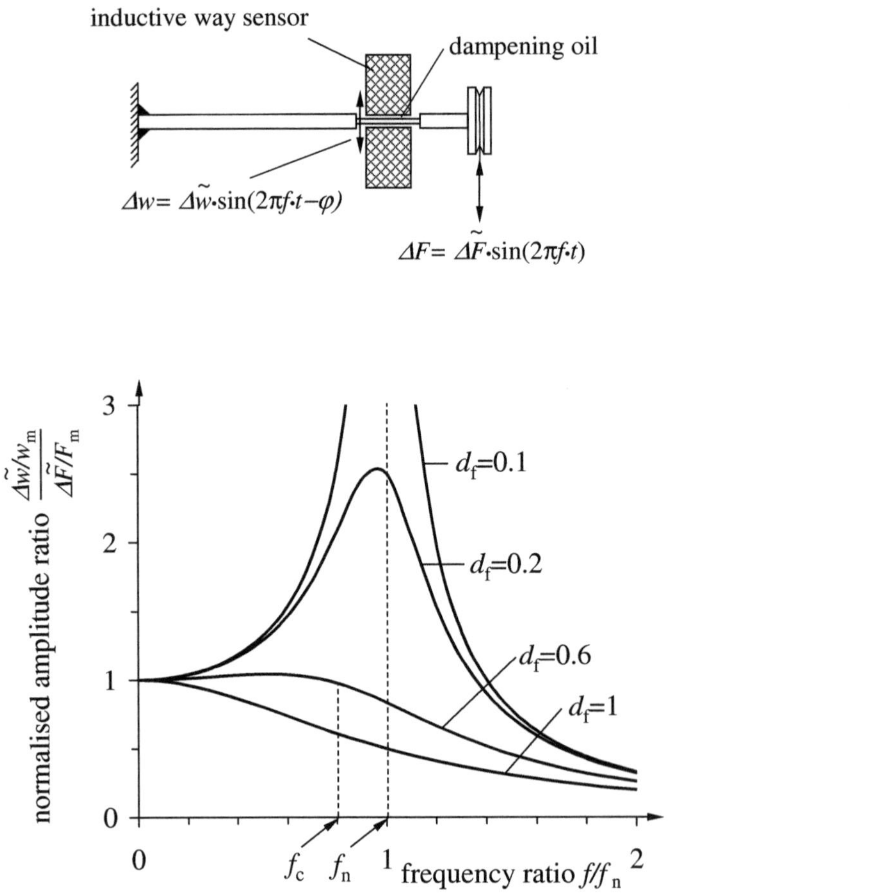

Fig. 6.4. Bending tongue of a yarn tensile force measuring sensor and normalised amplitude frequency responses depending on different dampening factors

mission without mistakes. Strong raised shifts of the bending tongue (caused always by steady exciting force amplitude at the bending tongue) will be recorded in the first case in a large frequency range. These already start near the steady state load ($f/f_\mathrm{n} \to 0$) and they decrease quickly to insignificant shifts above the natural frequency of the bending tongue. Such a measuring sensor would only be suitable for the measuring of quasi steady state or very slow yarn tensile force changes. All other frequencies are transmitted with amplitudes that are either too small or too large.

If the dampening is chosen too large then an exact amplitude reproduction is possible only likewise for quasi steady state changes of the yarn tensile force whereas amplitudes that are too small are measured in whole other frequency range.

A nearly exact amplitude measuring in the frequency range $0 \leq f \leq f_{\mathrm{n}}$ is possible if the dampening is optimally adjusted to $d_{\mathrm{f}} = 0.6$. The amplitude frequency response curve is nearly constant in the named frequency range. About 0.8 times of the natural frequency f_{n} of the freely oscillating, undampened bending tongue can be stated as critical frequency f_{c} for such a measurement system which should satisfy higher exactness pretensions. If the measuring sensor is used beyond this, then 20% of the decreasing amplitude (referring to the imprinted signal) appears already for $f = f_{\mathrm{n}}$.

Today offered yarn tensile force measuring sensors have critical frequencies in the range of $f_{\mathrm{c}} \leq 500$ Hz (in special cases also above it) and they fulfill virtually all wishes regarding the dynamic transmission properties. It is indeed problematic to fulfill the demands of a small reaction to the running yarn for narrow yarn tensile forces and for high yarn transport velocities. The latter is evident in high-speed spinning processes of the man-made fibre industry. Yarn guide rolls with light-motionable ball bearings are recommended for such tensile force measuring sensors. The named relations and regularities (collected here for the special case of the yarn tensile force measuring sensors) are of course valid for measuring sensors which measure signal courses of other process and product variables if the signal-recording element is a mass-spring-dampening-system.

6.4 Evaluation of the Tensile Force Time Function

6.4.1 Stationary Evaluation

It is surely correct that still today most recorded yarn tensile force courses are submitted to a steady state evaluation only, although in most cases powerful electronic computers are available. This is a contradiction in so far as just the time function of any process and product variables is an expression of the process dynamic. A pure steady state evaluation of a naturally dynamic measuring quantity gives away a considerable part of the information substance which it contains.

An assumed time course according to Fig. 6.5 is estimated normally to the following quantities:

a) mean value F_{m} results from

$$F_{\mathrm{m}} = \frac{1}{n} \cdot \sum_{\mathrm{i}=1}^{n} F_{\mathrm{i}} \tag{6.14}$$

b) maximum amplitude shift $\Delta F_{\max}$

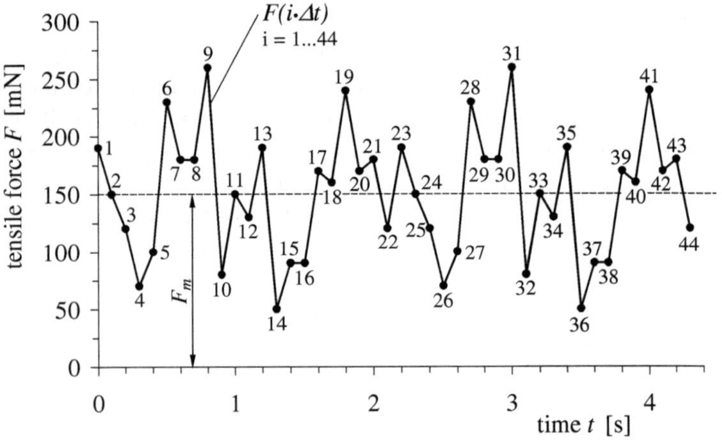

Fig. 6.5. Scanned time function of an assumed tensile force course; scan interval $\Delta t = 0.1$ s. $F_{\mathrm{m}} = 150$ mN, $d_{\mathrm{q}} = 56$ mN, $v_{\mathrm{c}} = 37\%$, $\Delta F_{\mathrm{max}} = 210$ mN

c) quadratic dispersion (variance), standard deviation or coefficient of variation as summary data for the fluctuation range of the time function course around the mean value. These quantities are calculated according to the following algorithms:

quadratic dispersion (variance) d_{q}^2

$$d_{\mathrm{q}}^2 = \frac{\sum_{\mathrm{i}=1}^{n}(F_{\mathrm{i}} - F_{\mathrm{m}})^2}{n-1} \tag{6.15}$$

standard deviation d_{q}

$$d_{\mathrm{q}} = \sqrt{\frac{\sum_{\mathrm{i}=1}^{n}(F_{\mathrm{i}} - F_{\mathrm{m}})^2}{n-1}} \tag{6.16}$$

coefficient of variation v_{c}

$$v_{\mathrm{c}} = \frac{1}{F_{\mathrm{m}}} \cdot \sqrt{\frac{\sum_{\mathrm{i}=1}^{n}(F_{\mathrm{i}} - F_{\mathrm{m}})^2}{n-1}} \cdot 100\% \tag{6.17}$$

The single signs mean:

- F_{i} discontinuous value of time function $F(t)$, taken in steps of Δt
- F_{m} mean value of the time function $F(t)$ in the range $F_1 \ldots F_{\mathrm{i}}$
- n maximum number of available single values F_{i}
- i running index

6.4.2 Dynamic Evaluation; Auto Correlation and Auto Power Density Spectrum Functions

Fundamentals

The steady state analysis does not give any data about the time behaviour of the recorded time functions, this means no data is given about the frequencies of a fluctuation course. Essentially three possibilities are given in order to get such related data:

a) Visual estimation of the time function. It is possible in exceptional cases to only recognise strong outstanding frequencies because many fluctuation courses are a mixture of frequencies.
b) Calculation of the auto-correlation function of the recorded time function.
c) Calculation of the auto-power density spectrum function of the recorded time function.

Calculation algorithms

The general calculation algorithms for the auto-correlation (ACF) and the auto-power density spectrum functions (APSF) have already been given in Sect. 2.4.4 (see Eqs. 2.16, 2.17, 2.20 to 2.22 and Fig. 2.9). The special equations can be read in the present case of the time function of the yarn tensile force as:

ACF, Integral representation:

$$K_{\mathrm{F}}(\tau) = \lim_{\mathrm{T}\to\infty} \frac{1}{T} \int_0^T [F(t) - F_{\mathrm{m}}][F(t+\tau) - F_{\mathrm{m}}]\mathrm{d}t \qquad (6.18)$$

ACF, Sum representation:

$$K_{\mathrm{F}}(k \cdot \Delta t) = \frac{1}{n-k} \sum_{\mathrm{i}=1}^{n-k} (F_{\mathrm{i}} - F_{\mathrm{m}})(F_{\mathrm{i+k}} - F_{\mathrm{m}}) \qquad (6.19)$$

Boundary condition: $k_{\mathrm{max}} \leq \frac{n}{5}$, with $k = 0, 1, 2, 3, \ldots k_{\mathrm{max}}$.

APSF, Integral representation:

$$S_{\mathrm{F}}(f) = \lim_{\mathrm{T}\to\infty} \frac{1}{T} \left\{ \left[\int_0^T (F(t) - F_{\mathrm{m}}) \cos(2\pi \cdot f \cdot t)\mathrm{d}t \right]^2 + \left[\int_0^T (F(t) - F_{\mathrm{m}}) \sin(2\pi \cdot f \cdot t)\mathrm{d}t \right]^2 \right\} \qquad (6.20)$$

APSF, Sum representation:

$$S_F(f) = \frac{1}{n}\left\{\left[\sum_{i=1}^{n}(F(t) - F_m)\cos(2\pi \cdot f \cdot i \cdot \Delta t)\right]^2 + \left[\sum_{i=1}^{n}(F(t) - F_m)\sin(2\pi \cdot f \cdot i \cdot \Delta t)\right]^2\right\} \tag{6.21}$$

Boundary condition: $\frac{5}{n \cdot \Delta t} \leq f \leq \frac{1}{2 \cdot \Delta t}$

APSF, calculated from $K_F(k \cdot \Delta t)$:

$$S_F(f) = 2\Delta t \sum_{k=1}^{m} K_F(k \cdot \Delta t) \cdot \cos(2\pi \cdot f \cdot k \cdot \Delta t) \tag{6.22}$$

Boundary condition: $\frac{1}{m \cdot \Delta t} \leq f \leq \frac{1}{2 \cdot \Delta t}$

$F(t)$	time function of the tensile force
F_m	mean value of the time function of tensile force, evaluation range $0 \leq t \leq T$ or $0 \leq k \cdot \Delta t \leq n - k$
F_i	discontinuous value of time function $F(t)$, taken in steps of Δt
T	time period of the integration range
$\tau, k \cdot \Delta t$	time shift
k, m, n	running (sequence) indices
$K_F(\tau), K_F(k \cdot \Delta t)$	single values of the ACF of time function $F(t)$ for τ or $k \cdot \Delta t$
$S_F(f)$	single values of the APSF of time function $F(t)$ for f

It has been previously referred to the use and the expanded assertion possibilities of these analysis procedures regarding the fineness unevenness analysis of threads and spun yarns [319]. Nevertheless, various applications have not been induced, because electronic computers were not yet available.

Calculation Example; Estimation Rules; Necessary Measurement and Evaluation Scopes

The formation of the ACF of an arbitrary given yarn tensile force course according to Eq. 6.19 will be demonstrated in the following. This time function is reduced to 44 equidistant values F_i which are keyed into the distance Δt (Fig. 6.5). The result are 10 single ACF-values $K_F(k \cdot \Delta t)$ for $k = 0, 1, \ldots 9$.

The calculation of these few values for the ACF (based on the underlied time function) is as you can see already very expensive. With this it is to be remarked, that an ACF-calculation, which fulfills a useful dynamic analysis requirement, supposes at least 1000 single values F_{i} (better more) from which 200 functional values $K_{\text{F}}(k \cdot \Delta t)$ can be calculated.

The expenses for the calculation of the APSF can be much larger because the according equations prescribe a continuous multiplication with sin- and cos-functions of changing frequency. Automatised measuring data recording of the time functions (and their processing) by means of computers are also an absolute prerequisite to an effective use of these methods.

From our simple, roughly divided function $F(t)$ (Fig.6.5) can be read, after the statistic estimation, as:
$F_{\text{m}} = 150$ mN; $\Delta F_{\text{max}} = 210$ mN; $d_{\text{q}}^2 = 3098$ mN^2; $d_{\text{q}} = 56$ mN; $v_{\text{c}} = 37\%$.

The 10 single values of the ACF according to Table 6.1 are drawn versus the related time shift $k \cdot \Delta t$ in the Fig. 6.6.

The question is now, how is the ACF and/or the APSF to be evaluated

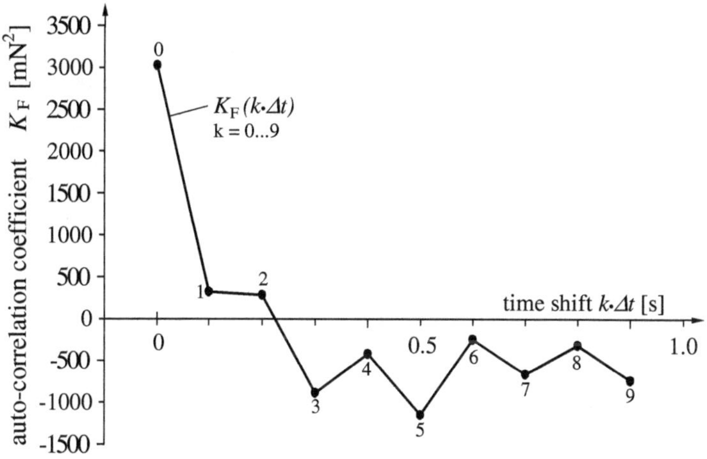

Fig. 6.6. Auto-correlation function (ACF) of the time function corresponding to Fig. 6.5 and Table 6.1

or to be discussed with respect to a deepened time function analysis. Some *simplified estimation rules* will be given for this, in the following, for different courses of the ACF and APSF which also allow a classification of our arbitraril chosen tensile force course according to Fig. 6.5.

a) Poriodical parts of the time function appear in the ACF as pure, unadulterated oscillations with the same cycle duration (measured in the units of the time shift τ or $k \cdot \Delta t$) which appears (mostly in a not exact discernible form) in the time function itself.

Table 6.1. Development of the ACF $K_F(k \cdot \Delta t)$ of the time function $F(t)$ corresponding to Fig. 6.5

i	F_i [mN]	$F_i - F_m$ [mN]	$(F_i - F_m)(F_{i+k} - F_m)$ for $k = 0$ [mN^2]	$k = 1$ [mN^2]	$k = 2$ [mN^2]	$k = 3$ [mN^2]	$k = 4$ [mN^2]	$k = 5$ [mN^2]	$k = 6$ [mN^2]	$k = 7$ [mN^2]	$k = 8$ [mN^2]	$k = 9$ [mN^2]
1	190	40	1600	0	-1200	-3200	-2000	3200	1200	1200	4400	-2800
2	150	0	0	0	0	0	0	0	0	0	0	0
3	120	-30	900	2400	1500	-2400	-900	-900	-3300	2100	0	600
4	70	-80	6400	4000	-6400	-2400	-2400	-8800	5600	0	1600	-3200
5	100	-50	2500	-4000	-1500	-1500	-5500	3500	0	1000	-2000	5000
6	230	80	6400	2400	2400	8800	-5600	0	-1600	3200	-8000	-4800
7	180	30	900	900	3300	-2100	0	-600	1200	-3000	-1800	-1800
8	180	30	900	3300	-2100	0	-600	1200	-3000	-1800	-1800	600
9	260	110	12100	-7700	0	-2200	4400	-11000	-6600	-6600	2200	1100
10	80	-70	4900	0	1400	-2800	7000	4200	4200	-1400	-700	-6300
11	150	0	0	0	0	0	0	0	0	0	0	0
12	130	-20	400	-800	2000	1200	1200	-400	-200	-1800	-400	-600
13	190	40	1600	-4000	-2400	-2400	800	400	3600	800	1200	-1200
14	50	-100	10000	6000	6000	-2000	-1000	-9000	-2000	-3000	3000	-4000
15	90	-60	3600	3600	-1200	-600	-5400	-1200	-1800	1800	-2400	0
16	90	-60	3600	-1200	-600	-5400	-1200	-1800	1800	-2400	0	1800
17	170	20	400	200	1800	400	600	-600	800	0	-600	-1600
18	160	10	100	900	200	300	-300	400	0	-300	-800	-500
19	240	90	8100	1800	2700	-2700	3600	0	-2700	-7200	-4500	7200
20	170	20	400	600	-600	800	0	-600	-1600	-1000	1600	600
21	180	30	900	-900	1200	0	-900	-2400	-1500	2400	900	900
22	120	-30	900	-1200	0	900	2400	1500	-2400	-900	-900	-3300
23	190	40	1600	0	-1200	-3200	-2000	3200	1200	1200	4400	-2800
24	150	0	0	0	0	0	0	0	0	0	0	0
25	120	-30	900	2400	1500	-2400	-900	-900	-3300	2100	0	600
26	70	-80	6400	4000	-6400	-2400	-2400	-8800	5600	0	1600	-3200
27	100	-50	2500	-4000	-1500	-1500	-5500	3500	0	1000	-2000	5000
28	230	80	6400	2400	2400	8800	-5600	0	-1600	3200	-8000	-4800
29	180	30	900	900	3300	-2100	0	-600	1200	-3000	-1800	-1800
30	180	30	900	3300	-2100	0	-600	1200	-3000	-1800	-1800	600
31	260	110	12100	-7700	0	-2200	4400	-11000	-6600	-6600	2200	1100
32	80	-70	4900	0	1400	-2800	7000	4200	4200	-1400	-700	-6300
33	150	0	0	0	0	0	0	0	0	0	0	0
34	130	-20	400	-800	2000	1200	1200	-400	-200	-1800	-400	-600
35	190	40	1600	-4000	-2400	-2400	800	400	3600	800	1200	-1200
36	50	-100	10000	6000	6000	-2000	-1000	-9000	-2000	-3000	3000	
37	90	-60	3600	3600	-1200	-600	-5400	-1200	-1800	1800		
38	90	-60	3600	-1200	-600	-5400	-1200	-1800	1800			
39	170	20	400	200	1800	400	600	-600				
40	160	10	100	900	200	300	-300					
41	240	90	8100	1800	2700	-2700						
42	170	20	400	600	-600							
43	180	30	900	-900								
44	120	-30	900									
Sum [mN^2]			133200	13800	11800	-36300	-16700	-44700	-9200	-24400	-11300	-25700
$n - k$			44	43	42	41	40	39	38	37	36	35
$K_F(k \cdot \Delta t)$ [mN^2]			3027	321	281	-885	-418	-1146	-242	-659	-314	-734

Periodical parts of the time function appear in the APSF as a maximum at these frequencies which the periodic parts themselfs possess in the time function.
The appropriate analysis situation is shown in Fig. 6.7. If the time function shows strict periodical oscillations then the ACF does not die-away to zero with increasing τ but it will monotonously oscillate like the time function itself. If the time function shows strict periodical oscillations then the APSF shows a narrow high maximum.

b) If the time function only has stochastic (statistic) parts then the ACF does continuously die-away like an exponential function and will be zero in

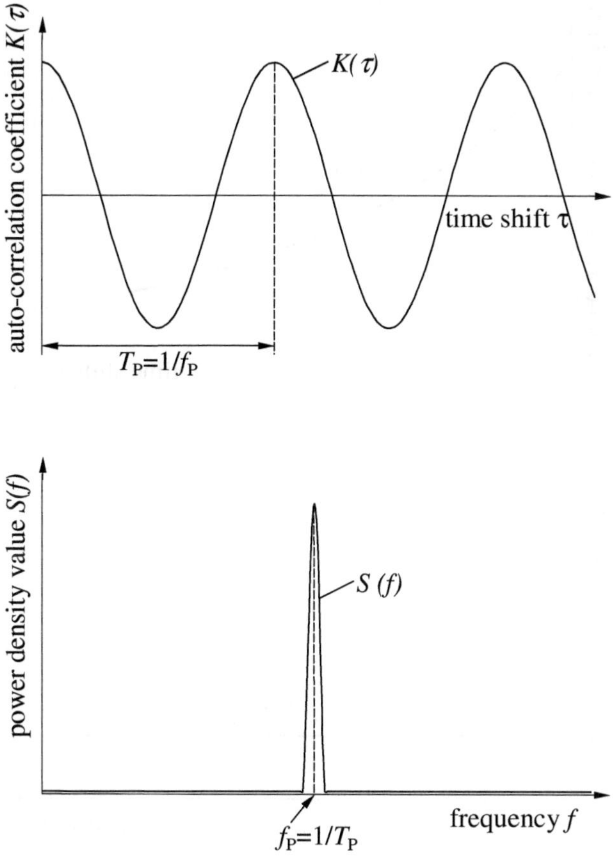

Fig. 6.7. Auto-correlation and auto-power density spectrum functions of a time function with a pure periodic (sinusoidal) change; T_P cycle duration of the change, f_P frequency of the change

the end. The APSF does not have any sort of maximum but it shows a more or less equalised (in the ideal case a constant) course (see Fig. 6.8).

c) If the time function has superimposed, periodic and statistic parts then these appear in good separated form in the ACF and APSF appropriate to the given criteria under a) and b). Figures 6.9 and 6.10 show two (separate) examples for the superposition of statistic and 1- or 2-times periodic parts in the basic time function.

The time function of the yarn tensile force Fig. 6.5 is to be characterised by means of its ACF (Fig. 6.6) as a function which consists of statistic parts and a one-time periodic part. Nevertheless, such an assertion would be supported more exactly by means of a longer analysis interval, an essentially greater

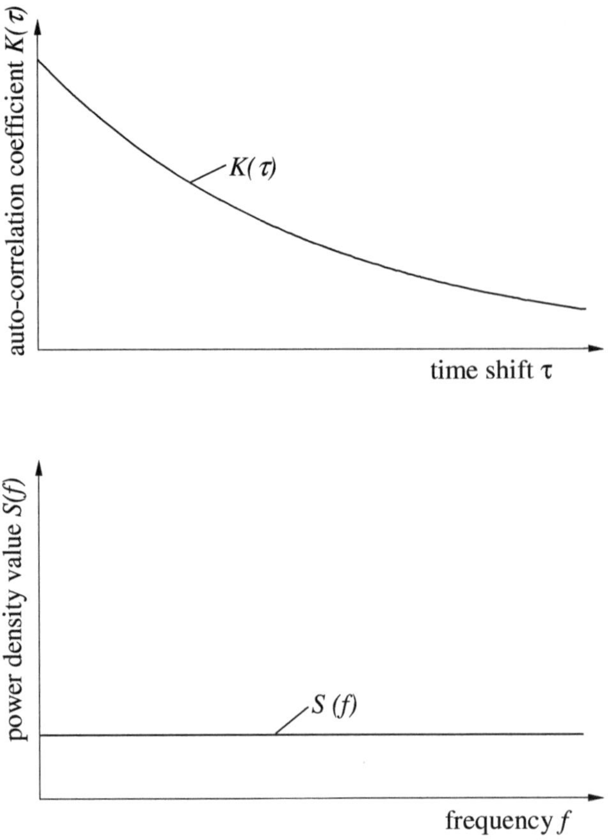

Fig. 6.8. Auto-correlation and auto-power density spectrum functions of a time function with pure stochastic changes

number of single tensile force values and a more extensive ACF-analysis. The necessary measuring and estimation volume to the calculation of the ACFs and APSFs (from more intensive analyzing time functions) can be fixed according to the following rules of thumb. Some likeness exists about the problem of the necessary measuring and gauge length of threads for the purpose of analysis of dynamic disturbances (see Sect. 4.5).

It is necessary to estimate the probable highest and lowest occurring frequencies $f_{\max}$ and $f_{\min}$ (by means of test records or a-priori knowledges) before appointing a measuring and analysis strategy, because the time functions of the product variable yarn tensile force can consist of high- as well as low-frequency fluctuation parts in proportion to the process step.

The necessary maximum analysis time T_{A} of the time function (and with it the measuring time for the complete recorded function) results from the condition that also a low frequency periodic change ($f_{\min}$) could be run at

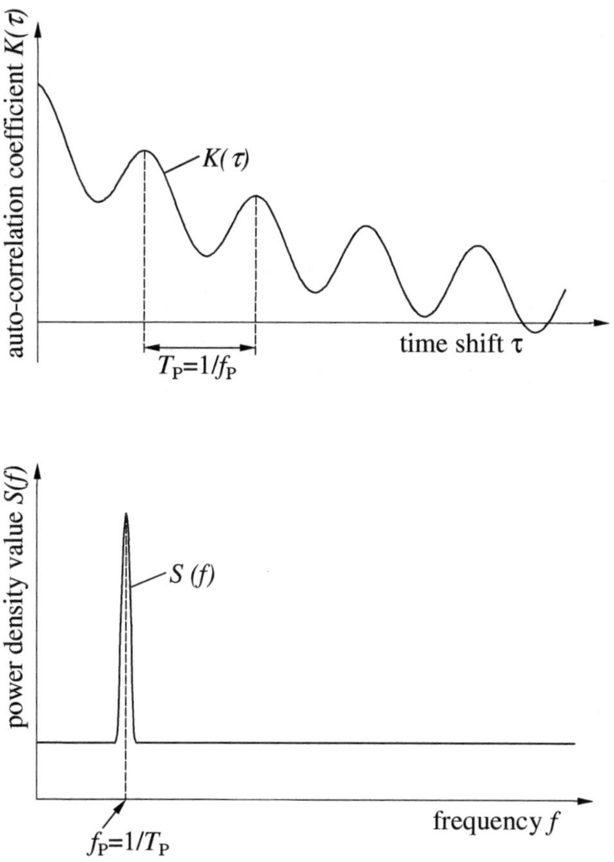

Fig. 6.9. Auto-correlation and auto-power density spectrum functions of a time function with (superposed) one periodic (sinusoidal) and stochastic changes; T_P cycle duration of the periodic change, f_P frequency of the periodic change

least 5-times. The condition for T_A is therefore:

$$T_\mathrm{A} \leq \frac{5}{f_\mathrm{min}} \tag{6.23}$$

If the calculation of the ACF and/or the APSF are realised according to the sum equations (Eqs. 6.19, 6.21) then the number n of equidistant taken single values F_i of the time function $F(t)$ can be derived from the following condition. A periodic change which runs with f_max should take at least 5 equidistant single values. This means that the taken interval Δt must be:

$$\Delta t \leq \frac{1}{5 \cdot f_\mathrm{max}} \tag{6.24}$$

The necessary number of single values n_min results with this to:

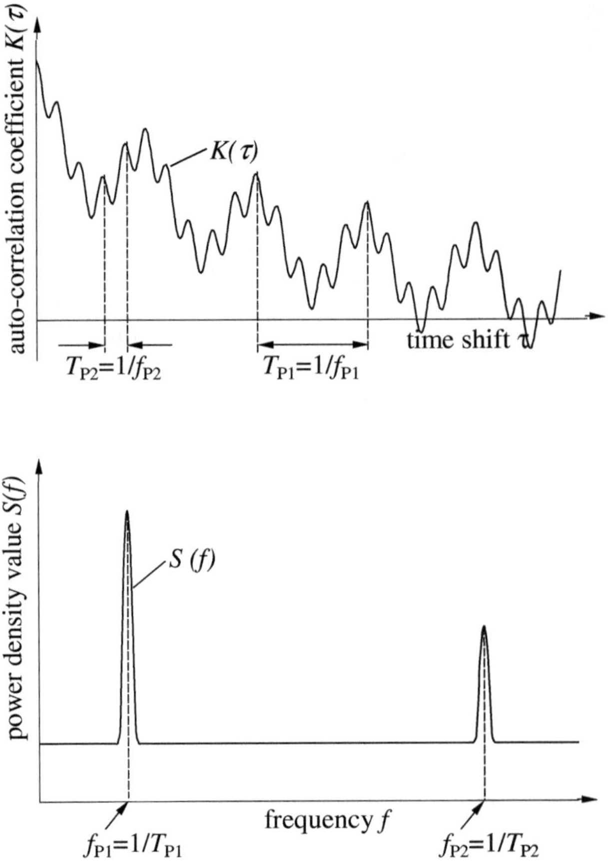

Fig. 6.10. Auto-correlation and auto-power density spectrum functions of a time function with two periodic (sinusoidal) and stochastic changes; T_{P1} and T_{P2} cycle durations of the periodic changes, f_{P1} and f_{P2} frequencies of the periodic changes

$$n_{\min} \geq \frac{T_{\mathrm{A}}}{\Delta t} = 25 \cdot \frac{f_{\max}}{f_{\min}} \tag{6.25}$$

A maximum possible time shift $\tau_{\max}$ can be recommended for the ACF-calculation as follows:

$$\tau_{\max} \leq \frac{T_{\mathrm{A}}}{5} = \frac{1}{f_{\min}} \tag{6.26}$$

If this condition is observed then the slowest can also occur.

It is easy to estimate according to Eq. 6.25 that the number of single values $n_{\min}$ must be more than 1000, if the ratio is $f_{\max}/f_{\min} = 40$ (for instance $f_{\max} = 4$ Hz, $f_{\min} = 0.1$ Hz).

Technological Example

The preceding statements will be demonstrated once more by means of the concentration of five yarn tensile force courses to their auto-correlation and auto-power density spectrum functions. We choose, in retrospect, five measured time functions of the yarn tensile force in the input line according to Fig. 5.15 in Sect. 5.1.5. The technological situation of this yarn input line has already been discussed in detail. It is a characteristic of this that extremely different, but overlookable time functions, can be submitted to the ACF- and APSF-calculation. The five time functions $F(t)$ have been characterised by an input line length $l_{\mathrm{m}} = 0.42$ m in the first case, an extended input line of $l_{\mathrm{ext}} = 0.45$ m in the second case, an extended input line of $l_{\mathrm{ext}} = 1.60$ m in the third case, an extended input line of $l_{\mathrm{ext}} = 4.80$ m in the fourth case, and an extended input line of $l_{\mathrm{ext}} = 6.40$ m in the fifth case, all according to the technological scheme of Fig. 5.12. The analyzed time functions and the, in each case attached, ACFs and APSFs are shown in Figs. 6.11 to 6.15.

The following conclud statements which confirm the discussion in Sect. 5.1.5 can be made:

a) The ACF of the undamped yarn tensile force course clearly has a strong periodic character with a main disturbance frequency of $f = 0.75$ Hz (equivalent yarn length of 1.3 m; see Fig. 6.11). It is equivalent with that to the basic type, outlined in Fig. 6.7. A second disturbance frequency ($f = 4.65$ Hz, equivalent yarn length 0.21 m) appears clearer with the increasing dampening of this main disturbance frequency. This can only be made visible by a drastic changed ordinate measure of the ACF-diagrams (see Figs. 6.12 to 6.15), because the main disturbance frequency in the undamped case (Fig. 6.11) dominates and covers all the other. This second disturbance frequency also correlated with the periphery of a small eccentric running input godet of the twister. The ACF passes over from the basic type of the one periodic disturbance to that of two periodic disturbances without discernible stochastic parts (compare Figs. 6.11 to 6.15 with Fig. 6.10).

The strong different ordinate measures of the ACFs (and APSFs) in the Figs. 6.11 to 6.15 demonstrate, in another way, the utmost effective calming of the tensile force course. It is to be remarked at this point that the start value of the ACF (for $\tau = 0$) corresponds to the well-known quadratic dispersion (variance).

b) The same interpretation is valid for the APSFs (*S(f)*-values) of the five time function courses. The Figs. 6.11 and 6.12 show the main disturbance frequency of 0.75 Hz clearly and independently, Fig. 6.13 also shows the second disturbance frequency of 4.65 Hz clearly, and in Figs. 6.14 to 6.15 (greatest dampening) show the second disturbance frequency only independently. One also notes here the different ordinate measures!

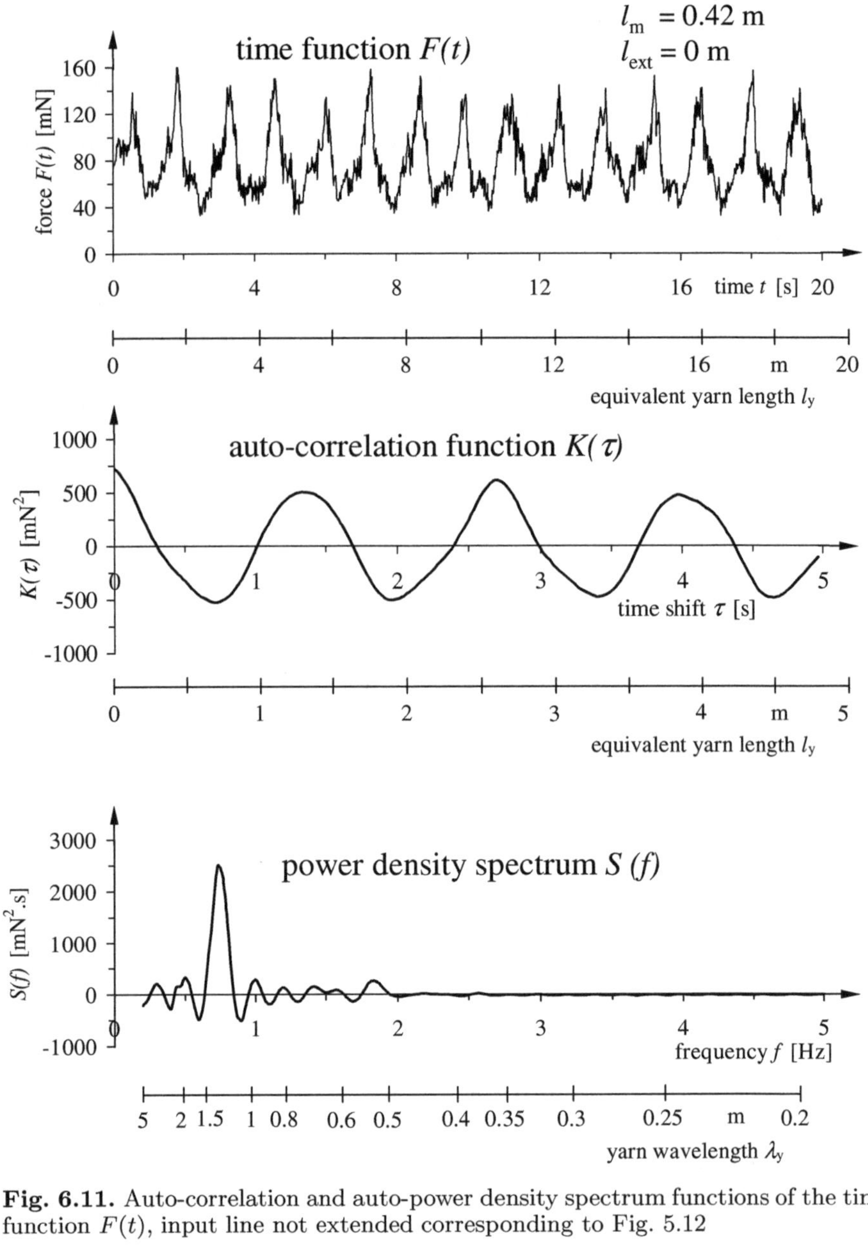

Fig. 6.11. Auto-correlation and auto-power density spectrum functions of the time function $F(t)$, input line not extended corresponding to Fig. 5.12

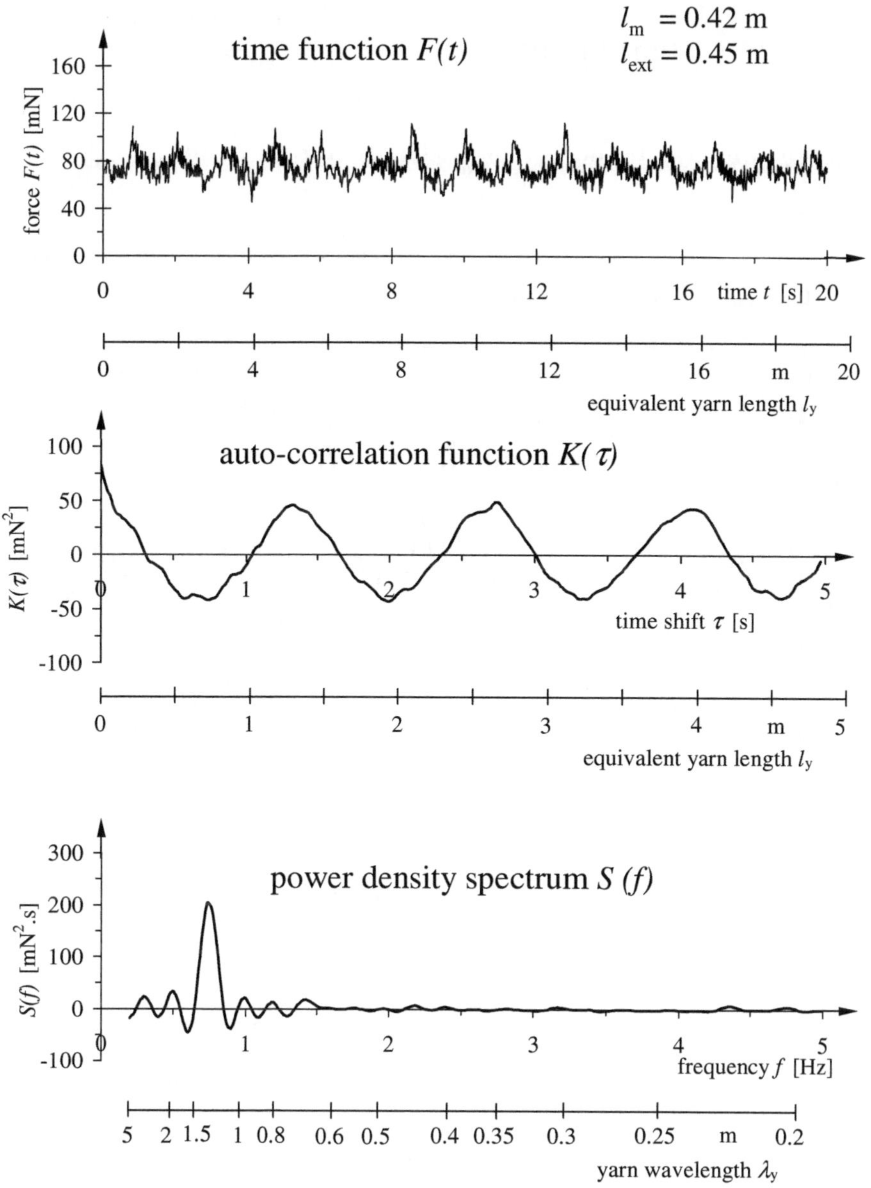

Fig. 6.12. Auto-correlation and auto-power density spectrum functions of the time function $F(t)$, extension of the input line $l_{\text{ext}} = 0.45$ m corresponding to Fig. 5.12

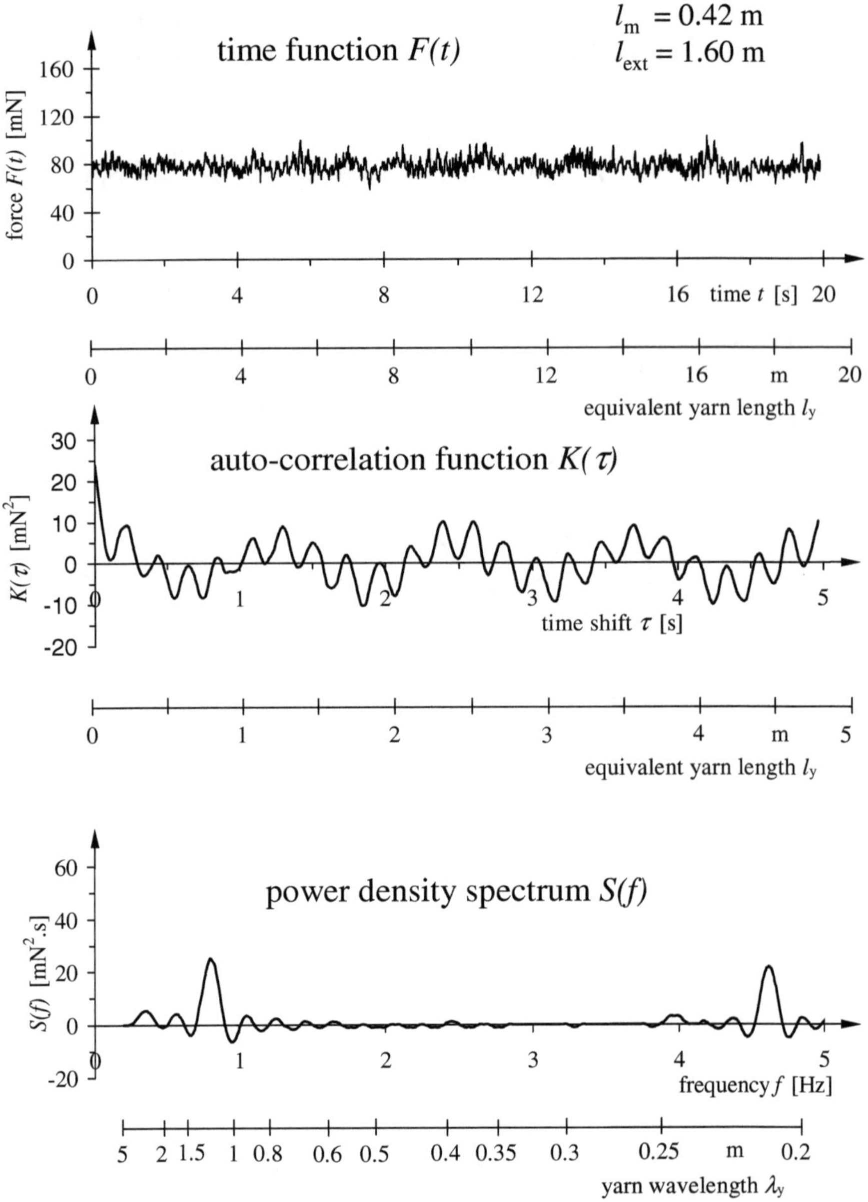

Fig. 6.13. Auto-correlation and auto-power density spectrum functions of the time function $F(t)$, extension of the input line $l_{ext} = 1.60$ m corresponding to Fig. 5.12

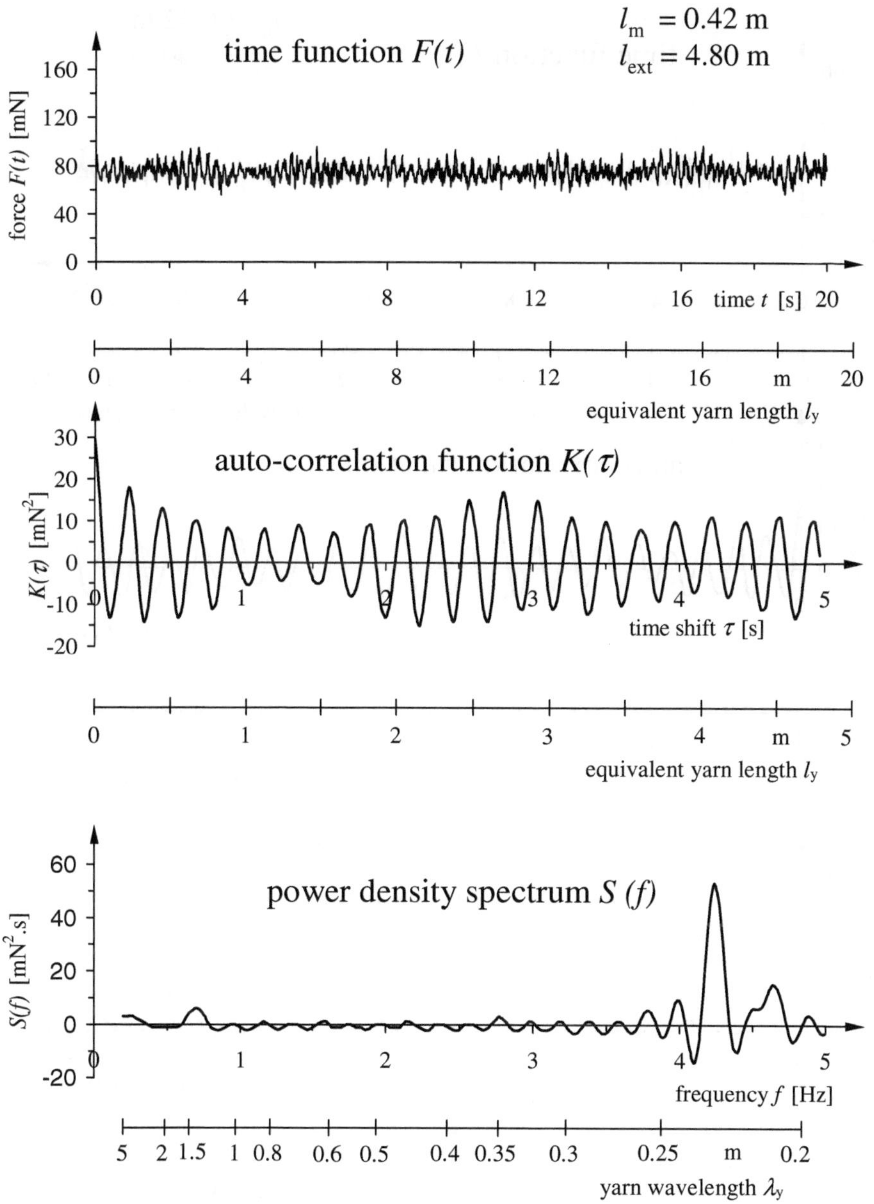

Fig. 6.14. Auto-correlation and auto-power density spectrum functions of the time function $F(t)$, extension of the input line $l_{\text{ext}} = 4.80$ m corresponding to Fig. 5.12

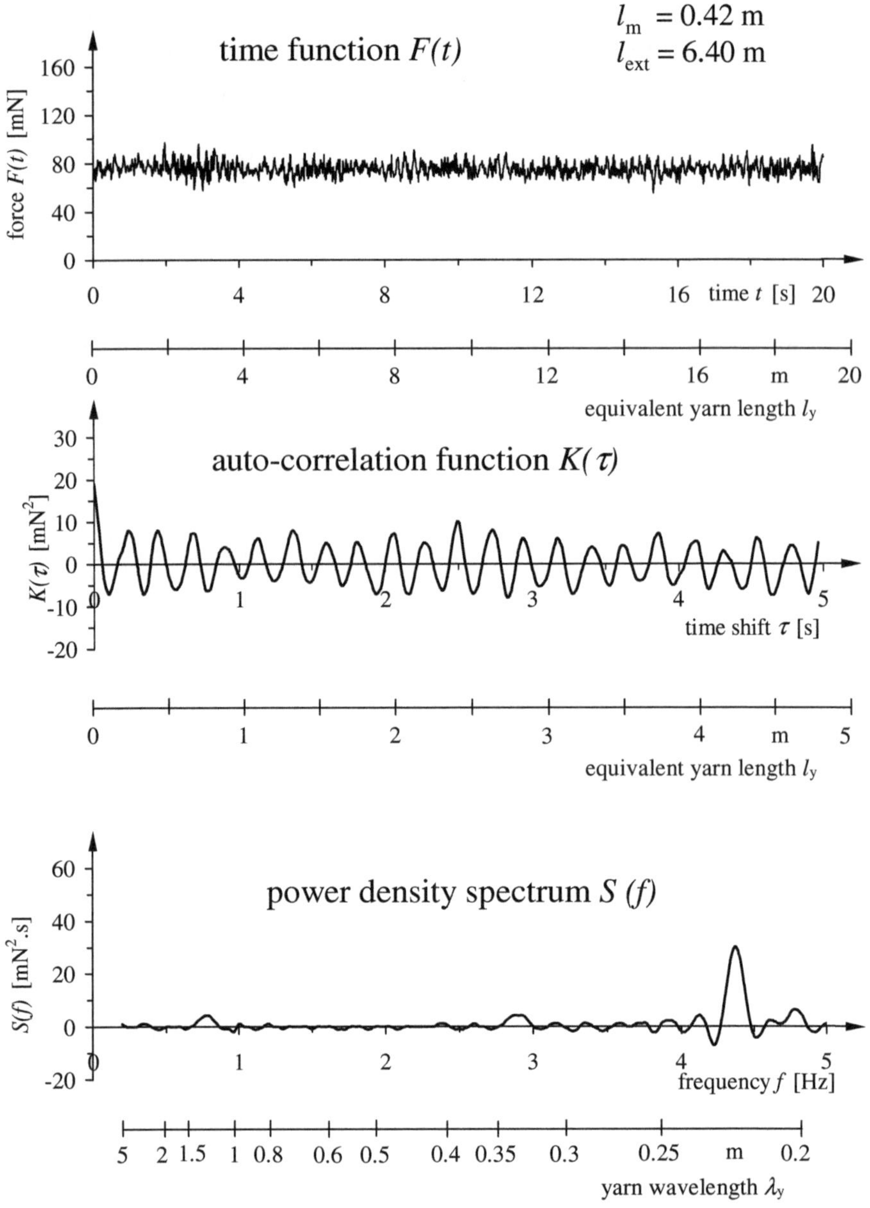

Fig. 6.15. Auto-correlation and auto-power density spectrum functions of the time function $F(t)$, extension of the input line $l_{\text{ext}} = 6.40$ m corresponding to Fig. 5.12

The overswing of the *S(f)*-curves to negative values near the main maximum (which theoretically should not occur) is caused by mistakes which arise from a too coarsely taken interval ($\Delta t = 1/30$ s), from the small number of single values for the time functions ($n = 600$), and from the small number of single values for the correlation function ($m = 150$). The limits of the used sum represent the APSF by means of Eq. 6.22. Nevertheless, the general statement is not even called into question in the presented case.

6.5 Combination Measurements and Evaluations

6.5.1 Task and Measurements

The dynamic behaviour of a process is normally distinguished not only by the time functions of *one* process or product variable. *Several* time functions and their mutual influence are to be analysed. The *combined* measurement and analysis of the time functions of

- the yarn tensile force $F(t)$,
- the yarn fineness $Tt_o(t)$,
- the tensile stress $\sigma(t) = F(t)/Tt_o(t)$

have hardly been used in the past with respect to the research of dynamic cause-effect relations in the beginning and the transmission of inner and outer yarn unevennesses and their relationships. A related analytical method has been presented until now only in just a few (own) papers [320, 321]. The following statements are essentially a brief conclusion of these.

The following two possibilities are practically given to the measurement recording of the named time functions:

a) Two channel synchronous measuring signal records of $F(t)$ and $Tt_o(t)$ in which the third time function $\sigma(t) = F(t)/Tt_o(t)$ is *simultaneously calculated* as the tensile stress (fineness related tensile force) by means of a quotient computer.

b) Two channel synchronous measuring signal records of $F(t)$ and $Tt_o(t)$ in which the time function of the tensile stress $\sigma(t)$ is calculated *after* the measuring signal records, point by point, by means of the division for each temporal related values. This calculation can be carried out either "by hand" or – better – also by means of a quotient computer.

It is an advantage for both methods, the primary records of the time functions of the yarn tensile force $F(t)$ and the yarn fineness $Tt_o(t)$, that they can be to stored by means of a two channel measuring magnetic tape device or of a quick external computer memory. It is then possible to temporally

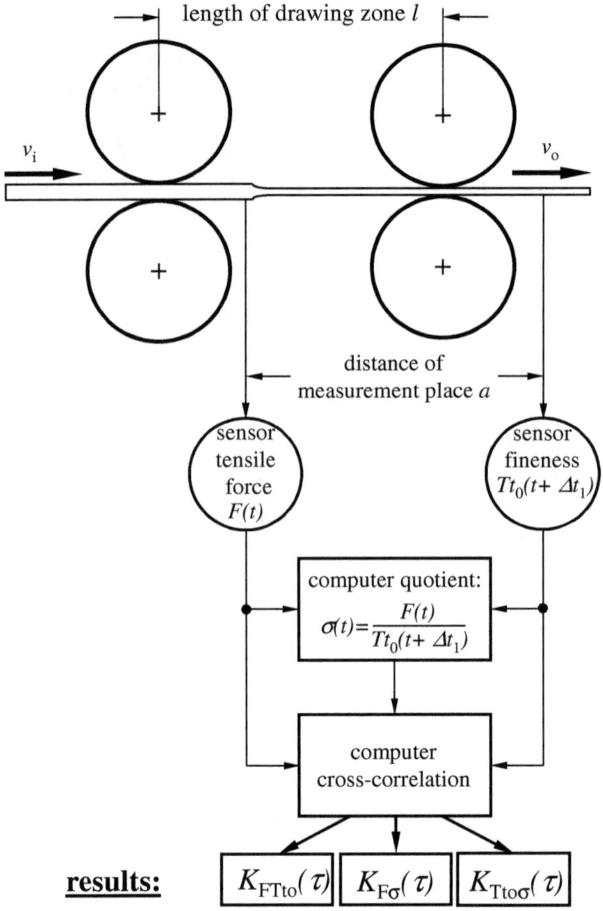

Fig. 6.16. Technological scheme of a combination measurement of yarn tensile force, yarn fineness and tensile stress in the drawing process with correlation evaluation

and locally separate the measuring signal records on the machine and the analysis in the lab. In this way, it is possible to carry out later the measuring process at any time to repeat in the slow or fast motion manner or the quotient calculation to the time function of the tensile stress $\sigma(t)$. It is possible in the same way to produce a data memory of the time functions which is available to the further described analysis as follows.

Figure 6.16 shows a conventionally designed variant which proposes a storage of the measuring values by means of a measuring magnetic tape device which the time function can be submitted to in a (cross-)correlation evaluation.

6.5.2 Dynamic Evaluation and Cross-Correlation Functions (CCF)

Fundamentals

First of all several synchronously determined time function courses of different process and product variables can also be evaluated visually, similarly as to Sect. 6.4 for the evaluation and analysis of the time function course of the yarn tensile force independently. This can take place, for instance, to observe whether changes of the $F(t)$-, $Tt_o(t)$- and $\sigma(t)$-courses correspond to each other and whether this correlation will be positive or negative.

Further analytical statements can give the cross-correlation function (CCF) or to the cross-power density spectrum function (CPSF) for the mutual dependences or the dynamic relationship of time functions of two process or product variables.

The calculation rules for the CCF of two time functions in the integral and sum representation has been already described in Sect. 2.4.4 (see Eqs. 2.18 and 2.19). We will only regard the CCF in the following. The interested reader can inquire as to the numerical more expensive formation rules and the (until now only little used) statement possibilities of the cross-power density spectrum function in [8] and [9].

Cross-correlation functions (CCF) of yarn tensile force, yarn fineness and tensile stress in the draw process

The following three cross-correlation functions can be formated (in each case in the integral and the sum representation) from the time functions of the yarn tensile force $F(t)$, the yarn fineness $Tt_o(t)$ and the tensile stress $\sigma(t)$:

CCF of the yarn tensile force and the yarn fineness

$$K_{\mathrm{FTt_o}}(\tau) = \lim_{\mathrm{T}\to\infty} \frac{1}{T} \int_0^T [F(t) - F_{\mathrm{m}}][Tt_{\mathrm{o}}(t+\tau) - Tt_{\mathrm{om}}]\mathrm{d}t \tag{6.27}$$

$$K_{\mathrm{FTt_o}}(k \cdot \Delta t) = \frac{1}{n-k} \sum_{\mathrm{i}=1}^{n-k} (F_{\mathrm{i}} - F_{\mathrm{m}})(Tt_{\mathrm{o(i+k)}} - Tt_{\mathrm{om}}) \tag{6.28}$$

Boundary condition: $m = k_{\max} \leq \frac{n}{5}$, which $k = 0, 1, 2, 3, \ldots m$

CCF of the yarn tensile force and the tensile stress

$$K_{\mathrm{F}\sigma}(\tau) = \lim_{\mathrm{T}\to\infty} \frac{1}{T} \int_0^T [F(t) - F_{\mathrm{m}}][\sigma(t+\tau) - \sigma_{\mathrm{m}}]\mathrm{d}t \tag{6.29}$$

$$K_{\mathrm{F}\sigma}(k \cdot \Delta t) = \frac{1}{n-k} \sum_{\mathrm{i}=1}^{n-k} (F_{\mathrm{i}} - F_{\mathrm{m}})(\sigma_{\mathrm{i+k}} - \sigma_{\mathrm{m}}) \tag{6.30}$$

Boundary condition: $m = k_{\mathrm{max}} \leq \frac{n}{5}$, which $k = 0, 1, 2, 3, ...m$

CCF of the yarn fineness and the tensile stress

$$K_{\mathrm{Tt_o}\sigma}(\tau) = \lim_{\mathrm{T}\to\infty} \frac{1}{T} \int_0^T [Tt_{\mathrm{o}}(t) - Tt_{\mathrm{om}}][\sigma(t+\tau) - \sigma_{\mathrm{m}}]\mathrm{d}t \tag{6.31}$$

$$K_{\mathrm{Tt_o}\sigma}(k \cdot \Delta t) = \frac{1}{n-k} \sum_{\mathrm{i}=1}^{n-k} (Tt_{\mathrm{oi}} - Tt_{\mathrm{om}})(\sigma_{\mathrm{i+k}} - \sigma_{\mathrm{m}}) \tag{6.32}$$

Boundary condition: $m = k_{\mathrm{max}} \leq \frac{n}{5}$, which $k = 0, 1, 2, 3, ...m$

The signs and symbols in Eqs. 6.27 to 6.30 mean:

$F(t), Tt_{\mathrm{o}}(t)$, $\sigma(t)$	time functions of the tensile force, yarn fineness, tensile stress
$F_{\mathrm{m}}, Tt_{\mathrm{om}}$, σ_{m}	mean values of the tensile force, yarn fineness, tensile stress in the range of $0 \leq t \leq T$ or $0 \leq k \cdot \Delta t \leq n-k$
$Tt_{\mathrm{oi}}, Tt_{\mathrm{o(i+k)}}$, $F_{\mathrm{i}}, \sigma_{\mathrm{i+k}}$	discontinuous values of the time functions $F(t), Tt_{\mathrm{o}}(t), \sigma(t)$, taken in steps of Δt
T	length of the integration range
$\tau, k \cdot \Delta t$	time shift
k, m, n	running (sequence) indices
$K_{\mathrm{FTt_o}}(\tau)$, $K_{\mathrm{F}\sigma}(\tau)$, $K_{\mathrm{Tt_o}\sigma}(\tau)$, $K_{\mathrm{FTt_o}}(k \cdot \Delta t)$, $K_{\mathrm{F}\sigma}(k \cdot \Delta t)$, $K_{\mathrm{Tt_o}\sigma}(k \cdot \Delta t)$	single values of the CCF of the time functions $F(t)$, $Tt_{\mathrm{o}}(t)$ and $\sigma(t)$ for τ or $k \cdot \Delta t$

The CCFs practically express the *interconnected dispersion* of the single time functions among one another and not only in the simple known manner for $\tau = 0$. This is enabled by means of the continuously or step-like realised time shift τ or $k \cdot \Delta t$ (change values of the one time function are related to the temporal with τ or $k \cdot \Delta t$ shifted change values of the other time function). The CCFs also enable an estimation of the *statistical relationship* between the values of the one time function and values of the other time function which are more or less distantly temporal.

The methods of combination measurements and evaluations of time functions have been applied extensively in the past to the draw process of polymer yarns. This process imprints on the yarn *drastic* fineness and E-modulus changes (consequently quality changes) which are effected by yarn tensile force. The following statements and ideas are not only applicable to the draw deformation but also to all fibre formation and fibre deformation processes which are similar to the draw process. Examples are the yarn deformation processes of classical melt spinning, of high speed spinning, but also of simultaneous or sequential draw texturing. It has already been hinted to the sharpened analysis situation at yarn deformation processes in a first extensive paper [321]. Through this it has been given that the effecting quantities for F and Tt_o can be correlated (temporal stable or instable) among one another in different manners with a more or less phase shift and that the changes of the secondary product variable tensile stress $\sigma(t)$ must be generated by an additional quotient calculation of $F(t)$ and $Tt_\mathrm{o}(t)$.

The following dependences on other process and product variables can be formally formulated for the primary quantities F and Tt_o which generate the tensile stress σ (see also Fig. 2.1 in Sect. 2.1.1):

$$Tt_\mathrm{o} = f(Tt_\mathrm{i}, v_\mathrm{o}/v_\mathrm{i}, l, E_\mathrm{yi})$$

$$F = f(Tt_\mathrm{i}, E_\mathrm{yi}, v_\mathrm{o}/v_\mathrm{i}, v_\mathrm{o}, l, T_\mathrm{y})$$

The difficulties from these mutual interweavings will be distinctly special if the characterising yarn quantity E_yo at the output of the drawing zone is included into the estimation of the inner yarn unevenness:

$$E_\mathrm{yo} = f(E_\mathrm{yi}, v_\mathrm{o}/v_\mathrm{i}, v_\mathrm{o}, l, Tt_\mathrm{i}, Tt_\mathrm{o}, F)$$

The combined measurement and estimation of time functions of yarn tensile force $F(t)$ and yarn fineness $Tt_\mathrm{o}(t)$ can give answers to the following questions:

a) Did induced fluctuate fineness changes to the yarn during the elongation or deformation process? Which quantity exists at the process input?
b) Did induced fluctuate substance property changes (E-modulus, tensile stress, plastic deformation part) on the yarn during the elongation or deformation process?
c) Which relation exists between form changes and reaction stresses? Which process influences are causally responsible for these changes?
d) Is a change of the yarn tensile force (which is marked often wrongly as yarn stress, though of no sort cross section relation exists) the effect of a change of the imprinted elongation or the effect of a change of the momentary yarn elasticity (which again depends on the cross section or the fineness as well as on the E-modulus of the yarn)?

Tendencies of disturbance transmission in the drawing process are compiled in Table 6.2. These come from theoretical considerations and experimental investigations for different combinations of possible process input disturbances. We should remember: Fineness changes $\Delta Tt_i/Tt_{im}$ are an expression of outer, of elastic modulus $\Delta E_{yi}/E_{yim}$ are an expression of inner unevennesses of the input yarn. The following conclusions can be deduced which emphasise the necessity of a detailed time function analysis:

a) Tensile force changes of the yarn in the drawing zone do not allow, in any case, for an unambiguous assertion about the situation of disturbance causes at the process input. The direction of the tensile force changes can be predicted when at least one input size or both input sizes change in the same direction. If fineness and E-modulus change turn out to contradict then the change dimensions of the one input size compared to the change dimensions of the other input size are responsible for the direction of the output disturbance. The absence of some reaction is imaginable, in the special case, if the contrary imprinted inner and outer unevennesses of the yarn at the drawing zone input are just canceled.

Table 6.2. Tendencies of disturbance transmission in the drawing process

Yarn disturbances at the process input: (length of the disturbance < length of drawing zone)		Yarn disturbances at the process output		
Fineness $\Delta Tt_i/Tt_{im}$	E-modulus $\Delta E_{yi}/E_{yim}$	Tensile force $\Delta F/F_m$	Fineness $\Delta Tt_o/Tt_{om}$	E-modulus $\Delta E_{yo}/E_{yom}$
0	+	+	↑	↓
0	−	−	↑	↓
+	0	+	↑	↑
−	0	−	↑	↑
+	+	+	↑	↓
−	−	−	↑	↓
+	−	+,−,0	↑,↓	↑,↓
−	+	+,−,0	↑,↓	↑,↓

↑ increasing unevenness
↓ decreasing unevenness
\+ positive deviation to the mean value
− negative deviation to the mean value
0 no deviation to the mean value

b) The calculation of the quotient of the (draw-)yarn tensile force to the yarn fineness (the real fineness related draw yarn tensile force) is a necessary but is not a sufficient prerequisite to the desirable separation of geometrical and material influences to the yarn reaction (outer and inner unevenness).

Identification Matrix to the Unevenness Analysis of Man Made Fibres in the Draw Process

An identification matrix of the unevenness analysis of man-made fibres in the draw process has been developed from the authors in the past which is based on physical-analytical and mathematical-logical ideas. This identification matrix allows a secure general valid assertion conclusion to the unevenness cause structure of the relative unoriented yarn at the drawing zone input and the oriented yarn at the drawing zone output. It has been tested in a row of simulation calculations for the start values (that means for $\tau = k \cdot \Delta t = 0$) of the three CCFs (Eqs. 6.27 to 6.32) which allow for a conclusion to the constellation of the unevenness causes of the investigated yarns.

These investigations have been carried out according to the signal scheme of Fig. 6.16 and included step-like (aperiodic) as well as rectangular and sinusoidal periodic changes of the fineness in the input yarn ΔTt_{i} and/or changes of its E-modulus ΔE_{yi}.[1]

The identification matrix is shown in Table 6.3. Disturbance constellations of the E-modulus and of the fineness of the yarn at the drawing zone input are indicated symbolically in the left columns, in which positive and negative disturbances of both product variables as well as positive and negative disturbances of only one product variable have been assumed. In the three right-hand columns the awaiting start value constellation of the three possible CCFs is inscribed according to the disturbance constellation. Complex composed and superimposed primary measurement signals of the yarn tensile force and the yarn fineness can be decoded with this (occurring for the first time) by means of the third measurement size tensile stress and with concentration on their three CCFs. A decoded disturbance cause description is similarly possible.

The identification matrix shows that disturbances and their combinations lead to the same start value constellations if they only differ in the sign of

[1] The following is to be remarked at this point: The calculation of the CCFs from the recorded time functions of the yarn tensile force, the yarn fineness and the tensile stress must be secure. This means only real, each other physical attached value of the yarn tensile force and the yarn fineness, will be related one upon another for the calculation of the tensile stress. This means (see Fig. 6.16) that the measuring value of the fineness at the drawing zone output correlates with the measuring value of the tensile force when this yarn piece has passed the position of the tensile force measuring sensor in the drawing zone. It is the moment $(t - \Delta t_1)$ according to the chosen draw conditions (output velocity v_{o}, draw ratio DR, distance of measurement place a), which Δt_1 can be signified as the necessary phase shift between the time functions of the tensile force and the fineness for the continuous quotient calculation to the tensile stress (see Fig. 6.16).

Table 6.3. Identification matrix to the unevenness analysis in the draw process

Variant of disturbance	Disturbance in the yarn at the process input (length of the disturbance <length of drawing zone)		Start value of the cross correlation function ($\tau = 0$)		
	Fineness ΔTt_{i}	E-modulus ΔE_{yi}	$K_{\mathrm{FTto}}(0)$	$K_{\mathrm{F}\sigma}(0)$	$K_{\mathrm{Tto}\sigma}(0)$
1	+	0	+	+	–
2	–	0	+	+	–
3	+	+	–	+	–
4	–	–	–	+	–
5	–	+	+	$-(0,+)$	–
6	+	–	+	$-(0,+)$	–
7	0	+	–	+	–
8	0	–	–	+	–

\+ positive deviation to the mean value; start value of the CCF positive
– negative deviation to the mean value; start value of the CCF negative

the disturbance (see Table 6.3, applicable to disturbance variants 1 and 2, 3 and 4, 5 and 6, 7 and 8). The following secure statements can be submitted from the named simulation calculations:

a) Disturbance variants 1 and 2 always result for disturbance lengths > 4 mm and any formed ΔTt_{i}

$$K_{\mathrm{FTto}}(0) > 0, K_{\mathrm{F}\sigma}(0) < 0, K_{\mathrm{Tto}\sigma}(0) < 0$$

A characteristic influence of the sign of the disturbance does not equate to the values of CCFs(0).

b) Disturbance variants 7 and 8 always result for any disturbances and any formed ΔE_{yi}

$$K_{\mathrm{FTto}}(0) < 0, K_{\mathrm{F}\sigma}(0) > 0, K_{\mathrm{Tto}\sigma}(0) < 0$$

The sign of the disturbance is also not provable here by the values of the CCFs(0).

c) The rectified synchronous disturbance of the E-modulus E_{yi} and the fineness Tt_{i} according to the disturbance variants 3 and 4 leads at the start values of the CCFs to the same results as with the disturbance variants 7 and 8. However, the variants 3 and 4 on the one hand and 7 and 8 on the other do further differ in the course of the CCFs by increasing τ-values. The sign of the $K_{\mathrm{F}\sigma}(\tau)$ changes from + to – with the unique E_{yi}-disturbance, whereas the function $K_{F\sigma}(\tau)$ is positive (and for a larger τ nearly zero) for rectified synchronous E_{yi}- and Tt_i-disturbances. This example shows that the whole function course of the CCFs (beyond the start values) can be nevertheless necessary for a detailed analysis of mistake causes for the draw process in the particular case.

d) Different effects superimpose themselves due to contrary synchronous E_{yi}- and Tt_i-disturbances depending on the disturbance parameters. $K_{FTto}(0) > 0$ and $K_{Tto\sigma}(0) < 0$ is valid when disturbances are contradictary. The sign changes for $K_{F\sigma}(0)$ depending on the length of the disturbance. $K_{F\sigma}(0)$ is negative for disturbances longer than 12 mm. If the disturbance is shorter then the $K_{F\sigma}(0)$ will be positive or zero.

The preceding informed investigations demonstrated problems of the signal analysis and their process-analytic importance contrary to most of the other dealt with questions of this book. It is obvious that the presented method of the qualitative unevenness analysis of man-made fibres in the classical draw process is not only valid for this process stage but in principle it is also transferable to analysis situations of the same kind on other continuous realised thread deformation processes during the thread formation and thread processing.

Finally, an idea for a device realisation for the described measuring and evaluation method should be developed. It is imaginable by means of the microelectronic tools available today that the electrical measuring signals (coming from both measuring sensors for the yarn fineness and the yarn tensile force) will be supplied by a specifically designed and appropriately programmed micro processor system, in which

- the calculation of the fineness related thread tensile force and the CCF-calculation repeatingly (appropriate selected integration times) can be realised,
- the appropriate results to the presented (or an expanded) identification matrix can be evaluated, and
- for instance the number of the identified disturbance variant (and if necessary, other intermediate results) displays as total result on a small digital screen.

Such a procedure of intelligent measuring value concentration would be serviceable not only for simplified handling but also for broader applications of the presented method [320].

References

1. G. Brack: *Dynamische Modelle verfahrenstechnischer Prozesse.* In: *Reihe Automatisierungstechnik* (Verlag Technik, Berlin 1971) Band 115
2. W. Weller: *Anwendung der Mikroelektronik in der Prozeßautomatisierung.* In: *Reihe Automatisierungstechnik* (Verlag Technik, Berlin 1981) Band 187
3. R. Beyreuther: *Dynamik von Fadenbildungs- und Fadenverarbeitungsprozessen*, 1. Aufl. (Fachbuchverlag, Leipzig 1986)
4. H. Strobel: *Experimentelle Systemanalyse* (Akademie-Verlag, Berlin 1975)
5. E.-G. Woschni: *Meßdynamik: Eine Einführung in die Theorie dynamischer Messungen* (S. Hirzel, Leipzig 1964)
6. F. H. Lange: *Spektrale Darstellung.* In: *Signale und Systeme* (Verlag Technik, Berlin 1969) Band I
7. F. H. Lange: *Regellose Vorgänge.* In: *Signale und Systeme* (Verlag Technik, Berlin 1971) Band III
8. H. Otto, M. Peschel: *Anwendung statistischer Methoden in der Regelungstechnik.* In: *Reihe Automatisierungstechnik* (Verlag Technik, Berlin 1970) Band 106
9. H. Schlitt: *Systemtheorie für regellose Vorgänge: Statistische Verfahren für die Nachrichten- und Systemtechnik* (Springer, Berlin 1960)
10. W. Oppelt: *Kleines Handbuch technischer Regelvorgänge* (esp. p. **55**) (Verlag Chemie, Weinheim 1960)
11. G. Doetsch: *Anleitung zum praktischen Gebrauch der Laplace-Transformation* (Verlag Oldenbourg, München 1961)
12. H. Dobesch: *Laplace-Transformation: Einführung, Berechnung von Einschwingvorgängen* (Verlag Technik, Berlin 1965)
13. R. C. Oldenbourg; H. Sartorius: *Dynamik selbsttätiger Regelungen: Band I: Allgemeine und mathematische Grundlagen, stetige und unstetige Regelungen, Nichtlinearitäten* (Verlag Oldenbourg, München 1951)
14. A. Ziabicki and K. Kedzierska: *Studies on the orientation phenomena by fiber formation from polymer melts. I. Preliminary investigations on polycapronamide.* J. Appl. Polym. Sci. **2**, 14 (1959)
15. A. Ziabicki: *Studies on the orientation phenomena by fiber formation from polymer melts. II. Theoretical considerations.* J. Appl. Polym. Sci. **2**, 24 (1959)
16. A. Ziabicki and K. Kedzierska: *Mechanical aspects of fibre spinning process in molten polymers. I. Stream diameter and velocity distribution along the spinning way.* Kolloid-Z. **171**, 51–61 (1960)
17. A. Ziabicki and K. Kedzierska: *Mechanical aspects of fibre spinning process in molten polymers. II. Stream broadening after the exit from the channel of spinneret.* Kolloid-Z. **171**, 111–119 (1960)
18. A. Ziabicki and K. Kedzierska: *Mechanical aspects of fibre spinning process in molten polymers. III. Tensile force and stress.* Kolloid-Z. **175**, 14–27 (1961)

19. A. Ziabicki: *Differentialgleichungen für Geschwindigkeitskomponenten beim Faserspinnen.* Kolloid-Z. **179** 2, 116–117 (1961)
20. A. Ziabicki and K. Kedzierska: *Studies on the orientation phenomena by fiber formation from polymer melts. III. Effects of structure on orientation. Condensation of polymers.* J. Appl. Polym. Sci. **6**, 111 (1962)
21. A. Ziabicki and K. Kedzierska: *Studies on the orientation phenomena by fiber formation from polymer melts. IV. Effects of structure on orientation. Polyethylene and polystyrene.* J. Appl. Polym. Sci. **6**, 361 (1962)
22. A. Ziabicki and R. Takserman-Krozer: *Effect on rheological factors on the length of liquid threads.* Kolloid-Z. **199**, 9–13 (1964)
23. A. Ziabicki and R. Takserman-Krozer: *Mechanism of breakage of liquid threads.* Kolloid-Z. **198**, 60–65 (1964)
24. E. H. Andrews: *Cooling a spinning thread-line.* Brit. J. Appl. Phys. **10**, 39–43 (1959)
25. S. Kase and T. Matsuo: *Studies on melt spinning. I. Fundamental equations on the dynamics on melt spinning.* J. Polym. Sci. **A3**, 2541–2554 (1965)
26. S. Kase and T. Matsuo: *Studies on melt spinning. II. Steady-state and transient solutions of fundamental equations compared with experimental results.* J. Appl. Polym. Sci. **11**, 251–287 (1967)
27. S. Kase: *Studies on melt spinning. III. Velocity field within the thread.* J. Appl. Polym. Sci. **18**, 3267–3278 (1974)
28. S. Kase: *Studies on melt spinning. IV. On the stability of melt spinning.* J. Appl. Polym. Sci. **18**, 3279–3304 (1974)
29. I. Hamana: *Der Verlauf der Fadenbildung beim Schmelzspinnen.* Lenzinger Berichte **26**, 118–132 (1968)
30. I. Hamana, M. Matsui, and S. Kato: *Der Verlauf der Fadenbildung beim Schmelzspinnen.* Melliand Textilberichte **50**, 382–388, 499–503 (1969)
31. C. D. Han: *A theoretical study on fiber spinnability.* Rheol. Acta **9**, 355–365 (1970)
32. C. D. Han and R. R. Lamonte: *Studies on melt spinning. I. Effect of molecular structure and molecular weight distribution on elongational viscosity.* Trans. Soc. Rheol. **16**, 447–472 (1972)
33. C. D. Han and R. R. Lamonte: *Studies on melt spinning. II. Analysis of the deformation and heat transfer processes.* J. Appl. Polym. Sci. **16**, 3285–3306 (1972)
34. C. D. Han and R. R. Lamonte: *Studies on melt spinning. III. Melt fracture and draw resonance.* J. Appl. Polym. Sci. **16**, 3307–3323 (1972)
35. C. D. Han and Y. W. Kim: *Studies on melt spinning. V. Elongational viscosity and spinnability of two-phase systems.* J. Appl. Polym. Sci. **18**, 2589–2603 (1974)
36. C. D. Han and Y. W. Kim: *Studies on melt spinning. VI. The effect of deformation history on elongational viscosity, spinnability, and thread instability.* J. Appl. Polym. Sci. **20**, 1555–1571 (1976)
37. C. D. Han and S. M. Apte: *Studies on melt spinning. VIII. The effect of molecular structure and cooling conditions on the severity of draw resonance.* J. Appl. Polym. Sci. **24**, 61–87 (1979)
38. H. H. George: *A steady state model model for melt spinning at intermediate winding speeds.* Polym. Eng. Sci. **22**, 292–299 (1982)
39. H. H. George, A. Holt, and A. Buckley: *A study of structural development in the high speed spinning of PET.* Polym. Eng. Sci. **23**, 95–99 (1983)
40. H. H. George: *Modeling the melt spinning process.* In: *Polymers for Fibers and Elastomers*, ACS Symp. **260**, 355–369 (1984)

41. J. Shimizu: *High-speed spinning – mechanism and fiber properties.* Kasen Geppo **30**, 42–51 (1977)
42. J. Shimizu, K. Toriumi, and Y. Imai: *High-speed melt spinning of isotactic polypropylene fibers.* Sen-i Gakkaishi **33**, 255–260 (1977)
43. J. Shimizu, K. Toriumi, and K. Tamai: *High-speed melt spinning of polyester filaments – effect of spinning velocity on the properties and molecular orientation.* Sen-i Gakkaishi **33**, 208–214 (1977)
44. J. Shimizu: *High-speed spinning and its industrial background.* Sen-i Gakkaishi **34**, P43–52 (1978)
45. J. Shimizu, N. Okui, A. Kaneko, and K. Toriumi: *High speed spinning of polyethylen terephthalate (PET), effects of meltdraw ratio and flow rate.* Sen-i Gakkaishi **34**, 64–71 (1978)
46. J. Shimizu, N. Okui, T. Kikutani, and K. Toriumi: *High-speed spinning of PET, effect of molecular weight.* Sen-i Gakkaishi **34**, T35–40, T93–98 (1978)
47. J. Shimizu, N. Okui, and Y. Imai: *High-speed melt spinning of isotactic polypropylene fibers – crystallization mechanism in the spinline and fiber structure and properties.* Sen-i Gakkaishi **35**, 405–412 (1979)
48. J. Shimizu, N. Okui, and Y. Imai: *High-speed spinning of isotactic polypropylene fibers – tensile properties of hard elastic fibers.* Sen-i Gakkaishi **36**, 166–174 (1980)
49. J. Shimizu, N. Okui, and T. Kikutani: *High speed melt spinning of PET – radial variation across fibers.* Sen-i Gakkaishi **37**, T135–T142 (1981)
50. J. Shimizu, N. Okui, T. Kikutani, A. Ono, and A. Takaku: *High speed melt spinning of nylon 6.* Sen-i Gakkaishi **37**, T143–T152 (1981)
51. J. Shimizu: *The fine structure of super high speed spun fibers and its structure development.* Sen-i Gakkaishi **38**, P499–P507 (1982)
52. J. Shimizu, N. Okui, T. Yamamoto, M. Ishii, and A. Takaku: *Melt spinning of polypropylene/polystyrene blends – morphology and ultra-fine fibril formation.* Sen-i Gakkaishi **38**, 1–9 (1982)
53. J. Shimizu, N. Okui, and K. Tamai: *Air drag in high-speed spinning.* Sen-i Gakkaishi **39**, T398–T407 (1983)
54. S. Kubo: *Air boundary layer on a filament in high-speed spinning.* In: A. Ziabicki, H. Kawai (Eds.): *High-Speed Fiber Spinning* (John Wiley & Sons, New York 1985)
55. C. Bos: *A sublayer description for the air friction acting on thin filaments during melt spinning.* Int. J. Heat Mass Transfer **31**, 167–76 (1988)
56. J. Shimizu and u. a.: *Fiber structure formation in high speed melt spinning of PET. Effect of take up velocity on unit cell parameters and superstructure of fibers.* Sen-i Gakkaishi **40**, T63–T71 (1984)
57. J. Shimizu: *High speed melt spinning of nylon 12, fiber structure and properties.* Sen-i Gakkaishi **41**, T513–T520 (1985)
58. J. Shimizu, N. Okui, and T. Kikutani: *Fine structures and physical properties of fibers melt-spun at high speeds from various polymers.* In: A. Ziabicki, H. Kawai (Eds.): *High Speed Fiber Spinning* (John Wiley & Sons, New York 1985)
59. J. Shimizu, N. Okui, and T. Kikutani: *Simulation of dynamics and structure formation in high-speed melt spinning.* In: A. Ziabicki, H. Kawai (Eds.): *High Speed Fiber Spinning* (John Wiley & Sons, New York 1985)
60. J. Shimizu, T. Kikutani, Y. Ohkoshi, and A. Takaku: *Melt-spinning of polyether-ether-ketone (PEEK) and the structure and properties of resulting fibers.* Sen-i Gakkaishi **43**, 507–519 (1987)
61. H. Yasuda: *Tenacity change and structural development in high-speed spinning of PET.* ACS/CSJ Chem. Conf. Honolulu, Hawaii, April 1-6, 19 (1979)

62. H. Yasuda, H. Sugiyama, and H. Yanagawa: *Simulation of the orientation and its distribution across a filament using the steady-state single filament melt spinning theory.* Sen-i Gakkaishi **35**, 370–375 (1979)
63. H. Yasuda: *Simulation of melt spinning by computer.* Sen-i Gakkaishi **33**, P11–P14 (1980)
64. H. Yasuda: *Structure development and physical properties of high-speed-spun PET fibres.* Sen-i Gakkaishi **38**, P514–P520 (1982)
65. H. Yasuda, H. Sugiyama, and S. Hayashi: *Theory of the steady-state multifilament melt spinning.* Sen-i Gakkaishi **40**, T227–T231 (1984)
66. H. Yasuda, H. Sugiyama, and S. Hayashi: *Applications of the steady-state multifilament melt spinning theory.* Sen-i Gakkaishi **40**, T488–T493 (1984)
67. H. Yasuda: *Structure development and physical properties of high-speed-spun PET fibers.* In: A. Ziabicki, H. Kawai (Eds.): *High Speed Fiber Spinning* (John Wiley & Sons, New York 1985)
68. A. Dutta and V. M. Nadkarni: *Identifying critical process variables in PET melt spinning.* Textile Res. J. **54**, 35–42 (1984)
69. C. L. Tucker: *Computer Modeling of Polymer Processing* (Carl Hanser Verlag, München 1989)
70. T. Nakajima: *Advanced Fibre Spinning Technology* (Woodhead Publishing Ltd. 1994)
71. D. R. Salem (Ed.): *Structure in Polymeric Fibers* (Carl Hanser Verlag, München 2001)
72. K. Higuchi, T. Katsu: Sen-i Kikai Gakkaishi **13**, 1 (1960)
73. Y. Sano, K. Orii: Sen-i Gakkaishi **24**, 212 (1968)
74. K. Fukuda: Sen-i Gakkaishi **22**, S-3 (1966)
75. A. B. Thompson, In: J. W. Hearle, R. H. Peters (Eds.), *Fibre Structure* (Butterworth, London 1953)
76. S. L.Anderson, R. Stubbs: J. Text. Inst. **49**, T-35 (1958)
77. A. Selwood: J. Text. Inst. **53**, T-576 (1962)
78. T. Ishibashi, K. Aoki, and T. Ishii: *Studies on melt spinning of nylon 6. I. Cooling and deformation behaviour and orientation of nylon 6 threadline.* J. Appl. Polym. Sci. **14**, 1597–1613 (1970)
79. T. Ishibashi and T. Ishii: *Studies on melt spinning of nylon 6. II. Effect of heating the threadline upon orientation and crystallization.* J. Appl. Polym. Sci. **20**, 335–344 (1976)
80. H. Ishihara and S. Kase: *Studies on melt spinning. V. Draw resonance as a limit cycle.* J. Appl. Polym. Sci. **19**, 557–565 (1975)
81. H. Ishihara and S. Kase: *Studies on melt spinning. VI. Simulation of draw resonance using Newtonian and power-law viscosities.* J. Appl. Polym. Sci. **20**, 169–191 (1976)
82. M. Matsui: *Air drag on a continuous filament in melt spinning.* Trans. Soc. Rheol. **20**, 465–474 (1976)
83. M. Matsui: *Fiber formation process in high speed spinning of PET.* Sen-i Gakkaishi **38**, P508–P513 (1982)
84. M. Matsui: *Fiber formation process in high-speed spinning of PET.* In: A. Ziabicki, H. Kawai (Eds.): *High-Speed Fiber Spinning* (John Wiley & Sons, New York 1985)
85. V. Pechoc: *Wärmetechnische Berechnungen zum Spinnprozeß synthetischer Fasern.* Faserforsch. u. Textiltechnik **10**, 62–67 (1959)
86. W. Roth and R. Schroth: *Über das Spinnen und Strecken von Polykondensatfäden. I. Einfluß der Spinnbedingungen auf den Streckvorgang.* Faserforsch. u. Textiltechnik **11**, 312–319 (1960)

87. W. Roth and R. Schroth: *Über das Spinnen und Strecken von Polykondensatfäden. II. Verstreckung unter verschiedenen Bedingungen.* Faserforsch. u. Textiltechnik **11**, 365–373 (1960)
88. W. Roth and R. Schroth: *Über das Spinnen und Strecken von Polykondensatfäden. III. Strukturveränderungen von Polyesterfäden durch Wärme- und Quellmitteleinfluß.* Faserforsch. u. Textiltechnik **12**, 361–369 (1961)
89. V. Gröbe and H. Versäumer: *Über die Fadenbildung beim Schmelzspinnen. I.* Faserforsch. u. Textiltechnik **14**, 249–256 (1963)
90. W. Roth and R. Schroth: *Über das Spinnen und Strecken von Polykondensatfäden. IV. Die kontinuierliche Verstreckung von PE und PA Fäden.* Faserforsch. u. Textiltechnik **14**, 507–516 (1963)
91. K. E. Perepelkin: *Einige Gesetzmäßigkeiten des Mechanismus des Spinnprozesses von Chemiefasern.* Khim. Volokna, 1–6 (1964)
92. G. Wilhelm: *Die Abkühlung eines aus der Schmelze gesponnenen Fadens im Spinnschacht.* Kolloid-Z. u. Z. Polymere **208**, 97–123 (1966)
93. K. E. Perepelkin: *Die wichtigsten dynamischen und kinematischen Gesetzmäßigkeiten der Spinnprozesse von Chemiefaserstoffen.* Sowj. Beitr. FFTT **5**, 281–292 (1968)
94. M. A. Matovich and J. R. A. Pearson: *Spinning a molten threadline. I. Steady state isothermal viscous flows.* Ind. Eng. Chem. Fundamentals **8**, 605–609 (1969)
95. J. M. Acierno, J. N. Dalton, J. M. Rodriguez, and J. L. White: *Rheological and heat transfer aspects on melt spinning of monofilament fibres of polyethylene and polystyrene.* J. Appl. Polym. Sci. **15**, 2395–2415 (1971)
96. K. Katayama, T. Amano, and K. Nakamura: *Structure formation during melt spinning.* Kolloid-Z. **226**, 125–134 (1971)
97. L. E. Abbott and J. L. White: *Melt spinning of high and low density polyethylene: The development of orientation and crystallinity and mechanical properties of spun fibers.* Appl. Polym. Symp. **20**, 247–268 (1973)
98. H. Graubner: *Untersuchungen über den Schmelzspinnprozeß und die Fadenbildung von Polyamid 6.* Angew. Makromol. Chem. **40/41**, 173–217 (1974)
99. M. M. Denn and J. S. Petrie: *Mechanics of a steady spinning of a viscoelastic liquid.* AIChE J. **21**, 791–799 (1975)
100. D. K. Gagon and M. M. Denn: *Simulation of the steady-state melt spinning of polymers with computer.* Polym. Eng. Sci. **21**, 844–853 (1981)
101. T. C. Papanastasiou, C. W. Macosco, L. E. Scriven, Z. Chen: *Fiber spinning of viscoelastic liquid.* AIChE J. **33**, 834–42 (1987)
102. T. S. Papanastasiou, S. M. Alai, Z. Chen: *High-speed, non-isothermal fiber spinning.* Int. Polym. Process. **9**, 148–158 (1994)
103. C. T. Kiang, J. A. Cuculo: *Influence of polymer characteristics and melt-spinning conditions on the production of fine denier PET fibers. I. Rheological characterization of PET polymer melt.* J. Appl. Polym. Sci. **46**, 55–65 (1992)
104. C. T. Kiang, J. A. Cuculo: *Influence of polymer characteristics and melt-spinning conditions on the production of fine denier PET fibers. II. melt spinning dynamics.* J. Appl. Polym. Sci. **46**, 67–82 (1992)
105. C. T. Kiang, J. A. Cuculo: *Influence of polymer characteristics and melt-spinning conditions on the production of fine denier PET fibers. III. Structure and properties of fine dernier as-spun PET fibers.* J. Appl. Polym. Sci. **46**, 83–97 (1992)
106. T. Kikutani, Y. Kawahara, T. Matsui, A. Takaku, J. Shimizu: *Measurement of filament temperature and analysis of orientation-induced crystallization behavior in high-speed melt spinning process.* Sen-i Gakkaishi **1**, 333–339 (1989)

107. T. Kikutani, T. Matsui, A. Takaku, J. Shimizu: *Diameter measurement in the vicinity of neck-like deformation in high-speed melt spinning process.* Sen-i Gakkaishi **45**, 441–6 (1989)
108. T. Kikutani, H. Morinaga, A. Takaku, J. Shimizu: *Effect of spinline quenching on structure development in high-speed melt spinning of poly(ethylene terephthalate). Diameter profiles in the spinline.* Int. Polym. Process. **5**, 20–24 (1990)
109. T. Kikutani, Y. Kawahara, N. Ogawa, N. Okui: *Capturing of real image of neck-like deformation from high-speed melt spinning line.* Sen-i Gakkaishi **50**, 561–566 (1994)
110. T. Kikutani, S. Arikawa, A. Takaku, N. Okui: *Fiber structure formation in high-speed melt spinning of sheath-core type bicomponent fibers.* Sen-i Gakkaishi **51**, 408–15 (1995)
111. T. Kikutani, K. Morohoshi, H. Y. Yoo: *Fiber structure development in high-speed melt spinning of copolyesters: poly(ethylene terephthalate-co-1, 4-cyclohexylene dimethylene terephthalate).* Polym. Eng. Sci. **35**, 942–9 (1995)
112. T. Kikutani, J. Radhakrishnan, S. Arikawa, A. Takaku, N. Okui, X. Jin, F. Niwa, Y. Kudo: *High-speed melt spinning of bicomponent fibers: mechanisms of fiber structure development in poly(ethylene terephthalate)/polypropylene system.* J. Appl. Polym. Sci. **62**, 1913–1924 (1996)
113. T. Kikutani, M. Radhakrishnan, M. Sato, N. Okui, A. Takaku: *High-speed melt spinning of PET. Mechanism of fiber structure formation using a liquid isothermal bath as modifier.* Int. Polym. Process. **11**, 42–49 (1996)
114. T. Kikutani: *Advances in fiber spinning technology.* Kobunshi **47**, 669–671 (1998)
115. T. Kikutani: *Fiber formation technology for polymers – accomplished technology and future direction of synthetic fibers.* Sen-i Gakkaishi **54**, P37–P42 (1998)
116. T. Kikutani: *High speed spinning of composite fibers.* Seikei Kako **10**, 78–85 (1998)
117. T. Kikutani: *High-speed melt spinning process and control of synthetic fiber structure.* Sen-i Gakkaishi **55**, P391–P396 (1999)
118. T. Kikutani: *High speed fiber spinning and fiber structure control.* Kobunshi Kako **48**, 162–168 (1999)
119. D. Salem: *Future view of structure formation in polymeric fibers.* Sen-i Gakkaishi **57**, 31 (2001)
120. G. Schmack, R. Beyreuther, H. Hofmann, U. Müller, D. Jehnichen: *Melt spinning of PA 4.6/6.* Chem. Fibers Int. **45** 6, 475–477 (1995)
121. G. Schmack, D. Jehnichen, R. Vogel, B. Tändler: *Biodegradable fibers of poly(3-hydroxybutyrate) produced by high-speed melt spinning and spin drawing.* J. Polym. Sci., Part B: Polym. Phys. **38**, 2841–2850 (2000)
122. G. Schmack, D. Jehnichen, R. Vogel, B. Tändler, R. Beyreuther, F. H. G. Jacobsen, S.: *Biodegradable fibers spun from polylactide generated by reactive extrusion.* J. Biotechnol. **86** 2, 151–160 (2001)
123. Y. Shimomura, H. Ito, T. Kikutani: *Fiber structure development in high-speed spinning of PP/PE sheath-core type bicomponent fibers.* ICAPP (Int. Conf. Adv. Polym. and Process.) Yonezawa, (2001)
124. J. E. Spruiell: *Structure and property development during the melt spinning of synthetic fibers.* NATO Sci. Ser., Ser. E **370**, 195–220 (2000)
125. J. E. Spruiell: *Structure formation during melt spinning.* In: D. R. Salem (Ed.): *Structure in Polymeric Fibers* (Carl Hanser Verlag, München 2001)
126. J. A. Cuculo, J. F. Hotter, Q. Zhou: *Advances in the control of spinline dynamics for enhanced properties.* In: D. R. Salem (Ed.): *Structure in Polymeric Fibers* (Carl Hanser Verlag, München 2001)

127. B. Tändler, G. Schmack, R. Vogel, D. Blechschmidt, R. Lindner: *Melt processing of a new biodegradable synthetic polymer in high-speed spinning and underpressure spunbonding process.* J. Polym. Env. **9**, 149–156 (2001)
128. W. Takarada, H. Ito, T. Kikutani, N. Okui: *Studies on high-speed melt spinning of noncircular cross-section fibers. I. Structural analysis of as-spun fibers.* J. Appl. Polym. Sci. **80**, 1575–1581 (2001)
129. W. Takarada, H. Ito, T. Kikutani, N. Okui: *Studies on high-speed melt spinning of noncircular cross-section fibers. II. On-line measurement of the spin line, including change in cross-sectional shape.* J. Appl. Polym. Sci. **80**, 1582–1588 (2001)
130. W. Takarada, H. Ito, T. Kikutani, N. Okui: *Studies on high-speed melt spinning of noncircular cross-section fibers. III. Modeling of melt spinnig process incorporating change in cross-sectional shape.* J. Appl. Polym. Sci. **80**, 1589–1600 (2001)
131. Z. Tang, N. Huang, X. Yu, P. Huo, X. Xu: *Computer simulation of PET high-speed thermal channel spinning.* Chem. Fibers Int. **51**, 437–441 (2001)
132. X. Wang, Z. Tang: *Necking point in PET high-speed fiber spinning.* Journal of China Textile University (English Edition) **18**, 16–20 (2001)
133. Y. Wu, G. Wu: *Structure and properties of high-speed melt-spun poly(trimethylene terephthalate) fibers.* Hecheng Xianwei Gongye **24**, 18–21 (2001)
134. M. Yoshimura, K. Iohara, H. Nagai, K. Koyama: *Fiber structure formation in high-speed melt spinning of poly(ethylene terephthalate) hybrid polymer.* IAPP (Int. Conf. Adv. Polym. and Process.) Yonezawa, 68–69 (2001)
135. E. B. Bond, J. E. Spruiell: *Melt spinning of metallocene catalyzed polypropylenes. II. As-spun filament structure and properties.* J. Appl. Polym. Sci. **82**, 3237–3247 (2001)
136. E. B. Bond, J. E. Spruiell: *Melt spinning of metallocene catalyzed polypropylenes. I. On-line measurements and their interpretation.* J. Appl. Polym. Sci. **82**, 3223–3236 (2001)
137. J. van Ruiten, R. Riedel, R. Deblieck, R. Brouwer, J. P. Penning: *Drawability and attainable mechanical properties of polyamide yarn using true stress-true strain curves.* J. Mater. Sci. **36**, 3119–3128 (2001)
138. R. Beyreuther, B. Tändler, M. Hoffmann, R. Vogel: *Reactive extrusion and melt spinning – a new technological route to special fibres.* J. Mater. Sci. **36** 7, 3103–3111 (2001)
139. M. Yoshimura, K. Iohara, H. Nagai, T. Takahashi, K. Koyama: *Fiber structure in high speed melt spinning of poly(ethylene terephthalate)/poly(methyl methacrylate).* Sen-i Gakkaishi **58**, 287–293 (2002)
140. R. J. Young, C. Riekel, M. M. Moran, R. J. Davies: *Deformation mechanisms in single polymer fibers.* Polym. Mater. Sci. Eng. [computer optical disk] **86**, 402–403 (2002)
141. J. Shimizu, T. Kikutani: *Dynamics and evolution of structure in fiber extrusion.* J. Appl. Polym. Sci. **83**, 539–558 (2002)
142. K. E. Perepelkin: *Polylactide fibers: Fabrication, properties, use, prospects. A review.* Fibre Chemistry (Translation of Khim. Vol.) **34**, 85–100 (2002)
143. L. K. Peterson: *Characterization of polylactic acid (PLA) fibers.* Microscope **50**, 37–43 (2002)
144. T. Kikutani: *Formation and structure of high mechanical performance fibers. II. flexible polymers.* J. Appl. Polym. Sci. **83**, 559–571 (2002)
145. S. J. Kim, S. S. Im: *Structure development and dynamic properties in high-speed spinning of high molecular weight PEN/PET copolyester fiber.* Fibers and Polymers **3**, 18–23 (2002)

146. H. Ito, T. Kikutani: *On-line measurement of orientation development in elongation process.* Seikei Kako **14**, 287–292 (2002)
147. M. H. Hong, J. L. a. White: *Birefringence and mechanical property development in melt spinning cyclopolyolefine filaments.* Int. Polym. Process. **17**, 53–59 (2002)
148. D. Choi, J. L. White: *Crystallization and orientation development in melt spinning isotactic PP of varying tacticity.* Int. Polym. Process. **17**, 233–243 (2002)
149. D. Choi, J. L. White: *Structure development in fiber extrusion of isotactic polypropylenes of varying tacticity.* Annual Technical Conference - Society of Plastics Engineers **60th**, 263–267 (2002)
150. R. Vogel, S. G. Hatzikiriakos, H. Brünig, B. Tändler, M. Golzar: *Improved spinnability of metallocene polyethylenes by using processing aids.* Int. Polym. Process. **18**, 67–73 (2003)
151. H. Brünig, R. Beyreuther, R. Vogel, B. Tändler: *Melt spinning of fine and ultra-fine PEEK-filaments.* J. Mater. Sci. **38**, 2149–2153 (2003)
152. S. Kase: *Simulation of melt spinning.* Sen-i Gakkaishi **38**, P418–P426 (1982)
153. K. Toriumi, A. Konda, J. Shimizu: *Dynamic simulation of melt spinning process.* Sen-i Gakkaishi **38**, 31–42, T49–T60 (1982)
154. J. C. Chang, M. M. Denn, S. Kase: *Dynamic simulation of melt spinning at low speed.* IEC Ind. Chem. Fundam. **21**, 13–17 (1982)
155. J. H. Bheda: *Mathematical modelling and experimental study of dynamics on structure development during high speed melt spinning of nylon-6.* PhD thesis, Univ. of Tennessee, Knoxville (1987)
156. J. H. Bheda, J. E. Spruiell: *Dynamics and structure development during high speed melt spinning of nylon 6. I. On-line experimental measurements.* J. Appl. Polym. Sci. **39**, 447–463 (1990)
157. Y. C. Bhuvanesh, V. B. Gupta: *Computer simulation of melt spinning of poly(ethylene terephthalate) using a steady-state model.* Indian J. Fibre Text. Res. **15**, 145–153 (1990)
158. Y. C. Bhuvanesh, V. B. Gupta: *Computer simulation of melt spinning of polypropylene fibers using a steady-state model.* J. Appl. Polym. Sci. **58**, 663–674 (1995)
159. Z. Ding, J. E. Spruiell: *An improved mathematical model of melt spinning process.* Polym. Mater. Sci. Eng. **73**, 550–551 (1995)
160. M.-G. Yoon: *Modeling of high speed spinning of nylon 6.* Han'guk Somyu Konghakhoechi **31**, 310–316 (1994)
161. V. Rauschenberger: *Recursive modeling of the fiber spinning process.* Chem. Fibers Int. **47**, 192–194 (1997)
162. V. Rauschenberger, H. M. Laun: *A recursive model for rheotens tests.* J. Rheol. **41**, 719–723 (1997)
163. A. J. McHugh, A. K. Doufas: *Simulation of fiber spinning including flow-induced crystallization.* Proc. Int. Congr. Rheol., 13th **3**, 386–388 (2000)
164. E. Mitsoulis, M. Beaulne: *Numerical simulation of rheological effects in fiber spinning.* Adv. Polym. Technol. **19**, 155–172 (2000)
165. H. Rave, H. Tiemeier, T. Gotz, D. Reinel-Bitzer, K. Steiner: *Simulation of the fiber spinning process.* Chem. Fibers Int. **51**, 427–428, 430–431 (2001)
166. L. Jarecki: *Computer modeling of fiber spinning from a crystallizing polymer melt. I. Mathematical model.* Polimery **46**, 335–343 (2001)
167. L. Jarecki: *Computer modeling of fiber spinning from a crystallization polymer melt. part II. Use of the model.* Polimery **46**, 420–427 (2001)

168. A. K. Doufas, A. J. McHugh: *Two-dimensional simulation of melt spinning with a microstructural model for flow-induced crystallization.* J. Rheol. **45**, 855–879 (2001)
169. A. K. Doufas, A. J. McHugh: *Simulation of melt spinning including flow-induced crystallization. Part III. Quantitative comparisons with PET spinline data.* J. Rheol. **45**, 403–420 (2001)
170. G. Allan, R. Yang, A. Fotheringham, R. Mather: *Neural modelling of polypropylene fibre processing: predicting the structure and properties and identifying the control parameters for specified fibres.* J. Mater. Sci. **36**, 3113–3118 (2001)
171. M. L. Ottone, J. A. Deiber: *A numerical method for the viscoelastic melt-spinning model with radial resolutions of temperature and stress fields.* Industrial & Engineering Chemistry Research **41**, 6345–6353 (2002)
172. Y. L. Joo, J. Sun, M. D. Smith, R. C. Armstrong, R. A. Brown, R. A. Ross: *Two-dimensional numerical analysis of non-isothermal melt spinning with and without phase transition.* J. Non-Newtonian Fluid Mech. **102**, 37–70 (2002)
173. P. H. Herrmans, P. Platzek: Kolloid. Z. **88**, 68 (1939)
174. R. S. Stein, F. H. Norris: J. Polym. Sci. **21**, 381 (1956)
175. P. Sajkiewicz, A. Ziabicki, L. Jarecki: *Effects of structure and formation conditions on mechanical properties of polyethylene terephthalate (PET) fibers.* Fibres & Textiles in Eastern Europe **6**, 44–51 (1998)
176. J. F. Spruiell and J. L. White: *Structural development during fiber processing. I. Kinetics and morphology of crystallization during melt spinning.* SPE 33rd Ann. Tech. Conf., ANTEC 75, Atlanta, 188–192 (1975)
177. J. L. White and Y. Ide: *Rheology and dynamic of fiber formation from polymer melts.* J. Appl. Polym. Sci.: Appl. Polym. Symp. **27**, 61–102 (1975)
178. W. Sattler, H. Dawcynski, W. Weiß, H. Henkel, R. Wagner, and H. Dittmann: *Untersuchungen zum Fadenbildungsvorgang von Polyamidseide bei hohen Spinngeschwindigkeiten.* Faserforsch. u. Textiltechnik **27**, 327–331 (1976)
179. V. G. Bankar, J. E. Spruiell, and J. L. White: *Melt spinning of nylon 6. I. Melt-spinning dynamics and rheological properties of nylon 6.* J. Appl. Polym. Sci. **21**, 2135–2155 (1977)
180. S. Kaufmann: *Zur Entstehung der Orientierung bei der Fadenbildung von Schmelzspinnpolymeren.* Faserforsch. u. Textiltechnik **28**, 83–84 (1977)
181. J. L. White: *Dynamic and structure development in melt spinning of fibers.* J. Rheol. **22**, 215 (1978)
182. Y. D. Kwon and D. C. Prevorsek: *Melt spinning of fibers: Effect of air drag.* J. Appl. Polym. Sci. **23**, 3105–3122 (1979)
183. A. Ziabicki: *Dynamics of multifilament vs. monofilament spinning.* In: F. Happey (Ed.): *Applied Fibre Science*, Vol 3 Chapt. 6, A. Ziabicki: *Fundamental studies of fibres.* (1979)
184. A. Ziabicki, H. Kawai (Eds.): *High Speed Fiber Spinning, Science and Engineering Aspects*, (John Wiley & Sons, New York 1985).
185. W. Kast, O. Kirscher, H. Reinecke, K. Wintermantel: *Konvektive Wärme- und Stoffübertragung* (Springer, Berlin 1974)
186. Y. Sano and S. Nishikawa, Kagaku Kogaku **28**, 275 (1964)
187. G. Bragato and G. Gianotti: *High speed spinning of PET. I. Steady state equations fundamental analysis. Structural property measurements along the spinning path. II. Orientation induced mechanism of cold crystallization in pre-oriented yarns.* Eur. Polym. J. **19**, 795–809 (1983)
188. H. Haberkorn, K. Hahn, H. Breuer, H. D. Dorrer, and P. Matthies: *On the neck-like deformation in high-speed spun polyamides.* J. Appl. Polym. Sci. **47**, 1551–79 (1993)

189. Y. Sano and N. Yamada, Kagaku Kogaku **30**, 997 (1966)
190. M. Copley and N. H. Chamberlain: *Filament attenuation in melt spinning and its effect on axial temperature gradient.* Appl. Polym. Symp., 27–50 (1967)
191. L. R. Glicksman: *The dynamics of a heated free jet of variable viscosity at low Reynold numbers.* J. Amer. Soc. Mechan. Eng. **90D**, 343–354 (1968)
192. L. R. Glicksman, Glass Technology **9**, 131 (1968)
193. W. Conti and E. Sorta: *Über die Abkühlung des spinnenden Fadens.* Faserforsch. u. Textiltechnik **21**, 509–513 (1970)
194. K. F. Zieminski and J. E. Spruiell: *A mathematical model of the melt spinning of synthetic fibers applicable to crystallizable polymers.* Int. Conf. Man Made Fibers, Beijing, 18.–22. Nov. (1985)
195. K. F. Zieminski: *Development and applicability of a mathematical model for the high-speed melt spinning of crystallizable polymers.* PhD thesis, Univ. of Tennessee, Knoxville (1986)
196. K. F. Zieminski and J. E. Spruiell: *On-line studies and computer simulation of the melt spinning of nylon-66 filaments.* J. Appl. Polym. Sci. **35**, 2223–45 (1988)
197. R. M. Patel, J. H. Bheda, and J. E. Spruiell: *Dynamics and structure development during high-speed melt spinning of nylon 6. II. Mathematical modeling.* J. Appl. Polym. Sci. **42**, 1671–82 (1991)
198. R. M. Patel and J. E. Spruiell: *Crystallization kinetics during polymer processing – analysis of available approaches for processing modeling.* Polym. Eng. Sci. **31**, 730–38 (1991)
199. K. Katayama, M.-G. Yoon: *Polymer crystallization in melt spinning.* Sen-i Gakkaishi **38**, P434–P441 (1982)
200. K. Katayama and M.-G. Yoon: *Polymer crystallization in melt spinning: Mathematical simulation.* In: A. Ziabicki, H. Kawai (Eds.): *High-Speed Fiber Spinning* (John Wiley & Sons, New York 1985)
201. A. Ziabicki: J. Non-Newtonian Fluid Mech. **30**, 157 (1988)
202. Y. Ohkoshi, T. Kikutani, A. Konda, and J. Shimizu: *Melt spinning of polyether-ether-ketone (PEEK) – cooling, thinning, and structure development on spin-line.* Sen-i Gakkaishi **49**, 211–219 (1993)
203. Y. Ohkoshi, C. Park, Y. Goto, M. Nagura, K. Toriumi, and T. Kikutani: *Cooling behavior of the spinning line of poly(ether ether ketone).* Sen-i Gakkaishi **56**, 340–347 (2000)
204. M. Golzar: *Melt spinning of fine PEEK filaments*, PhD thesis, Univ. of Technology Dresden, Dresden (2004)
205. R. Hilpert: *Wärmeabgabe von geheizten Drähten und Rohren im Luftstrom.* Forsch. Ing.-Wes. **4**, 215–224 (1933)
206. McAdams, In: V. Gnielinski: *Berechnung mittlerer Wärme- und Stoffübergangskoeffizienten.* Forsch. Ing.-Wes. **41**, 145–153 (1975)
207. H. Brünig, R. Beyreuther, H. Hofmann: *The influence of quench air on fibre formation and properties in the melt spinning process.* Intern. Fiber J. **14** 4, 104–107 (1999)
208. B. C. Sakiadis: *Boundary-layer behavior on continuous solid sufaces: I. boundary-layer equations for two-dimensional and axisymmetric flow.* AIChE J. **7**, 26–28 (1961)
209. B. C. Sakiadis: *Boundary-layer behavior on continuous solid sufaces: II. The boundary-layer on a continuous flat surface.* AIChE J. **7**, 221–225 (1961)
210. B. C. Sakiadis: *Boundary-layer behavior on continuous solid sufaces: III. The boundary-layer on a continuous cylindrical surface.* AIChE J. **7**, 467–472 (1961)

211. J. Gould and F. S. Smith: *Der Luftwiderstand monofiler Chemiefasern bei axialer Bewegung mit Geschwindigkeiten bis 100 m/s.* J. Text. Inst. **71**, 38–49 (1980)
212. N. Phan-Tien: Trans. Soc. Rheol. **22**, 259 (1978)
213. J. S. Denton, J. A. Cuculo, and P. A. Tucker: *Computer simulation of high-speed spinning of PET.* J. Appl. Polym. Sci. **57**, 939–951 (1995)
214. A. Ziabicki, A. Wasiak: *Crystallization of polymers in melt spinning.* Int. Symp. Man Made Fibers II, Kalinin 1977, Prepr. **1**, 214–218 (1977)
215. A. Ziabicki: *Development of orientation and crystallinity in high-speed spun fibers.* Fiber Producer Conf., 23.-25.10.1984, Greensville, 2-1–2-3 (1984)
216. A. Ziabicki, L. Jarecki: *The theory of molecular orientation and oriented crystallization in high-speed spinning.* In: A. Ziabicki, H. Kawai (Eds.): *High-Speed Fiber Spinning, Science and Engineering Aspects.* (John Wiley & Sons, New York 1985)
217. J. M. Schultz: *Theory of crystallization in high-speed spinning.* Polym. Eng. Sci. **31**, 661–6 (1991)
218. A. Ziabicki: *Crystallization of polymers in variable external conditions. Part 1. General equations.* Colloid Polym. Sci. **274**, 209–217 (1996)
219. A. Ziabicki: *Crystallization of polymers in variable external conditions. Part 2. Effects of cooling in the absence of stress and orientation.* Colloid Polym. Sci. **274**, 705–716 (1996)
220. A. Ziabicki, L. Jarecki: *Effect of crystallization on spinning fibers from polymer melts.* Polimery **43**, 293–298 (1998)
221. A. Ziabicki, P. Sajkiewicz: *Crystallization of polymers in variable external conditions. Part 3. Experimental determination of kinetic characteristics.* Colloid Polym. Sci. **276**, 680–689 (1998)
222. A. Ziabicki: *Crystallization of polymers in variable external conditions. Part 4. Isothermal crystallization in the presence of variable tensile stress or hydrostatic pressure.* Colloid Polym. Sci. **277**, 752–761 (1999)
223. A. Ziabicki: *Theory of crystal nucleation in polymers. Part I.* Polimery **45**, 520–527 (2000)
224. A. Ziabicki: *Theory of nucleation of polymer crystallization. Part II. Multidimensional theory and examples of its use.* Polimery **45**, 581–591 (2000)
225. A. Wasiak, A. Ziabicki: *Influence of spinning conditions and orientation on crystallinity of PA and PET fibres.* J. Appl. Polym. Sci.: Appl. Polym. Symp. **27**, 111–119 (1975)
226. A. Wasiak: *Crystallization of polymers in oriented state.* Colloid Polym. Sci. **259**, 135–145 (1981)
227. A. Wasiak: *A numerical method for the evaluation af data on the kinetics of polymer crystallization.* Chemtracts: Macromol. Chem. **2**, 348–349 (1991)
228. A. Wasiak: *Relationship between crystallization temperature and melting temperature in crystalline materials.* Chemtracts: Macromol. Chem. **3**, 47–49 (1992)
229. A. Wasiak: *Maximum crystal growth rate and its corresponding state.* Chemtracts: Macromol. Chem. **3**, 44–46 (1992)
230. A. Wasiak, M. Saiu, V. Brucato, G. Titomanlio: *Crystallization of polymer melts under fast cooling. II. High purity iPP.* Chemtracts: Macromol. Chem. **4**, 101–103 (1993)
231. A. Wasiak: *X-ray diffraction and calorimetric studies on polypropylene crystallization in non-isothermal conditions.* Mater., Funct. Des., Proc. Eur. Conf. Adv. Mater. Process. Appl., 5th **2**, 2/31–2/34 (1997)

232. A. Wasiak, P. Sajkiewicz, A. Wozniak: *Effects of cooling rate on crystallinity of i-polypropylene and polyethylene terephthalate crystallized in nonisothermal condititons.* J. Polym. Sci., Part B: Polym. Phys. **37**, 2821–2827 (1999)
233. P. Sajkiewicz: *Kinetics of crystallization of polymers.* Polimery **46**, 768–776 (2001)
234. P. Sajkiewicz, L. Carpaneto, A. Wasiak: *Application of the Ozawa model to non-isothermal crystallization of poly(ethylene terephthalate).* Polymer **42**, 5365–5370 (2001)
235. I. J. Rao, K. R. Rajagopal: *A continuum model for flow induced crystallization in polymers.* Annual Technical Conference - Society of Plastics Engineers **59th**, 1785–1789 (2001)
236. A. Ziabicki, G. C. Alfonso: *A simple model of flow-induced crystallization memory.* Macromol. Symp. **185**, 211–231 (2002)
237. P. Sajkiewicz: *Kinetics of crystallisation of polymers – a review.* Progress in Rubber, Plastics and Recycling Technology **18**, 195–215 (2002)
238. P. Sajkiewicz: *Transient and athermal effects in the crystallization of polymers. Part I. Isothermal crystallization.* J. Polym. Sci., Part B: Polymer Physics **40**, 1835–1849 (2002)
239. P. Sajkiewicz: *Transient and athermal effects in the crystallization of polymers. II. Nonisothermal crystallization.* J. Polym. Sci., Part B: Polym. Phys. **41**, 68–79 (2002)
240. J. M. Samon, J. M. Schultz, B. S. Hsiao: *Structure development in the early stages of crystallization during melt spinning.* Polymer **43**, 1873–1875 (2002)
241. B. Falkai, W. Gießler, G. Spilgies, H. Wilsing: *Strukturvergleich von schnellgesponnenen Polyesterfäden auf Basis Ethylenglykol und Dimethylolcyclohexan.* Int. Symp. Man Made Fibers III, Kalinin 1981, Prepr. **1/1**, 137–172 (1981)
242. T. Kawaguchi: *Industrial aspects of high-speed spinning.* In: A. Ziabicki, H. Kawai (Eds.): *High-Speed Fiber Spinning* (John Wiley & Sons, New York 1985)
243. H. M. Heuvel, R. Huisman: *Effect of winding speed and other spinning conditions on the physical structure of as-spun PET and nylon 6 yarns.* In: A. Ziabicki, H. Kawai (Eds.): *High-Speed Fiber Spinning* (John Wiley & Sons, New York 1985)
244. G. Vassilatos, B. H. Knox, H. R. E. Frankfort: *Dynamics, structure development and fiber properties in high-speed spinning of PET.* In A. Ziabicki, H. Kawai (Eds.): *High-Speed Fiber Spinning* (John Wiley & Sons, New York 1985)
245. H. G. Zachmann, M. Bark, A. Teitge, S. Röber: *Untersuchungen zur Entstehung der Feinstruktur beim Schnellspinnen und beim Spinnen unter normalen Bedingungen.* Intern. Conf. Man-made Fibres, Dornbirn (1990)
246. C. D. Bechev: *Crystallization behaviour of oriented poly(ethylene terephthalate).* NATO ASI Ser., C **405**, 515–520 (1993)
247. V. R. Borovskii, N. F. Klochko, A. M. Pankeev and L. M. Polovets: Khim. Volokna, **3**, 23 (1974)
248. A. M. Kiritchenko, *et al.*: Khim. Volokna **6**, 56 (1976)
249. A. Ziabicki, L. Jarecki and A. Wasiak: *Dynamic modeling of melt spinning.* Comput. Theor. Polym. Sci. **8**, 143–147 (1998)
250. A. Szaniawski and A. Zachara: *Die Durchdringung des Bades bis zur Fadenbildungszone der Kunstfaserbündel.* Polimery **19**, 143–146 (1974)
251. A. Szaniawski: *Über die Strömung innerhalb eines Kunstfaserbündels.* Polimery **20**, 350 (1975).

252. A. Szaniawski and A. Zachara: *Der Einfluß des Druckunterschiedes zwischen beiden Seiten eines ebenen Chemiefaserbündels auf die Strömungen in seinem Inneren.* Polimery **20**, 87–90 (1975)
253. A. Szaniawski and A. Zachara: *Hydrodynamic problems of multifilament spinning.* Chem. Vlakna **26**, 96–103 (1976)
254. A. Zachara: *How to estimate the flow characteristics in multifilament spinning. Part I and II.* Fiber World, pp. 13–18 and 52–60 (1987)
255. T. Matsuo, H. Yasuda, and H. Sugiyama: *Phenomenological theories for the melt spinning process and its applications.* Int. Symp. Man Made Fibres, Kalinin 1977, Prepr. **2**, 206–222 (1977)
256. H. Ishihara, S. Hayashi, and H. Ikeuchi: *Computer simulation of multifilament air jet spinning.* Intern. Polym. Process. **7**, 91–95 (1989)
257. H. Ishihara, S. Hayashi, Intern. Symp.Man Made Fibres, Kalinin 1990, Prepr. **1**, 64–82 (1990)
258. A. Dutta: *Melt spinning of (multifilament) poly(ethylene terephthalate) fibers: A simulation approach.* Polym. Eng. Sci. **27**, 1050–1058 (1987)
259. A. Dutta: *Role of quench air profiles in multifilament melt spinning of PET fibers.* Text. Res. J. **57**, 13–19 (1987)
260. A. Schöne and H. Brünig: *Modelling of multifilament spinning.* Arch. Mech. **42**, 571–582 (1990).
261. H. Brünig: *Computer system FABIL – application of the filament formation model in multifilament melt spinning,* Khim. Volokna, 35–40 (1994)
262. R. Beyreuther, H. Brünig, A. Schöne, and W. Müller: *Spinndüsen-Spinnschacht-Bereich als Quelle von Ungleichmäßigkeiten beim hochkapillarigen PES-Stapelfaserspinnen.* Melliand Textilberichte **72** 6, 401–406 (1991)
263. N. Hajji, J. E. Spruiell, F. M. Lu, S. Malkan, and G. C. Richardson: *Modeling of the 'REICOFIL' spunbonding process.* INDA J. Nonwoven Res. **4**, 16–21 (1991)
264. S. Misra, J. E. Spruiell, and G. C. Richeson: *Investigation of the spunbonding process via mathematical modeling.* INDA J. Nonwoven Res. **5**, 13–19 (1992)
265. R. Beyreuther, H. Brünig, D. Blechschmidt: *High filament velocities in the underpressure spunbonding nonwoven process – realized with the help of a fiber spinning model.* Int. Fibre J. **12** 6, 129–134 (1997)
266. R. Beyreuther and A. Schöne: *Zum Problem des Fadenbruchs beim Schmelzspinnprozeß organischer Hochpolymere unter Berücksichtigung von makroskopischen Störstellen im Elementarfaden.* Acta Polymerica **34** 2, 68–72 (1983)
267. R. Beyreuther: *Grenzen der Elementarfadenfeinheit beim Schmelzspinnen.* Melliand Textilberichte **72**, 795–799 (1991)
268. D. De Kee and K. F. Wissbrun: *Polymer rheology.* Physics Today **51**, 24–29 (1998)
269. R. Beyreuther and R. Vogel: *Spinnability of polymer melts. A complex problem in basic research.* Int. Polym. Process. **11**, 154–158 (1996).
270. A. Ghijsels, C. H. Massardier, R. M. Brandley: *Brittle melt rupture phenomena in polymer processing.* Int. Polym. Process. **12**, 147–154 (1997)
271. A. Ghijsels, J. De Clippeleir: *Melt strength behaviour of polypropylenes.* Int. Polym. Process. **9**, 252–257 (1994)
272. A. Ghijsels, J. L. Thomason, A. Miedema: *A melt-rupture model for predicting the spinnability of polypropylenes.* Annual Meeting of Polymer Process. Soc., Stuttgart (1995)
273. V. Rauschenberger, H. M. Laun: *Spinnbarkeit und Versagensverhalten thermoplastischer Polymere aus rheologischer Sicht.* Melt Spinning of Polymers

and Glass in Polymer Research, Symposium 2002, May 13-14, Institute of Polymer Research Dresden (2002)

274. A. Ziabicki: *Fundamentals of Fiber Formation*, 1st edn. (Springer, New York 1976)
275. R. Vogel, G. Schauer, R. Beyreuther: *Melt Spinning of Ultra-Low-Density Polyethylene Engage to Elastic Fibers.* Chem. Fibers Int. **45** 10, 268–269 (1995)
276. M. Hoffmann, R. Beyreuther, R. Vogel, B. Tändler: *Reactive Extrusion and Melt Spinning: Results from Polyolefine elastomer.* Chem. Fibers Int. **49** 14, 410–412 (1999)
277. R. Vogel, G. Schauer, G. Schmack, R. Beyreuther, E. Meyer: *Generation of a Polyolefine Fiber with Higher Tenacity by Favourable Blending.* Chem. Fibers Int. **46** 11, 421–432 (1996)
278. R. Vogel, H. Brünig, R. Beyreuther, B. Tändler, D. Voigt: *Rheological and Theoretical Estimation of the Spinnability of Polyolefines, Part 1: Rheological Study.* Int. Polym. Process. **XIV** 1, 69–74 (1999)
279. Hütte: *Des Ingenieurs Taschenbuch, theoretische Grundlagen* (Wilhelm Ernst und Sohn, Berlin 1955)
280. A. Schöne: *Physikalisch-analytisches Modell des Fadenbildungs- und -deformationsprozesses von Polyamid-6 unter klassischen und Schnellspinnbedingungen.* Dissertation, Akademie der Wissenschaften der DDR, Berlin (1982)
281. A. Schöne: *Modellbeschreibung des Fadenbildungsvorganges beim Schmelzspinnen unter den Bedingungen klassischer und hoher Spinngeschwindigkeiten.* Chem. Vlakna **32** 1, 25–41 (1982)
282. R. Beyreuther: *Theoretische und experimentelle Untersuchungen über die technologischen Ursachen und über die Möglichkeiten zur Verminderung der Feinheitsungleichmäßigkeit von düsengezogener Glasseide.* Dissertation, Technische Universität Dresden, Fakultät für Technologie, Dresden (1968)
283. R. Beyreuther: *Dynamisches Verhalten einer Glasseidenspinnanlage: Teile I bis IV.* Messen, Steuern, Regeln (Automatisierungspraxis) **14** 12, 232–237, **15** 1, 11–14, **15** 7, 144–147, **15** 9, 195–197 (1971/1972)
284. W. Wegener, H. Bechlenberg: *Regelanlage zum Vergleichmäßigen von Faserbändern.* Textilpraxis **14** 1, 35–41, **14** 2, 129–132, **14** 3, 245–249, **14** 4, 359–364, **14** 5, 456–463, **14** 6, 552–556, **14** 7, 672–678, **14** 8, 774–779, **14** 9, 877–882, **14** 10, 990–995, **14** 11, 1107–1114, **14** 12, 1212–1218 (1959); **15** 1, 10–16, **15** 2, 136–140, **15** 3, 259–266, **15** 4, 378–383 (1960)
285. V. P. Chavkin, L. N. Ginzburg: *Dynamik des Vergleichmäßigungsprozesses auf der Walzenkarde.* Textil. Prom. (Textil-Ind.) **26**, 4–12, (1966); translated in Germ. in: Sowj. Beiträge zur Faserforschung und Textiltechnik **4**, 81–90, (1967)
286. A. G. Sevostjanov: *Bewerten der Vergleichmäßigungsfähigkeit auf der Deckelkarde.* Textil. Prom. (Textil-Ind.) **28**, 23–25, (1968); translated in Germ. in: Sowj. Beiträge zur Faserforschung und Textiltechnik **5**, 418–421 (1968)
287. V. S. Anikin, S. D. Siganov, A. G. Sevostjanov: *Untersuchung zur vergleichmäßigenden Wirkung der Deckelkarde*, Izv. vyssich uč. Zav. Technol. textil. Prom. (Hochschulnachr. Technol. Textil-Ind.) **6**, 40–44, (1971); translated in Germ. in: Sowj. Beiträge zur Faserforschung und Textiltechnik **9**, 306–309 (1972)
288. R. Wächter: *Theoretische und experimentelle Untersuchungen spezieller Probleme des Übertragungsverhaltens der Schlagmaschine mit Vergleichmäßigungseinrichtungen.* Dissertation, Technische Hochschule Karl-Marx-Stadt, Karl-Marx-Stadt (1964)

289. R. Wächter, R. Beyreuther: *Ein pneumatischer Regler zur Vergleichmäßigung von Vorgarnen an der Ringspinnmaschine.* Faserforschung und Textiltechnik **16** 5, 225–233 (1965)
290. V. P. Chavkin, A. S. Molcanov: *Übertragungsfunktionen der Streckwerke.* Izv. vyssich uč. Zav. Technol. textil. Prom. (Hochschulnachr. Technol. Textil-Ind.) **5**, 119–126 (1963)
291. R. Beyreuther: *Dynamische Modellierung des Transportvorganges laufender Fäden. Teil 1: Entwicklung der Modellvorstellung, Teil 2: Quantitative Auswertung der Modellgleichungen.* Faserforschung und Textiltechnik/Zeitschrift für Polymerforschung **27** 5, 253–260, **27** 8, 389–395 (1976)
292. G. Kessler: *Das zeitliche Verhalten einer kontinuierlichen elastischen Bahn zwischen aufeinanderfolgenden Walzenpaaren. Teile I und II.*, Regelungstechnik **8** 12, 436–439 (1960), **9** 4, 154–159 (1961)
293. G. Brandenburg: *Ein mathematisches Modell für eine durchlaufende elastische Stoffbahn in einem System angetriebener, umschlungener Walzen. Teile I bis III.*, Regelungstechnik **21** 3, 69–77, **21** 4, 125–129, **21** 5, 157–162 (1973)
294. G. Brandenburg: *Über das dynamische Verhalten durchlaufender elastischer Stoffbahnen bei Kraftübertragung durch Coulombsche Reibung in einem System angetriebener und umschlungener Walzen.* Dissertation, Technische Hochschule München, München (1971)
295. R. Backmann: *Modellierung der Förderung elastischer Stoffe in Verarbeitungsmaschinen.* Dissertation, Technische Hochschule Karl-Marx-Stadt, Karl-Marx-Stadt (1981)
296. R. Beyreuther, H. Hofmann, G. Schauer: *Untersuchungen zur Dämpfung von Fadenzugkraft- und -dehnungsschwankungen in Fadeneinlaufstrecken.*, Textiltechnik **31** 6, 341–345 (1981)
297. DD-PS 148 241: *Verfahren zur Verminderung von Dehnungsschwankungen laufender Fäden.* Anm.: AdW der DDR, Inst. f. Technologie der Fasern Dresden. - Erf.: R. Beyreuther, H. Hofmann, G. Schauer, K. Pohl, R. Bittner. - Anm.-Tag: 27.12.1979. - Ausg.-Tag: 13.5.1981
298. R. Beyreuther, H. Henkel, O. Lochmüller: *Beeinflussung der Feinheitsungleichmäßigkeit von Chemieseiden durch die Dynamik der Fadenchangierung an Wicklern.* Textiltechnik **30** 9, 554–558 (1980)
299. V. N. Nuždin, V. M. Korolov, A. V. Choljavin: *Berechnung der durch die Fadenführerbewegung hervorgerufenen Schwankungen der Fadenspannung.* Izv. vyssich uč. Zav. Technol. textil. Prom. (Hochschulnachr. Technol. Textil-Ind.) **1**, 133–136 (1977)
300. G. A. Spescha: *Titerschwankungen als Folge der Drehzahlschwankungen von Streckrollen und Spulantrieben.* Chemiefasern/Textilindustrie **27** 9, 776–788 (1977)
301. J. Lünenschloß, H. Weinsdörfer: *Drehungsverteilung im Faden während des Falschdraht-Texturierens.* Chemiefasern/Textilindustrie **20** 11, 980–984, (1970)
302. H. W. Krause: *Mechanik der Torsionskräuselung.* Textilveredlung **5** 8, 643–647 (1970)
303. J. Lünenschloß, E. Kirschbaum: *Ein neues Verfahren zur Messung des Torsionsmomentes beim FD-Texturieren mit konventionellen und Friktionsdrallelementen.* Textil-Praxis intern. **29** 4, 434–436 (1974)
304. J. Lünenschloß, L. Coll-Tortosa, K. Fischer: *Die Messung der Fadendrehung und ihrer Schwankungen am laufenden Faden während des Texturiervorganges.* Chemiefasern/Textilindustrie **23/75** 4, 297–301 (1973)

305. J. Lünenschloß: *Einfluß von Drallgeberart und Texturierbedingungen auf den Drehungsdichte- und Fadenzugkraftverlauf beim FD-Texturieren.* Chemiefasern/Textilindustrie **24/76**, 267–271 (1974)
306. W. Wegener, L. Coll-Tortosa: *Die Wirkungsweise von Friktionsbüchsen auf die Drallerteilung bei verschiedenen Einstellungen der Falschdrahtmaschine.* Textil-Praxis intern. **29** 6, 773–775, **29** 8, 1043–1045 (1974)
307. J. Marko: *Die Struktur von im Torsionsverfahren texturierten Garnen.* Melliand Textilberichte **54** 10, 1019–1024 (1973)
308. K. Greenwood: *Kontrolle der Garnqualität bei der Friktionstexturierung.* Chemiefasern/Textilindustrie **26/78**, 401–407 (1976)
309. H. Weinsdörfer: *Drallschlupf oder Drallstau an Falschdrahtspindeln?* Textil-Praxis intern. **31** 1, 34 (1976)
310. G. Mutschler: *Entscheidende Parameter für die Erzielung einer konstanten Fadenqualität bei der Friktionstexturierung.* Chemiefasern/Textilindustrie **26/78** 10, 884–891 (1976)
311. V. Pechoč, J. Staš: *Ohrev vláken Stykem s chrevnou plochou.* Chem. Vlakna **22** 5, 75–77 (1972)
312. J. Staš, V. Pechoč: *Tepelný výpočet galety.* Chem. Vlakna **23** 1, 8–10 (1973)
313. G. Rottig: *Anwendung des Differenzenverfahrens bei der Bestimmung des Temperaturverlaufes im Faden.* Textiltechnik **23** 1, 24–27 (1973)
314. A. Recknagel: *Physik: Schwingungen und Wellen, Wärmelehre* (Verlag Technik, Berlin 1966)
315. S. Kaufmann, R. Beyreuther: *Experimentelle Bestimmung der Wärmeübergangszahlen von laufenden Polyamidfäden bei Abkühlung in Luft.* Faserforschung und Textiltechnik/Zeitschrift für Polymerforschung **28** 6, 293–301 (1977)
316. H. Faltin: *Technische Wärmelehre: Anhang Stoffwerte und Formelsammlung* (Akademie-Verlag, Berlin 1961)
317. R. Beyreuther: *Untersuchungen zur Dynamik des Wärmeüberganges von Chemieseidenfäden an beheizten Galetten.* Faserforschung und Textiltechnik/Zeitschrift für Polymerforschung **25** 11, 483–489 (1974)
318. R. Barthel, H. Hofmann: *Fadenzugkraftmessung in der Webereivorbereitung und Weberei.* Deutsche Textiltechnik **14** 5, 267–273 (1964)
319. H. Giesekus: *Die statistische Analyse der Garn- und Fadenungleichmäßigkeit: Teil I: Grundlagen und praktische Durchführung, Teil II: Ideales Garn, reales Garn, Fäden.* Faserforschung und Textiltechnik **10** 6, 275–282, **10** 7, 338–345, **10** 8, 358–368, **10** 9, 420–429 (1959)
320. DD-PS 208 674: *Verfahren zur Bestimmung der Ursachenstruktur von Ungleichmäßigkeiten in Fäden aus organischen Hochpolymeren.* Anm.: AdW der DDR, Inst. f. Technologie der Fasern Dresden. - Erf.: R. Beyreuther, W. Nicht. - Anm.-Tag: 20.7.1982. - Ausg.-Tag: 4.4.1984
321. R. Beyreuther, W. Nicht: *Die Korrelationsanalyse als eine Methode zur Erforschung der Ursachen von inneren und äußeren Fadenungleichmäßigkeiten im Reckprozeß.* Acta Polymerica **34** 2, 80–85 (1983)

List of Symbols

Symbol	*Description*
a	distance of measurement place
c	specific heat capacity of thread material
c_{air}	specific heat capacity of air
c_{f}	air friction coefficient
c_{p}	specific heat capacity of polymer
d	diameter of capillary hole of spinneret
d_{f}	dampening factor
d_{q}	standard deviation
d_{q}^2	quadratic dispersion (variance)
ddr	draw down ratio
$\frac{\mathrm{d}}{\mathrm{d}x}, \frac{\mathrm{d}}{\mathrm{d}t}$	derivatives in ordinary DEs
$\frac{\mathrm{d}^u}{\mathrm{d}t^u}$	u^{th} derivative to the time, LAPLACIAN p
$\mathrm{d}x,\ \mathrm{d}t,\ \mathrm{d}t^u,$	infinitesimal quantity of the appropriate size
$\mathrm{d}l,\ \mathrm{d}m,\ \mathrm{d}p,$	infinitesimal quantity of the appropriate size
$\mathrm{d}Tt_{\mathrm{o}},\ \mathrm{d}T,\ \mathrm{d}T_{\mathrm{y}}$	infinitesimal quantity of the appropriate size
$\mathrm{d}M,\ \mathrm{d}S_{\mathrm{y}}$	infinitesimal quantity of the appropriate size
$\mathrm{d}\alpha$	infinitesimal quantity of the appropriate size
e	asymmetry parameter
$\mathrm{e}^{\mu\alpha},\ \mathrm{e}^{\mu\alpha_{\mathrm{m}}},\ \mathrm{e}^{\mu_{\mathrm{m}}\alpha_{\mathrm{m}}}$	rope friction factor
$\mathrm{e}^{\mathrm{j}\varphi(\omega)}$	factor of the phase shift between cause- and effect oscillation in the complex plane
f	HERRMAN's orientation factor (Chap. 3)
f	disturbance frequency, frequency in the dimension Hz
f_{am}	amorphous orientation factor

continuation

Symbol	*Description*
f_{cr}	crystalline orientation factor
f_{c}	critical frequency
f_{ch}	critical frequency of heat transfer dynamic
f_{c1}, f_{c2}	critical frequencies of drafting zones 1 and 2
f_{m}	mains frequency
Δf_{m}	mains frequency change
f_{P}, f_{P1}, f_{P2}	frequencies of the changes
f_{ts}	frequency of traverse motion thread guide
$\widetilde{\Delta f_{\mathrm{m}}}$	sinusoidal mains frequency change
$f(t)$	time dependent function
f_{x}, f_{y}	components of external force density, cartesian coordinates
g	gravitational acceleration
$g(y)$	reduced stream function, cartesian coordinates
h	glass level
h_{m}	glass level, mean value
Δh	glass level change
$h(r)$	reduced stream function, radial coordinates
i	running (sequence) index
$i \cdot \Delta t$	time shift (power density spectrum)
j	imaginary unit $\mathrm{j}^2 = -1$
k	running (sequence) index
k	BOLTZMANN constant (Chap. 3)
k	transfer factor of heating transformer (Chap. 4)
$k(y)$	force density function
$k \cdot \Delta t$	time shift (correlation function)
l	length of the appropriate zone or roll setting
l_{m}	length of the appropriate zone or roll setting, mean value
Δl	length change of the appropriate zone or roll setting
$\widetilde{\Delta l}$	sinusoidal length change of the appropriate zone or roll setting
l_{b}	length of bobbin
l_{ext}	input line extension
$l_{\mathrm{ext}}/l_{\mathrm{m}}$	extension factor
l_{f}	middle staple length of fibres
l_{h}	high of traverse motion triangle

continuation

Symbol	*Description*
l_s	distance from spinneret
l_t	thread length of one twist
l_u	streched, untwisted length of one twist
l_0	length of unloaded fibre/yarn/thread
l_1	length of F loaded fibre/yarn/thread
$l_1,\ l_2,\ l_3$	lengths of heated godet systems
l_1	length of the roll setting zone 1
$l_{1\mathrm{m}}$	length of the roll setting zone 1, mean value
Δl_1	length change of the roll setting zone 1
l_2	length of the roll setting zone 2
$l_{2\mathrm{m}}$	length of the roll setting zone 2, mean value
Δl_2	length change of the roll setting zone 2
$l_{\mathrm{T_g}}$	distance spinneret-glass transition point
m	running (sequence) index
n	running (sequence) index
n	AVRAMI exponent (crystallisation rate, Chap. 3)
n_b	revolution number of bobbin motor
n_{bm}	revolution number of bobbin motor, mean value
$\Delta n_\mathrm{b},\ \Delta n_{\mathrm{b1}},\ \Delta n_{\mathrm{b2}}$	revolution number changes of bobbin motor
$\widetilde{\Delta n}$	sinusoidal revolution number change of bobbin motor
n_b	revolution number of bobbin
n_i	revolution number of input drawing godet
n_o	revolution number of output drawing godet
n_p	revolution number of motor spinning pump
n_s	revolution number of false twist spindle
n_{sm}	revolution number of false twist spindle, mean value
Δn_s	revolution number change of false twist spindle
$\widetilde{\Delta n_\mathrm{s}}$	sinusoidal revolution number change of false twist spindle
p_e	pressure of melt from extruder
p_s	pressure of melt before the spinneret
$p^u = \dfrac{\mathrm{d}^u}{\mathrm{d}t^u}$	LAPLACIAN
p_ν	zero value of the integrand of Eq. 2.42
$p_{,\mathrm{x}},\ p_{,\mathrm{y}}$	components of pressure gradient, cartesian coordinates
q	number of worker-angle-stripper pairs (roller top card, Chap. 4)

continuation

Symbol	*Description*
q_{i}	cross section of capillary hole
q_{im}	cross section of capillary hole, mean value
Δq_{i}	cross section change of capillary hole
$\widetilde{\Delta q_{\mathrm{i}}}$	sinusoidal cross section change of capillary hole
q_{o}	throughput of one single filament
q_{s}	cross section of ready formated fibre
q_{sm}	cross section of ready formated fibre, mean value
Δq_{s}	cross section change of ready formated fibre
$\widetilde{\Delta q_{\mathrm{s}}}$	sinusoidal cross section change of of ready formated fibre
$\dot{q}_{\mathrm{s}}$	derivative of q_{s} to t
$q_{\mathrm{x}},\ q_{\mathrm{y}}$	components of heat flow
r_{g}	gear ratio
s	length of capillary hole of spinneret
t	running time
Δt	time interval
Δt_1	necessary time shift between time functions of tensile force and fineness to continuous quotient calculation to fineness related yarn tensile force (Fig. 6.16)
t_0	time to the start point zero
tm_{a}	amplitude of traverse motion at winder
tm_{f}	frequency of traverse motion at winder
u	order of the LAPLACIAN
v	fibre/yarn/thread velocity
v_0	extrusion velocity
v_{a}	velocity of quenching air
Δv_{a}	velocity change of quenching air
Δv_{a}	velocity change of quenching air
v_{air}	air velocity
v_{c}	coefficient of variation (Chap. 6)
v_{i}	velocity of input fibre/yarn/thread or godet or present velocity of twisted thread in texturing zone
v_{input}	velocity of input fibre mass
v_{im}	velocity of input fibre/yarn/thread or godet or present velocity of twisted thread in texturing zone, mean values

continuation

Symbol	*Description*
Δv_{i}	velocity change of input fibre/yarn/thread or godet or present velocity change of twisted thread the texturing zone
$\widetilde{\Delta v_{\mathrm{i}}}$	sinusoidal velocity change of input fibre/yarn/thread
$v_{\mathrm{i}}(t)$	velocity of input fibre/yarn/thread or godet, time function
$v_{\mathrm{i1}}(t)$	velocity of input fibre/yarn/thread, level 1, time function
$v_{\mathrm{i2}}(t)$	velocity of input fibre/yarn/thread, level 2, time function
v_{o}	velocity of output or take-down velocity of fibre/yarn/thread or godet
v_{output}	velocity of output fibre mass
v_{om}	velocity of output fibre/yarn/thread or godet, mean value
Δv_{o}	velocity change of output fibre/yarn/thread or godet
$\widetilde{\Delta v_{\mathrm{o}}}$	sinusoidal velocity change of ouput fibre/yarn/thread or godet
v_{o}'	velocity of twisted thread in the peel off moment
$v_{\mathrm{o}}(t)$	velocity of output fibre/yarn/thread or godet, time function
$v_{\mathrm{o1}}(t)$	velocity of output fibre/yarn/thread, level 1, time function
$v_{\mathrm{o2}}(t)$	velocity of output fibre/yarn/thread, level 2, time function
v_{s}	take-down (spinning) velocity of formated fibre
v_{sm}	take-down (spinning) velocity of formated fibre, mean value
Δv_{s}, Δv_{s1}, Δv_{s2}	take-down (spinning) velocity changes of formated fibre
$\widetilde{\Delta v_{\mathrm{s}}}$	sinusoidal take-down (spinning) velocity change of formated fibre
v_{tm}	linear velocity of traverse motion thread guide, mean value
v_0, v_{B}	(initial) quenching air velocity
v_{L}	take-up velocity
v_{x}, v_{y}	velocity components of air, cartesian coordinates
v_{x}, v_{r}	velocity components of air, radial coordinates
v_{z}	velocity of middle drafting godet or input velocity of untwisted thread in texturing zone

continuation

Symbol	*Description*
v_{zm}	velocity of middle drafting godet or velocity of input dead time line or velocity of the tensionless fibre/yarn/thread or input velocity of untwisted thread in texturing zone, mean values
Δv_{z}	velocity change of middle drafting godet or velocity change of input dead time line or input velocity change of untwisted thread in texturing zone
$\widetilde{\Delta v_{\mathrm{z}}}$	sinusoidal velocity change of untwisted thread in texturing zone
v'_{z}	velocity of twisted thread in texturing zone
$v_{\|\|},\ v_{\perp}$	axial and cross air velocity
w	bending tongue shift
w_{m}	bending tongue shift, mean value
Δw	bending tongue shift change
$\widetilde{\Delta w}$	sinusoidal bending tongue shift change
x	coordinate in fibre direction, distance from spinneret (Chap. 3)
$x(t)$	time function (Chap. 2)
$\bar{x}$	mean value of the time function (Chap. 2)
Δx	cause variable (Chap. 2)
x_0	length measurement corresponding to Eqs. 5.96, 5.97
$x_1, ..., x_{\mathrm{n}},$ $x_{\mathrm{i}},\ x_{\mathrm{i+k}}$	discrete values of the time function $x(t)$ (Chap. 2)
x_{i}	input fibre mass/time (carding engines, Figs. 4.15, 4.18)
x_{im}	input fibre mass/time (carding engines), mean value
Δx_{i}	input fibre mass/time change (carding engines, Figs. 4.16, 4.18)
$\widetilde{\Delta x_{\mathrm{i}}}$	sinusoidal input fibre mass/time change (carding engines)
$\Delta x,\ \Delta x_1, ..., \Delta x_{\mathrm{n}},$ $\Delta x_{\mathrm{i1}},\ \Delta x_{\mathrm{i2}}$	partial fibre mass/time changes (roller top card, Figs. 4.16, 4.17)
x_{o}	output fibre mass/time (carding engines, Figs. 4.15, 4.18)
x_{om}	output fibre mass/time (carding engines), mean value
Δx_{o}	output fibre mass/time change (carding engines, Figs. 4.16, 4.18)
$\widetilde{\Delta x_{\mathrm{o}}}$	sinusoidal output fibre mass/time change (carding engines)

continuation

Symbol	*Description*
Δx_{o2}	partial fibre mass/time change (roller top card, Fig. 4.16)
x_{s}	distance to solidification point
$\widetilde{\Delta x}(\omega)$	vector of the sinusoidal disturbance of the independent cause variable
$y(t)$	time function (Chap. 2)
$y_1, \ldots, y_{\mathrm{n}},\ y_{\mathrm{i+k}}$	discrete values of the time function $y(t)$ (Chap. 2)
$\bar{y}$	mean value of the time function (Chap. 2)
Δy	effect variable (Chap. 2)
$\Delta y_1,\ \Delta y_2, \ldots, \Delta y_{\mathrm{n}}$	partial fibre mass/time changes (roller top card, Figs. 4.16, 4.17)
$\widetilde{\Delta y}(\omega)$	vector of the sinusoidal disturbance of the dependent response (effect) variable
$\widetilde{\Delta y}(\omega_0)$	vector of the sinusoidal disturbance of the dependent response (effect) variable for ω_0
$\widetilde{\Delta y}(\omega_1)$	vector of the sinusoidal disturbance of the dependent response (effect) variable for ω_1
$\widetilde{\Delta y}(\omega_2)$	vector of the sinusoidal disturbance of the dependent response (effect) variable for ω_2
$\widetilde{\Delta y}(\omega_3)$	vector of the sinusoidal disturbance of the dependent response (effect) variable for ω_3
z	number of capillary holes of spinneret or number of total twists in the zone l_{m} or number of thread wraps around heated godets
z_{i}	number of fibres in cross section of coming in sliver
z_{im}	number of fibres in cross section of coming in sliver, mean value
Δz_{i}	change of number of fibres in cross section of coming in sliver
$\widetilde{\Delta z_{\mathrm{i}}}$	sinusoidal change of number of fibres in cross section of coming in sliver
z_{o}	number of fibres in cross section of coming out sliver
z_{om}	number of fibres in cross section of coming out sliver, mean value
z_{z}	number of fibres in cross section of coming out/in sliver from/to drafting zones 1/2
z_{zm}	number of fibres in cross section of coming out/in sliver from/to drafting zones 1/2, mean value

continuation

Symbol	*Description*
Δz_{z}	change of number of fibres in cross section of coming out/in sliver from/to drafting zones 1/2
A	actuator device
A	filament cross section
$A_{\mathrm{a}},\ B_{\mathrm{a}},\ C_{\mathrm{a}},\ E_{\mathrm{a}}$	abbreviation factors in the Eqs. 5.23, 5.24
$A_{\mathrm{b}},\ B_{\mathrm{b}},\ C_{\mathrm{b}}$	abbreviation factors in the Eqs. 5.38, 5.39
A_{s}	surface of glass melt
A_{zm}	to the fineness related rise of the force-elongation-curve of fibre/yarn/thread
ACF	auto-correlation function
APSF	auto-power density spectrum function
B	width of filament bundle in multifilament spinning, quenching air direction
$C_1,\ C_2,\ C_3$	constants
CAC	coordination automatic controller
CC	coordination controller
CCF	cross-correlation function
CPCD	connection programmed controller device
CV	coefficient of variation (Chap. 3)
D	filament diameter (Chap. 3)
D	wind-up/godet diameter
D_{m}	wind-up/godet/friction element diameter, mean value
ΔD	wind-up/godet diameter change
$\widetilde{\Delta D}$	sinusoidal wind-up diameter change
D_0	diameter of capillary hole
D_{i}	diameter of input godet
D_{o}	diameter of output godet
D_{L}	filament diameter at take-up distance L (Chap. 3)
D_{y}	fibre/yarn/thread diameter
De	*Deborah* number
DE	differential equation
DR	draw ratio
DMCC	digital multi channel controller
DPP	data processing peripherals
E	elongational elastic modulus (melt)
E_{y}	elastic modulus of fibre/yarn/thread
E_{yi}	elastic modulus of input fibre/yarn/thread

continuation

Symbol	*Description*
E_{yim}	elastic modulus of input fibre/yarn/thread, mean value
E_{yo}	elastic modulus of output fibre/yarn/thread
E_{yom}	elastic modulus of output fibre/yarn/thread, mean value
E_{a}	activation energy (ARRHENIUS)
EDPS	electronic data processing system
$F,\ F_1,\ F_2$	tensile forces of fibre/yarn/thread
F_{m}	tensile force of fibre/yarn/thread, mean value
ΔF	tensile force change of fibre/yarn/thread
$\widetilde{\Delta F}$	sinusoidal tensile force change of fibre/yarn/thread
ΔF_1	tensile force change (amplitude) of fibre/yarn/thread
$F(t)$	tensile force of fibre/yarn/thread, time function
F_{m}	mean value of $F(t)$
F_{i}	discontinuous value of time function $F(t)$
$F_1(t)$	tensile force of fibre/yarn/thread, level 1, time function
$F_2(t)$	tensile force of fibre/yarn/thread, level 2, time function
F_{b}	tensile force at break of fibre/yarn/thread
$F_{\mathrm{T_g}}$	tensile force of fibre/yarn/thread at the glass transition temperature T_{g}
F_{i}	discontinuous value of time function $F(t)$ (Chap. 6)
F_{i}	tensile force of fibre/yarn/thread at friction thread line input
F_{o}	tensile force of fibre/yarn/thread at friction thread line output
F_0	initial force (at capillary)
F_{drag}	air drag tensile force
F_{inert}	inertial tensile force
F_{grav}	gravitational tensile force
F_{surf}	surface tensile force
F_{rheo}	(rheological) fibre force
F_{L}	take-up force at distance L
F_{y}	yarn tensile force (drawing process)
FT	false twist
G	modulus (upper convected MAXWELL model)
$G(p)$	dynamic transfer function
$G_1(p)$	dynamic transfer function drafting zone 1
$G_2(p)$	dynamic transfer function drafting zone 2

continuation

Symbol	*Description*
$G_{\mathrm{D}}(p)$	dynamic transfer function of two steps drafting process
$G_1(p)...G_5(p)$	dynamic transfer functions of FT-texturing process
$\|G_1(\mathrm{j}f)\|...\|G_5(\mathrm{j}f)\|$	normalized amplitude frequency responses of FT-texturing process
$G_{\mathrm{fc}}(p)$	dynamic transfer function of stationary flat card
$G_{\mathrm{fc}}(\mathrm{j}\omega)$	complex frequency response of stationary flat card
$\|G_{\mathrm{fc}}(\mathrm{j}\omega)\|$	amplitude frequency response of stationary flat card
$\|G_{\mathrm{fc}}(\mathrm{j}\lambda_{\mathrm{o}})\|$	amplitude frequency response of stationary flat card (Fig. 4.19)
$G_{\mathrm{rc}}(p)$	dynamic transfer function of roller top card
$G_{\mathrm{W1}}(p)...G_{\mathrm{Wq}}(p)$	dynamic transfer functions of worker-angle-stripper-pairs
$G(\mathrm{j}\omega)$	complex frequency response
$\|G(\mathrm{j}\omega)\|$	amplitude frequency response
$G(\mathrm{j}f)$	complex frequency response
$\|G(\mathrm{j}f)\|$	amplitude frequency response
$\|G[\mathrm{j}(f/f_{\mathrm{c}})]\|$	normalized amplitude frequency response
$G_{\mathrm{D}}(\mathrm{j}\omega)$	complex frequency response of two steps drafting process
$\|G_{\mathrm{D}}(\mathrm{j}\omega)\|$	amplitude frequency response of two steps drafting process
G_{i}	abbreviation for dynamic transfer function $G_{\mathrm{i}}(p)$
$G_1...G_{18}$	single transfer elements of functional block diagram according to Fig. 4.12 and Table 4.2, abbrev. for dynamic transfer functions $G_1(p)...G_{18}(p)$
$G_{\mathrm{z1}}...G_{\mathrm{z9}}$	disturbance transfer functions (Table 4.3), abbrev. for dynamic transfer functions $G_{\mathrm{z1}}(p)...G_{\mathrm{z9}}(p)$
$\|G_{\mathrm{z1}}\|...\|G_{\mathrm{z9}}\|$	disturbance amplitude frequency responses (Table 4.3, Fig. 4.13), abbrev. for amplitude frequency responses $\|G_{\mathrm{z1}}(\mathrm{j}f)\|...\|G_{\mathrm{z9}}(\mathrm{j}f)\|$
Gr	GRASHOF number
ΔH	heat of fusion
I_{h}	heating current
I_{hm}	heating current, mean value
ΔI_{h}	heating current change
$\mathrm{Im}(\omega)$	imaginary part of complex frequency response
K	fibre mass distribution coefficient (roller top card, Fig. 4.17)

continuation

Symbol	*Description*
K_a	fibre mass distribution coefficient (roller top card, Fig. 4.16)
$K_F(\tau)$, $K_F(k \cdot \Delta t)$	single values of ACF of time function $F(t)$ for τ or $k \cdot \Delta t$
K_h	polymer specific constant for the heat transfer
K_p	polymer specific constant for conversion fibre/yarn/ thread fineness to diameter
$K(T, \sigma)$	crystallisation rate
K_{max}	crystallisation rate constant
$K(\tau)$	auto- or cross-correlation function, integral representation
$K(k \cdot \Delta t)$	auto- or cross-correlation function, sum representation
$K_{FTt_o}(\tau)$	cross-correlation function of $F(t)$ and $Tt_o(t)$
$K_{FTt_o}(0)$	start value of $K_{FTt_o}(\tau)$ for $\tau = 0$
$K_{F\sigma}(\tau)$	cross-correlation function of $F(t)$ and $\sigma(t)$
$K_{F\sigma}(0)$	start value of $K_{F\sigma}(\tau)$ for $\tau = 0$
$K_{Tt_o\sigma}(\tau)$	cross-correlation function of $Tt_o(t)$ and $\sigma(t)$
$K_{Tt_o\sigma}(0)$	start value of $K_{Tt_o\sigma}(\tau)$ for $\tau = 0$
K_s	shortening factor (false twist texturing process)
K_S	amplification factor of transfer element G_4
$K_{\overline{S}}$	amplification factor of transfer element G_8
$K_{K(T_O)}$	amplification factor of transfer element G_9
K_K	amplification factor of transfer element G_{12}
K_U	amplification factor of transfer element G_{13}
K_{W1}, K_{W2}	polymer specific constants in Eqs. 5.96, 5.97
$K_{\overline{U}}$	amplification factor of transfer element G_{14}
L	LAPLACE-transformation
L_{-1}	LAPLACE-retransformation
L	take-up distance, length of take-up channel (Chap. 3)
L_c	cooling length
L_{cl}	necessary cutting length of fibre/yarn/thread
L_d	delay thread length
L_{dv}	delay thread length for a velocity disturbance Δv_i
$L_{d\mu}$	delay thread length for a velocity disturbance $\Delta\mu$
L_{gl}	necessary gauge length of fibre/yarn/thread
M	measuring device
M	molecular weight (Chap. 3)
M	fibre/yarn/thread mass at the friction element (Chap. 5)
MC	micro computer
MP	micro processor

continuation

Symbol	*Description*
MPC	micro processor controller
MVC	measuring value computer
MVCO	measuring value concentrator
N	number of filaments in fibre bundle
$N(\Theta)$	orientational distribution function
N_a	abbreviation factor in Eq. 5.56
N_h	heating power
N_hm	heating power, mean value
ΔN_h	heating power change
Nu	NUSSELT number
PA	poly(amide)
PET	poly(ethylene terephthalate)
PP	poly(propylene)
PCD	programmable controller device
P_i	LEGENDRE polynomials
Pr	PRANDTL number
Q	mass throughput of polymer
Q_air	(mass) flow rate of air
Q_i	glass mass inflow
ΔQ_i, ΔQ_i1, ΔQ_i2	glass mass inflow changes
$\widetilde{\Delta Q_\mathrm{i1}}$	sinusoidal glass mass inflow change
Q_s	throughput through the spinneret
ΔQ_s	throughput change through the spinneret
$\widetilde{\Delta Q_\mathrm{s}}$	sinusoidal throughput change through the spinneret
Q_sm	throughput through the spinneret, mean value
ΔQ_s, ΔQ_s1, ΔQ_s2	throughput changes through the spinneret
Q_p	throughput through the spinning pump
R	gas constant (Chap. 3)
R	filament (fibre) radius (Chap. 3)
R	reduction factor (Chap. 5)
R_y	fineness related tensile force F/Tt_1
R_0	radius of capillary hole, initial filament radius
R_1, R_2	inner and outer radius of radially symmetric filament bundle
Re, $Re_{\|\|}$, $Re_\perp$	REYNOLDS number, related to $v_{\|\|}, v_\perp$
$R_{\mathrm{T_g}}$	fineness related tensile force of fibre/yarn/thread at the glass transition temperature T_g

continuation

Symbol	*Description*
$\Delta R_{\mathrm{T_g}}$	fineness related tensile force change of fibre/yarn/ thread at the glass transition temperature T_{g}
$\mathrm{Re}(\omega)$	real part of frequency response
R_{h}	heating resistance
R_{hm}	heating resistance, mean value
ΔR_{h}	heating resistance change
R_{s}	flow resistance of spinneret holes
$\mathrm{Res}[S(p)]$	residue of $S(p)$
S_{y}	surface of disc-shaped pice
$S(\omega),\ S(f)$	power density spectrum function
$S_{\mathrm{F}}(f)$	single value of the APSF of time function $F(t)$ for f
$S(p)$	abbreviation of the integrand of Eq. 2.42
T	temperature or time period of the integration range
T_{A}	necessary maximum analysis time
$T_1,\ T_2, ..., T_{\mathrm{n}}$	thread temperatures T_{y} after pass of single heating and cooling lines
T_0	extrusion temperature, initial temperature
$T_{\mathrm{a}},\ T_{\mathrm{air}}$	temperature of air
T_{b}	temperature at beginning of heat transfer
T_{bf}	bobbin formation time
T_{c}	time constant
T_{ch}	time constant of heat transfer dynamic
$T_{\mathrm{c1}},\ T_{\mathrm{c2}}$	time constants (stationary flat card)
$T_{\mathrm{cd1}},\ T_{\mathrm{cd2}}$	time constants of drafting zones 1 and 2
$T_{\mathrm{d}},\ T_{\mathrm{d1}},\ T_{\mathrm{d2}}$	dead (transport) times
T_{H}	time constant of transfer element G_4
$T_{\overline{\mathrm{H}}}$	time constant of transfer element G_7
T_{h}	time constant of transfer element G_9
$T_{\mathrm{K1}},\ T_{\mathrm{K2}}$	time constants of transfer element G_{12}
T_{U}	time constant of transfer elements $G_{13},\ G_{14}$
T_{e}	temperature of melt from extruder
T_{f}	temperature of filament
T_{g}	glass transition temperature
T_{g}	temperature of glass melt
T_{gm}	temperature of glass melt, mean value
ΔT_{g}	temperature change of glass melt
$\widetilde{\Delta T_{\mathrm{g}}}$	sinusoidal temperature change of glass melt
T_{h}	temperature of heat medium
T_{hg}	temperature of heated godet
T_{m}	melt temperature

continuation

Symbol	*Description*
T_{D}	twist density
T_{Dm}	twist density, mean value
ΔT_{D}	twist density change
$\widetilde{\Delta T_{\mathrm{D}}}$	sinusoidal twist density change
$\dot{T}_{\mathrm{D}}$	derivative of T_{D} to t
T_{P}	cycle duration
T_{P1}	cycle duration, correlating to circuit frequency ω_1
T_{P2}	cycle duration, correlating to circuit frequency ω_2
T_{r}	temperature of cool medium
T_{s}	temperature of the spinneret
T_{sm}	temperature of the spinneret, mean value
ΔT_{s}, ΔT_{s1}, ΔT_{s2}	temperature changes of the spinneret
$\widetilde{\Delta T_{\mathrm{s}}}$	sinusoidal temperature change of the spinneret
T_{y}	temperature of fibre/yarn/thread
T_{ym}	temperature of fibre/yarn/thread, mean value
ΔT_{y}	temperature change of fibre/yarn/thread
TN	tenacity of fibre/yarn/thread
Tt	fineness (titre) of fibre/yarn/thread
Tt_{m}	fineness (titre) of fibre/yarn/thread, mean value
ΔTt	fineness (titre) change of fibre/yarn/thread
$\widetilde{\Delta Tt}$	sinusoidal fineness (titre) change of fibre/yarn/thread
Tt_{f}	fineness of single fibre
Tt_{fm}	fineness of single fibre, mean value
ΔTt_{f}	fineness change of single fibre
$\widetilde{\Delta Tt_{\mathrm{f}}}$	sinusoidal fineness change of single fibre
Tt_{i}	fineness of input fibre/yarn/thread
Tt_{ii}	fineness of input fibre/yarn/thread before the ith elongation step
Tt_{im}	fineness of input fibre/yarn/thread, mean value
Tt_{o}	fineness of output fibre/yarn/thread
Tt_{om}	fineness of output fibre/yarn/thread, mean value
ΔTt_{o}	fineness change of output fibre/yarn/thread
$\widetilde{\Delta Tt_{\mathrm{o}}}$	sinusoidal fineness change of output fibre/yarn/thread
$\dot{T}t_{\mathrm{o}}$	derivative of Tt_{o} to t
$\dot{T}t_{\mathrm{om}}$	derivative of Tt_{om} to t
$\Delta\dot{T}t_{\mathrm{o}}$	derivative of ΔTt_{o} to t
$\widetilde{\Delta Tt_{\mathrm{o}}(\omega)}$	vector of sinusoidal disturbances of the effect variable Tt_{o} for ω
$Tt_{\mathrm{o}}(t)$	fineness of fibre/yarn/thread, time function
Tt_{om}	mean value of $Tt_{\mathrm{o}}(t)$
Tt_{oi}, $Tt_{\mathrm{o(i+k)}}$	discontinuous values of time function $Tt_{\mathrm{o}}(t)$

continuation

Symbol	*Description*
Tt_{oi}	fineness of output fibre/yarn/thread after the ith elongation step
Tt_0	fineness of unloaded fibre/yarn/thread
$Tt_{0\mathrm{m}}$	fineness of unloaded fibre/yarn/thread, mean value
ΔTt_0	fineness change of unloaded fibre/yarn/thread
Tt_1	fineness of F loaded fibre/yarn/thread
$Tt_{1\mathrm{m}}$	fineness of loaded fibre/yarn/thread, mean value
ΔTt_1	fineness change of loaded fibre/yarn/thread
Tt_{L}	filament fineness at take-up distance L
Tt_{s}	fineness of ready formated fibre
Tt_{sm}	fineness of ready formated fibre, mean value
ΔTt_{s}, ΔTt_{s1}, ΔTt_{s2}	fineness changes of ready formated fibre
$\widetilde{\Delta Tt_{\mathrm{s}}}$	sinusoidal fineness change of ready formated fibre
$\dot{T}t_{\mathrm{s}}$	derivative of Tt_{s} to t
Tt_{z}	sliver fineness at output/input of drafting zones 1/2
Tt_{zm}	sliver fineness at output/input of drafting zones 1/2 or fineness of the tensionless fibre/yarn/thread or fibre/yarn/thread fineness of dead time line input, mean values
ΔTt_{z}	sliver fineness change at output/input of drafting zones 1/2 or fibre/yarn/thread fineness change of dead time line input
U^*	activation energy for segment motion
U_{h}	heating voltage
U_{hm}	heating voltage, mean value
ΔU_{h}	heating voltage change
$\widetilde{\Delta U_{\mathrm{h}}}$	sinusoidal heating voltage change
U_{m}	mains voltage
ΔU_{m}	mains voltage change
$\widetilde{\Delta U_{\mathrm{m}}}$	sinusoidal mains voltage change
V	draft of sliver
V_1	draft of sliver drafting zone 1
V_2	draft of sliver drafting zone 2
W	width of filament bundle in multifilament spinning, perpendicular to quenching direction
W_{a}	abbreviation factor in Eq. 5.56
X_{c}	crystallinity

continuation

Symbol	*Description*
α	heat transfer coefficient
α_1	heat transfer coefficient thread/metallic surface
α_2	heat transfer coefficient thread/surrounding air
α_{rt}	resistance-temperature coefficient of the Pt-Rh-spinneret oven
α	angle of wrap
α_{m}	angle of wrap, mean value
$\Delta\alpha$	angle of wrap change
β	parameter (to force density) in multifilament theory
γ	parameter (to force density) in multifilament theory
Δn	(total) birefringence
Δn_{am}	amorphous birefringence
Δn_{cr}	crystalline birefringence
$\varDelta_{\mathrm{i}}$	birefringence of input fibre/yarn/thread
$\varDelta_{\mathrm{o}}$	birefringence of output fibre/yarn/thread
$\frac{\partial}{\partial x}, \frac{\partial}{\partial y}$	derivatives in partial DEs
$\frac{\partial}{\partial z}, \frac{\partial}{\partial t}$	derivatives in partial DEs
$\frac{\partial \varPhi}{\partial v_{\mathrm{s}}}, \frac{\partial \varPhi}{\partial v_{\mathrm{i}}}, \frac{\partial \varPhi}{\partial l}$	partial derivatives from $\varPhi$ to the appropriate sizes
$\frac{\partial \varPhi}{\partial q_{\mathrm{s}}}, \frac{\partial \varPhi}{\partial q_{\mathrm{i}}}, \frac{\partial \varPhi}{\partial \varrho}$	partial derivatives from $\varPhi$ to the appropriate sizes
$\frac{\partial \varPhi}{\partial T_{\mathrm{D}}}, \frac{\partial \varPhi}{\partial n_{\mathrm{s}}}, \frac{\partial \varPhi}{\partial \mu}$	partial derivatives from $\varPhi$ to the appropriate sizes
$\frac{\partial \varPhi}{\partial v_{\mathrm{z}}}, \frac{\partial \varPhi}{\partial v_{\mathrm{o}}}, \frac{\partial \varPhi}{\partial Tt}$	partial derivatives from $\varPhi$ to the appropriate sizes
ϵ, ε	elongation of fibre/yarn/thread (in %, resp. logarithmic (HENCKY) measure)
$\Delta\epsilon$	elongation change of fibre/yarn/thread (in %)
ϵ_{b}	elongation at break of fibre/yarn/thread (in %)
ε_{o}	elastic (orientational) part of elongational deformation (HENCKY measure)

continuation

Symbol	*Description*
$\varepsilon_{\mathrm{oi}}$	reached orientation elongation of fibre/yarn/thread in the ith elongation step (HENCKY measure)
$\varepsilon_{\mathrm{o1}}$	reached orientation elongation of fibre/yarn/thread at the glass transition point (HENCKY measure)
$\Delta\varepsilon_{\mathrm{o1}}$	reached orientation elongation change of fibre/yarn/thread at the glass transition point (HENCKY measure)
$\varepsilon_{\mathrm{omax}}$	maximum orientation elongation of fibre/yarn/thread (HENCKY measure)
ε_{v}	viscous part of elongational deformation (HENCKY measure)
$\dot{\varepsilon}_{\mathrm{v}}$	viscous elongational deformation rate (HENCKY measure)
$\varepsilon(t)$	elongation of fibre/yarn/thread, time function
$\epsilon_1,\ \epsilon_2$	elongations (in %) correlated with F_1 and F_2
ε_{m}	emissivity (heat radiation)
η	elongational viscosity
η_{app}	apparent elongational viscosity
η_{air}	dynamic viscosity of air
η_{e}	viscosity of melt from extruder
η_{s}	viscosity of melt in the spinneret
η_{h}	heating yield
θ	angle
λ	relaxation time
λ_{air}	heat conductivity of air
λ_{f}	wavelength of disturbance
λ_{i}	wavelength of disturbance effect in an input web
λ_{o}	wavelength of disturbance effect in an output web
μ	coefficient of friction between fibre/yarn/thread and fixed friction guide or false twist spindle
μ_{m}	coefficient of friction between fibre/yarn/thread and fixed friction guide or false twist spindle, mean values
$\Delta\mu$	change of coefficient of friction between fibre/yarn/thread and fixed friction guide or false twist spindle
$\widetilde{\Delta\mu}$	sinusoidal change of coefficient of friction between fibre/yarn/thread and fixed friction guide or false twist spindle

continuation

Symbol	*Description*
ν_{air}	kinematic viscosity of air
ϱ_{air}	density of air
ϱ	density of polymer or thread material
ϱ_{m}	density of polymer, mean value
$\Delta\varrho$	density change of polymer
$\widetilde{\Delta\varrho}$	sinusoidal density change of polymer
ϱ_{am}	mass density of amorphous polymer
ϱ_{cr}	mass density of crystalline polymer
ϱ_{m}	mass density of glass
ϱ_{p}	mass density of polymer
σ	filament stress
$\sigma(t)$	fineness related tensile force of fibre/yarn/thread, time function
σ_{m}	mean value of $\sigma(t)$
$\sigma_{\mathrm{i+k}}$	discontinuous value of time function $\sigma(t)$
$\sigma_{\mathrm{T_g}}$	tension $F_{\mathrm{T_g}}/Tt_{\mathrm{s}}$ of fibre/yarn/thread at the glass transition point
σ_{SB}	STEPHAN-BOLTZMANN constant
σ_{surf}	surface tension (specific surface energy)
τ	time shift (correlation function)
τ_{f}	shear stress at filament surface
φ	phase shift angle
φ_{a}	humidity of quenching air
φ_1	phase shift angle, correlating to circuit frequency ω_1
φ_2	phase shift angle, correlating to circuit frequency ω_2
φ_3	phase shift angle, correlating to circuit frequency ω_3
$\varphi(\omega)$, $\varphi(f)$	phase frequency responses
Φ	fluidity
Φ	symbol for a nonlinear differential equation
$\Phi(x,r)$	potential flow function, radial coordinates
$\Psi(x,y)$	potential flow function, cartesian coordinates
ω	circular/excitation frequency
ω_0	circular/excitation frequency, level 0
ω_1	circular/excitation frequency, level 1
ω_2	circular/excitation frequency, level 2

continuation

Symbol	*Description*
ω_3	circular/excitation frequency, level 3
ω_{c}	critical (circular) frequency
$\lrcorner\!\ulcorner$	step
$\perp$	impulse

Index

MIX
Papier aus verantwortungsvollen Quellen
Paper from responsible sources
FSC® C105338

If you have any concerns about our products,
you can contact us on
ProductSafety@springernature.com

In case Publisher is established outside the EU,
the EU authorized representative is:
Springer Nature Customer Service Center GmbH
Europaplatz 3, 69115 Heidelberg, Germany

Printed by Libri Plureos GmbH
in Hamburg, Germany